The Slater
Field Guide
to
AUSTRALIAN
BIRDS

D0668877

The Slater
Field Guide
to
AUSTRALIAN
BIRDS

Peter Slater • Pat Slater • Raoul Slater

WELDON
PUBLISHING
SYDNEY · HONG KONG · CHICAGO · LONDON

To our parents, Sam and Nell Slater
Aubrey and Judy Moore

Distributed by Gary Allen Pty Ltd

9 Cooper Street, Smithfield, NSW 2164
A Kevin Weldon Production
Published by Weldon Publishing
a division of Kevin Weldon & Associates Pty Limited
372 Eastern Valley Way, Willoughby, NSW 2068, Australia

First published 1986
Reprinted 1988
Revised edition 1989
Reprinted 1990, 1991, 1992
Copyright Paintings © 1986 Peter Slater
            Text © 1986 Pat Slater
            Maps © 1986 Raoul Slater
Wholly designed in Australia
© Design Kevin Weldon & Associates Pty Ltd
Typeset in Australia by Caxtons, Adelaide
Printed in Australia by The Griffin Press, Adelaide

National Library of Australia
Cataloguing-in-Publication Data

Slater, Peter, 1932–
    The Slater Field Guide to Australian Birds

    Includes index.
    ISBN 0 947116 99 0

    1. Birds — Australia — Identification. I. Slater,
    Pat, 1937– . II. Slater, Raoul, 1966– . III.
    Title. IV. Title: Field guide to Australian birds.

598.2994

Designer: Lynne Tracey
Managing Editor: Lisa Berryman
Editor: Beverley Barnes

Contents

Introduction

A field guide has most value when it contains sufficient information for *every* species in the area it covers to be identified by anyone with a modicum of intelligence and a pair of binoculars. This guide seeks to provide such information in the most compact and portable form possible for the 756 birds so far recorded in Australia and Tasmania and at sea over the continental shelf. The illustrations and text are arranged in order from the birds considered to be the most primitive to those considered the most modern. Most bird books are arranged in the same way, so knowledgeable birdwatchers should have no trouble finding the birds they want. For those unfamiliar with bird books, who may have trouble knowing which page to turn to, the following guide may help.

Water birds

1. Aerial birds, usually seen flying over water: (a) salt water: albatrosses, petrels, shearwaters, storm petrels, gannets, frigatebirds pp 18–47; (b) salt or fresh water: gulls, terns pp 136–49.
2. Swimming birds, usually seen swimming or perched near water; all have short legs with some sort of webbing between the toes: (a) salt water: penguins pp 14–17; (b) salt or fresh water: cormorants and shags p 48, pelicans p 50, phalaropes p 130; (c) mainly fresh water: grebes p 12, ducks, swans and geese pp 60–71.
3. Large wading birds: egrets, herons, bitterns pp 52–7; storks and cranes p 50; oystercatchers, stilts p 132; curlews, godwits pp 120–3.
4. Small wading birds, usually seen on beaches, mudflats, islands and around swamps and lakes: plovers, snipe pp 98–105, 118; curlews p 122.
5. Reed birds, usually secretive although often vocal: crakes, rails and gallinules pp 92–7; reed-warbler p 242.

Land birds

1. Ground birds that feed on ground, and never or very seldom perch in trees: (a) large flightless birds: ostriches, emu, cassowary p 10; (b) medium-sized, long-legged ground birds: mound-builders, stone-curlews, bustards p 86, lapwings p 98; (c) small, short-legged ground birds: quail, button-quail pp 88–91; logrunners p 236.
2. Ground birds that feed and spend much time on ground but also perch in bushes and trees: pittas, thrushes, scrub-birds, bristlebirds, wagtails, larks, pipits pp 202–9; whipbirds, wedgebills, babblers, quail-thrush pp 236–41; grasswrens, scrubwrens pp 252–9; whitefaces p 264; chats p 302; starlings, mud-nest builders pp 314–17; magpies p 324; lyrebirds p 332.
3. Grass birds, usually seen in long grass: quail p 88; emu-wrens, cisticolas, grassbirds, fairy-wrens, grasswrens pp 244–53; finches pp 306–13.
4. Aerial birds: swifts, swallows pp 198–201.
5. Birds of prey: eagles, hawks, falcons, kites pp 72–85.
6. Night birds: owls, frogmouths, nightjars pp 190–7.
7. Medium-sized 'bush birds', usually seen perching in trees; some feed on ground but spend most of their time in trees: pigeons and doves, cockatoos, lorikeets, parrots, cuckoos, kingfishers pp 150–89; cuckoo-shrikes p 210; wattlebirds, friarbirds pp 274–7; orioles p 314; drongoes p 316; currawongs p 322; bowerbirds pp 326–9.
8. Small 'bush birds': trillers p 212; flycatchers, wrens, robins, whistlers, monarchs, fantails pp 214–35; warblers (gerygones, thornbills) pp 260–9; honeyeaters, silvereyes pp 280–301; pardalotes p 304.
9. Tree-trunk birds, usually seen searching for insects in tree bark: sittellas, treecreepers pp 270–3.

Having decided where to begin looking in trying to identify a bird, it may be necessary to examine several successive plates of illustrations, because some groups of similar birds are large, eg sandpipers, honeyeaters and flycatchers. With this in mind, I grouped together the birds most similar in appearance to make it easier to compare them. The page title may be of some help – for instance, 'teetering sandpipers', 'white-gaped honeyeaters', 'reddish grasswrens' or 'scrubwrens with prominent eyebrows'.

Most pages are set out in the same way. At the top is a brief discussion of the birds described, with some hints that help identification, followed by more detailed information about each bird.

Common name: in most cases the name used is that recommended by the Royal Australasian Ornithologists Union. We have made exceptions in line with other popular field guides, (a) where academic names are recommended for common names, eg Hylacola for heathwren and Gerygone for warbler; (b) where unnecessary changes to old-established names have been recommended, eg Australian Kestrel for Nankeen Kestrel (in these cases, both names are included).

Scientific names: the academic names used are those recommended by the Royal Australasian Ornithologists Union. When their new, long-awaited checklist is issued in the near future there will be a number of changes to these names, but while these changes are very exciting to ornithologists I would not advocate that birdwatchers get too upset about them, as their effect on birdwatching will be minimal – a bird will still be a bird, no matter what pigeonhole the academics stuff it in.

The scientific name of each bird is given in two parts. The first, always with a capital letter, is the generic name, which tells what **genus** (group of closely related species) it belongs to. It is this name that may change, as different ideas about relationships emerge. The second name, which always begins with a small letter, is the specific name, which tells what **species** the bird is. This name is changed only extremely rarely, although its ending may alter when it is changed from one genus to another; thus when the Wompoo Pigeon was classified in the genus *Megaloprepia* its specific name was *magnifica;* when it was moved to the genus *Ptilinopus* its specific name became *magnificus.* In the body of the text, reference is sometimes made to a third academic name, the **subspecies** (also called **race** or **form**). Most subspecies of birds differ from the basic stock in minor ways and are of no relevance here, but some can be identified easily in the field, so these are referred to thus: 'short-tailed form (race *modestus*) with underparts pale fawn only faintly streaked white', enabling one to determine which subspecies one is looking at. It certainly makes birdwatching more interesting, and at least some of them will prove to be full species with further investigation.

Size: the measurement given is the length from bill tip to tail tip and is a rough guide only. Sometimes wingspread is given also. Occasionally other more precise measurements are given in the text, usually in millimetres. These are generally for very similar birds that are likely to come to hand, eg beach-washed seabirds. A suitable ruler marked in millimetres is printed at the back of the book.

Key sentence: in the first sentence are given the characteristics that in combination differentiate this species from all others. If female, juvenile and immature plumages are different from the male, they are given in that order. If it is not possible to distinguish the species in the field, as in the case of reed-warblers, this is stated plainly. Nothing is gained by guessing.

Description: the text has been prepared on the assumption that the illustrations preclude the need for detailed descriptions of the plumage. If a bird is shown in the illustration with a blue head, there is little point in writing 'the head is blue' in the text opposite. However, where such colours are relevant in identification, mention is made. Similar species are referred to with distinguishing features spelled out. Phases of plumage and subspecies that are identifiable in the field are discussed. Any habits that may be used in identification are added, eg flight characteristics, methods of feeding, etc.

Voice: the various calls made by birds can often help identification, eg the two very similar tattlers have very different calls. However, describing many calls in words is unsatisfactory; once the call is known, the words can be seen to fit, but it is at the wrong end of the identification process. There are many recordings of bird songs available which may help. For aesthetic and technical reasons, I recommend two: (a) John Hutchinson's *Index of Australian Bird Calls*, available in a series of discs or cassettes from John Hutchinson, Brockman St, Balingup, WA 6253; (b) *Field Guide to Australian Bird Song*, arranged by Rex Buckingham and available from the Bird Observers' Club, PO Box 185, Nunawading, Vic 3131.

Nesting: in the space available it is not possible to describe nests in detail, but a brief description is given together with the number and colour of eggs. On the plate opposite, a typical egg is illustrated, with approximate measurements in millimetres. These illustrations are based on the Cameron Collection in the Queensland Museum – any eggs not available there were painted from the Cornwall Collection, also housed in the Queensland Museum. Where space was available, a drawing of a nest was added to the plate as well.

Range: given under this heading are (in order) status, habitat and distribution. The **status** of each bird is rather subjective; such terms as common, rare, abundant, etc, are relative and have different meaning for different birds. However, they give some idea of whether a species may be readily found or may have to be searched for. There are a number of categories into which birds may fall: (a) *resident:* lives in the area all year, eg Yellow Robin. (b) *nomad:* moves from area to area with no seasonal regularity, eg Crimson Chat. (c) *migrant:* moves from area to area on a regular basis. There are several forms of migration: (i) summer breeding migrant, a species that comes into Australia from the north to breed in summer, eg White-tailed Kingfisher; (ii) summer non-breeding migrant, a species that breeds in the northern hemisphere and spends summer in Australia, eg Sharp-tailed Sandpiper; (iii) winter breeding migrant, a species that breeds in Australia in winter and moves elsewhere in summer, eg Great-winged Petrel; (iv) winter non-breeding migrant, a species that breeds to the south in summer and visits Australia in winter, eg Cape Petrel; (v) internal migrant, a species that breeds in southern Australia in summer and moves to northern Australia in winter, eg White-winged Triller; (vi) local migrant, a species that migrates over a short distance, usually breeding at higher altitudes in summer and moving down to lower altitudes in winter, eg Rose Robin; (vii) trans-Bassian migrant, a species that breeds in Tasmania and migrates into south-eastern mainland Australia, eg Swift Parrot; (viii) trans-Tasman migrant, a species that breeds in New Zealand and migrates to Australia, eg White-fronted Tern.

Habitat: the main habitats are shown here in eight maps. Their names are mostly self-explanatory.

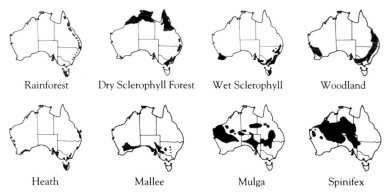

| Rainforest | Dry Sclerophyll Forest | Wet Sclerophyll | Woodland |
| Heath | Mallee | Mulga | Spinifex |

Distribution: finally, the distribution is indicated briefly in words and is shown on a small map. The main aim of these maps is to aid in identification; they are not large enough to be regarded as precise, but if a bird is seen *well* outside the area indicated, there are two possibilities: either an incorrect identification has been made; or the bird occupies an area previously unreported, so the matter should be taken further if checks on better maps confirm the position. Accurate maps on a larger scale may be consulted in the RAOU *Atlas of Australian Birds,* and *Tasmanian Bird Atlas* by David Thomas.

Illustrations: the coloured plates were commenced in April 1978, and most were finished late in 1985. As well as typical examples of each species, illustration of various plumage forms, juveniles, immatures and subspecies were included. The following symbols were used:

    male ♂   female ♀

ad–adult

juv–juvenile, a bird in its first set of feathers, illustrated only if it differs from the adults.

imm–immature, a bird that after its first moult, still differs from adults. There are two sorts of immatures: (a) those that moult *during* their first year, usually retaining juvenile wing and tail feathers (exceptions are some male wrens, which moult completely at six months); (b) those that moult at the *end* of their first year, usually having a complete change of feathers, eg birds of prey.

sub-ad–sub-adult, a bird that after its second moult has a plumage that is different from both immature and adult; only a few examples, such as Wedge-tailed Eagle, Wandering Albatross.

br–breeding, a bird that assumes fresh, usually decorative plumage before the breeding season.

non-br–non-breeding, a bird that assumes a dull plumage after breeding.

Identification is only the first step in bird study; this book is aimed at helping you take that first step. The next stage is to find out more about birds. There are many books available, but those I recommend are: firstly, the *Reader's Digest Book of Australian Birds* (2nd edition), and *The Australian Bird Atlas*, produced by the Royal Australasian Ornithologists Union, or *The Tasmanian Bird Atlas* by David Thomas if you live in Tasmania. These summarise what is known about Australian birds and where they occur. Secondly, *Bird Life* by Ian Rowley, which gives more detailed accounts of a small number of very interesting birds, an ideal introduction to the fascinating world of bird behaviour; and *Understanding Australian Birds*, by J. D. Macdonald, an authoritative but easy-to-read discussion on bird behaviour and biology. Thirdly, *The Australian Birdfinder* by Michael Morcombe, which gives details about the best places to see birds. There are very few inexpensive popular books which add much to the information found in these six. Best among the exceptions are: *The Kookaburra* by Veronica Parry, *Waterfowl in Australia* by H. J. Frith, *Pigeons and Doves of Australia* by H. J. Frith, *The Mallee Fowl* by H. J. Frith, *Nightwatchmen of Bush and Plain* by David Fleay, and *Hawks in Focus* by Jack and Lindsay Cupper. For those interested in photographing birds, we suggest *The Technique of Bird Photography* (4th edition) by John Warham, or *Masterpieces of Australian Bird Photography* by Peter Slater. For bird artists, none is better than *Wildlife Painting: Techniques of Modern Masters* by Susan Rayfield. For home gardeners, *Birdscaping Your Garden* by George Adams is first rate, giving numerous hints on attracting birds to your immediate surroundings.

Acknowledgements: among the hundreds of people who have given us assistance, we must single out for special mention the ornithological staff of the Queensland Museum, Glen Ingram, Wayne Longmore and, in the early stages, Lenore Wedgewood and Don Vernon. Without their cooperation in making available suitable specimens, this book could never have been completed. To the curators of other museums who loaned specimens, our grateful thanks. Chris Corben and John McKean kept us up to date on the 'new birds' that are constantly being added to the Australian list by a growing band of dedicated 'super twitchers'. (They usually managed to wait until a plate was finished before telling us about a new bird that should be included on it, meaning that the painting had to be done again.) Chris Corben was most generous in his assistance, particularly with seabirds and waders. Stephen Debus made available his considerable knowledge of corvids. Rob Elvish checked parts of the manuscript and suggested a tighter, more logical format. John Hutchinson, as always, gave us access to his tapes of bird calls and songs.

Ruth Berry painted six of the plates. They are identified by her initials in the bottom right hand corner (honeyeaters, finches and lorikeets). Our thanks to her for helping out when we were trying to speed up the delivery time of the manuscript.

Our thanks also to those who helped along the way: Eric Sedgwick, Angus Robinson, D. L. Serventy, Vincent Serventy, Eric Lindgren, Eric McCrum, David Ride, Kevn Griffiths, John Halse, Ray Garstone, H. J. Frith, John Calaby, Peter Fullagar, Gerry Van Tets, Joseph Forshaw, Graeme Chapman, Tony D'Andria, Wim Vestjens, Julian Ford, Fred T. H. Smith, Stanley Breeden, Joanna and Barry Morgan, Billie Gill, J. D. Macdonald, Noel Cusa, Ali and Col Lloyd, Mr and Mrs Monty Schrader, Herb Rabig, H. O. Officer, Glen Threlfo, Brian Coates, Len Robinson, Roy Wheeler, Richard Weatherley, Shane Parker, Glen Storr, Ron Johnstone, J. S. Robinson, Bob Lovell, Graham Pizzey, Joyce and Ivan Fien, Jim Bravery, Cecil Cameron, Peter O'Reilly, Alister McKissock, Larry Bryant, Gary Norwood, Don Shivas, D. D. Dow, R. Schodde, A. R. McEvey, Lawrie Muller, Henry and Cathy Nix and Michael Morcombe.

Grateful thanks to Bob Brown and to many of those already mentioned above for suggested improvements to the second edition.

LARGE FLIGHTLESS BIRDS

Two large species of flightless birds remain in Australia from a probably large number of species that became extinct during or since Pleistocene times, most recent being the Dwarf Emu, gone from King I and Kangaroo I by 1830, and the Tasmanian Emu, exterminated by 1870. The familiar Emu of the mainland still occurs in large numbers in unsettled areas, has ability to exploit good seasons in interior to maintain populations. During droughts may move in numbers into agricultural areas, giving rise to infamous 'emu war' of 1932, and construction of emu-proof fences. Feeds on fruit, leaves, insects. The Cassowary, being confined to rainforest and antipathetic to disturbance, is vulnerable unless large areas of tropical rainforest remain intact. For such a large, brightly-coloured bird it is surprisingly hard to see in rainforest, unless it is startled, when it crashes off through undergrowth with head down. Feeds mainly on fallen fruit, also fungi, insects, etc. The Ostrich was introduced from South Africa during 1870s for feather farming, a long-defunct industry, and only a few feral flocks remain in SA. Egg shell remnants at Port Augusta approximated those of northern race *camelus*, so birds there may be hybrids between *camelus* and South African race *australis*.

Cassowary   *Casuarius casuarius* 1·5–2·0 m
Large black flightless bird with coloured wattles, found in north-eastern rainforest. Heavy casque on head aids progress through vines. *Imm:* brown with rudimentary casque, attains full colour at three years. *Juv:* striped buff-yellow and black or brown, stays with male for about nine months. *Voice:* loud hollow booming. *Nesting:* 3–4 light brown eggs laid on ground, incubated by male; female polyandrous. *Range:* uncommon to rare in tropical rainforest from Pascoe R south to Hinchinbrook I and Paluma Ra (Mt Spec), Qld.

Emu   *Dromaius novaehollandiae* 2 m
Large brown flightless bird of open country ranging from woodland to desert plains. Western birds are darker and lack white ruff when breeding. *Juv:* striped cream and brown, stay with male up to 18 months. *Voice:* hollow booming (female); guttural ratcheting (male). *Nesting:* 6–12 heavily granulated dark green eggs, incubated by male. *Range:* common in unsettled areas, uncommon in settled areas, mainly in open woodland to plains.

Ostrich   *Struthio camelus* 2·5 m
Large flightless bird of African plains introduced to Australia for feather farming. *Male:* enormous, black and white with long naked neck and thighs, and white feathers in tail and vestigial wings used for display. *Female:* brown, distinguished from Emu by larger size, white plumes on wings and tail, large naked thighs and unfeathered neck. *Nesting:* male polygamous; small group of females lay up to 20 white eggs in common nest, incubated by male and females. *Range:* small feral flocks at 'Redcliffe', near Burra, SA, and near Port Augusta, SA.

DROMAIIDAE, CASUARIIDAE, STRUTHIONIDAE

LARGE FLIGHTLESS BIRDS

ad

juv

CASSOWARY

EMU

♂

♀

OSTRICH

150 x 125

140 x 95

130 x 90

Cassowary

Emu

Ostrich

GREBES

Grebes are aquatic birds mainly observed on fresh water and less frequently on saline lakes and estuaries; occurrence on dry land is accidental. Feathers are densely packed and waterproof, legs are set well back on body, making walking ungainly, but are well adapted for swimming with lobed toes. Food is taken underwater. Usually fly at night. Nest a heap of floating vegetation; eggs soon become stained and are covered with waterweed when nest is left. Chicks (downies) can swim soon after hatching but often ride on parent's back and return to nest to roost for about a fortnight. Identification of breeding birds is easy, but non-breeding plumages of Hoary-headed and Australasian are very similar. Eurasian Little Grebe may occur occasionally on some coastal tropical wetlands.

Hoary-headed Grebe   *Poliocephalus poliocephalus*   27 cm
Small greyish grebe with white streaks on head. *Non-breeding:* very similar to Australasian, but paler and with dark cap extending below eye and often with black patch on the hindneck, only visible when bird is swimming directly away. *Juv:* dusky markings on head and neck, more obscure than Australasian. In all plumages except downy (chick) a broad white wing stripe shows in flight. *Nesting:* 3–7 eggs, white at first, becoming nest-stained. *Range:* less common than Australasian in most areas, more likely to be seen on inland lakes and more partial to salt water. May turn up almost anywhere in the interior after rain has filled claypans, etc, where they may breed in colonies.

Australasian Grebe   *Tachybaptus novaehollandiae*   25 cm
Small brownish grebe with dark head, narrow chestnut streak on side of head, yellow eye and broad white wing stripe. Eurasian Little has broader chestnut patch on side of head and neck, dark eye and only a small patch of white in wing. *Non-breeding:* very similar to Hoary-headed but browner and with dark cap passing through eye (in incomplete moult this is not obvious). *Juv:* head dusky grey with pale patches. *Downy:* striped black and white with red bill and chestnut patch over eye. *Call:* vociferous chittering – a very noisy bird, calling day and night. *Nesting:* 3–7 eggs white at first, becoming nest-stained, sometimes almost black. *Range:* common on fresh water ponds, dams and lakes all over Australia, rarely on salt water or estuaries. (Look for Eurasian Little Grebe *Tachybaptus ruficollis* which is similar but has dark eye, much more rufous on head and neck and shows little white on wing in flight. Probably non-breeding and immature are indistinguishable from Australasian except for amount of white in wing; possibly occurred at Derby, WA, on ephemeral swamp after wet season, 1961.)

Great Crested Grebe   *Podiceps cristatus*   50 cm
Large brown grebe with prominent ear tufts; chestnut flanks shown are often hidden by wings while swimming. During display ear tufts expand into large ruff. Does not appear to have distinct non-breeding plumage as do northern hemisphere races. *Juv:* lacks tufts, has brownish cap, much larger than other immature grebes. *Downy:* striped black and white. *Call:* loud chittering. *Nesting:* 3–7 eggs, white at first, becoming nest-stained. *Range:* uncommon to rare on lakes and pond in south-west and south-east; sometimes numbers congregate on a particular lake where they may breed in loose colony.

GREBES

HOARY-HEADED

br

non-br

AUSTRALASIAN

br

non-br

juv

LITTLE

ad

GREAT CRESTED

juv

50 x 35

35 x 25

40 x 28

Great Crested   (fresh)   Australasian   (stained)   Hoary-headed

CRESTLESS PENGUINS

Penguins are flightless seabirds living on fish caught underwater. They swim effortlessly, low in the water, using large flippers for propulsion, and can survive at sea for long periods. On land they stand upright and walk or hop awkwardly. They must come ashore to moult. New feathers grow quickly. Most nest on islands of southern oceans but a few breed in Antarctica. Only the Little nests in Australia and is the only commonly seen species in Australian waters. Most of the others on this page have been seen only once or twice. On the page overleaf are species that have prominent yellow crests in adult plumage. Any penguin is likely to be seen anywhere along southern coast, so ranges given are approximate only.

King Penguin *Aptenodytes patagonicus* 95 cm
Very large, grey-backed penguin with black face and orange 'ear' patches. *Juv:* similar but with greyer face and pale yellow ear patches. (Emperor Penguin *A. forsteri* is larger with comparatively smaller bill with down-drooping tip; juvenile similar to juvenile King but has white ear patches.) *Range:* very rare vagrant from Antarctica to south-east, mainly Tas and Vic.

Gentoo Penguin *Pygoscelis papua* 80 cm
Large penguin with white patch over top of head and orange or yellow sides to bill and feet. *Juv:* similar but with duller markings on head. *Range:* very rare vagrant from subantarctic islands to Tas.

Adelie Penguin *Pygoscelis adeliae* 70 cm
Medium-sized penguin with characteristic head silhouette, stubby bill and white ring around eye giving staring appearance. *Juv:* similar but throat whitish and eye-ring duller. *Range:* very rare vagrant from Antarctica to south-east and south-west (one record each).

Chinstrap Penguin *Pygoscelis antarctica* 68 cm
Medium-sized penguin with white face and black line around chin. *Juv:* similar but with scattered dark feathers on face. *Range:* very rare vagrant from Antarctic and subantarctic islands to Tas.

Magellanic Penguin *Spheniscus magellanicus* 70 cm
Medium-sized penguin with black face and two black breastbands. *Juv and imm:* greyish face and single dusky breastband. (Some Jackass Penguins *S. demersus* from South Africa have two breastbands, but most have only one and juv and imm have dark face.) *Range:* a very rare vagrant from southern South America and Falklands to south-east.

Little Penguin *Eudyptula minor* 35 cm
Small penguin with blue-grey to blackish-grey and whitish underparts; worn plumage more brownish. While moulting, usually under coastal bushes, overhanging rocks or in burrows, covered with untidy tufts of old down and feathers. *Juv:* similar to adult but before fledging covered with brownish down. *Nesting:* in burrows, under overhanging rocks or under dense vegetation in spring and summer on islands of southern coast from Carnac I, WA to Broughton I, NSW. *Range:* common in southern waters from south Qld to Sharks Bay, WA.

PETER SLATER

PENGUINS

KING

GENTOO

CHINSTRAP

ADELIE

MAGELLANIC

LITTLE

56 x 42

Little

CRESTED PENGUINS

All are uncommon visitors, usually found on southern beaches in winter. Swimming birds are very difficult to identify, but beached individuals are easier. Moulting or immature birds may lack plumes or crests, and identification will depend on bill shape, size and underwing pattern. Immature of Rockhopper, Fiordland and Snares are most difficult to separate, and evidence (specimen or photograph) should accompany records. Beachdrifts are tame but may bite savagely if handled.

Erect-crested Penguin *Eudyptes sclateri* 70 cm
Large crested penguin with black back and throat and distinctive upright, bristly crest starting near gape but not meeting on forehead; pattern under flipper bolder than other crested penguins. *Juv:* throat white, crest vestigial. *Imm:* throat greyish, crest small, similar to other immatures, best told by underwing pattern. *Range:* very rare visitor to south-east coast, one record WA.

Rockhopper Penguin *Eudyptes chrysocome* 50 cm
Small crested penguin with blue-black back and throat, and with black and yellow drooping crest starting over eye, not meeting on forehead. *Juv:* throat whitish, crest undeveloped. *Imm:* throat greyish. Two forms occur, separated by underflipper pattern: (a) race *chrysocome* most likely in east, and (b) race *mosleyi* in west; watch for race *filholi* with underwing pattern similar to Fiordland. *Range:* uncommon visitor to south coast, mainly south-east and south-west.

Fiordland Penguin *Eudyptes pachyrhynchus* 55–60 cm
Only crested penguin without pale naked skin at base of bill; back and throat blue-black with white streaks on cheeks; crest compact with short feathers not as bright as Snares but wider and not as bushy at end; bill when viewed from above has bowed edges. *Imm:* less obvious crest and greyish throat, told by lack of bare skin at base of beak. *Range:* uncommon visitor to south-east.

Snares Penguin *Eudyptes robustus* 55–60 cm
Middle-sized penguin with blackish back and throat, compact, drooping crest starting above gape. Similar to Fiordland but darker above, has pale pink naked skin at base of bill, crest narrower and bushier at end and lacks whitish streaks on cheeks. *Imm:* less obvious crest and greyish throat. *Range:* very rare visitor to south-east.

Royal Penguin *Eudyptes chrysolophus schlegeli* 73 cm
Large crested penguin with white face and throat, long drooping crest meeting on forehead. *Imm:* small crest. Some individuals have face and throat dusky, and a form known as Macaroni Penguin (race *chrysolophus*) has forehead and throat black. *Range:* rare visitor to south-east, with most records in Tas.

PENGUINS

ERECT-CRESTED

a   b

juv

ROCKHOPPER   ad

FIORDLAND
juv   ad

SNARES
juv   ad

ROYAL
juv   ad

BLACK-BACKED ALBATROSSES (MOLLYMAWKS)

Albatrosses are large oceanic flying birds with long narrow wings, capable of remaining at sea for years. Feed on plankton, fish and squid. Nest on islands, either on ground or on dense platform of grass, earth and droppings. One egg is laid; incubation and fledging period extended, longest of all birds. Shy Albatross is only species nesting in Australia. **Mollymawks** have black upperwings with varying amounts of white on underwings. Soar for long periods, often following ships, settle on water near fishing boats in anticipation of food. Range southern waters, rarely north of tropic. Underwing pattern, amount of grey on head and neck, colour of bill most important field characters. Immatures more difficult to identify; careful notes should accompany records.

Black-browed Albatross *Diomedea melanophrys* 85 cm (ws 2·2 m)
Adult: only mollymawk with yellow bill, white head and broad black leading edge to underwing (broadest of all adult mollymawks), bull-necked shape (Yellow-nosed more slender). *Imm:* bill dark greyish horn with black tip, greyish collar, less white on underwing (very similar to young Grey-headed, which has uniformly dark or yellow-tipped bill). Dark-eyed form (race *melanophrys*) from southern islands more common than pale-eyed form (race *impavida*) from NZ which has more black on underwing. *Range:* common mostly May–Nov in southern waters, less common towards tropic.

Grey-headed Albatross *Diomedea chrysostoma* 90 cm (ws 2·1 m)
Adult: grey-headed mollymawk with bicoloured bill (black with yellow top and bottom), broad black underwing margins (narrower, more clear-cut than Black-browed, broader than Yellow-nosed, Buller's and Shy). *Imm:* less white in underwing, bill dark horn with yellowish tip (imm Black-browed has black tip). *Juv:* underwing dark like darkest Black-browed, uniformly dark bill. *Range:* uncommon to rare mostly May–Sep in southern waters north to tropic.

Yellow-nosed Albatross *Diomedea chlororhynchos* 80 cm (ws 2.1 m)
Small slender mollymawk with bicoloured bill (black with yellow top), *either* with head white with greyish cheeks and little or no 'eyebrow', narrow yellow 'nose' (Indian Ocean form race *bassi*, breeding in Indian Ocean but ranging well into Pacific) *or* with head and neck grey with prominent eyebrow, broader yellow 'nose' (Atlantic form *chlororhynchos*, breeding in Atlantic Ocean); narrow black underwing margins with leading edge broader (Shy has both margins narrower, same width, black patch in armpit), narrower wing-margins than Grey-headed, more slender and with more black in wingtips than Buller's. *Juv:* only mollymawk with white head and black bill. *Range:* common mostly May–Sept in southern waters north to Tropic — Atlantic form rare vagrant to south coast.

Buller's Albatross *Diomedea bulleri* 85 cm(ws 2 m)
Adult: grey-headed mollymawk with white forehead, bicoloured bill (black with broad yellow top, narrower yellow bottom), underwing with broad black leading margin like Yellow-nosed but with less dark at tip; Black-browed and Grey-headed have broader leading margin. *Juv:* similar with bill dark horn (young Grey-headed, Black-browed have much less white in underwing). *Range:* rare in south-eastern waters north to Byron Bay, NSW.

Shy Albatross *Diomedea cauta* 1 m (ws 2–2·75 m)
Adult: large mollymawk with narrow black margin to underwing, diagnostic small black patch in armpit, greenish-grey to yellow bill, white cap, greyish lores and cheeks. *Imm:* dull greyish black-tipped bill, grey head and neck. Australian form (race *cauta*) mainly white head in adult, bill greenish-grey; Grey backed form (race *salvini*) more black in wingtip, bill always with dark tip to mandible; Chatham Islands form (race *eremita*) rich yellow bill, dark grey head and neck, *Nesting:* Sep–Apr on Albatross Rock in Bass Strait; Mewstone and Pedra Branca off southern Tas; 1 white egg. *Range:* common in south-eastern waters north to tropic, less common farther west to Perth, WA.

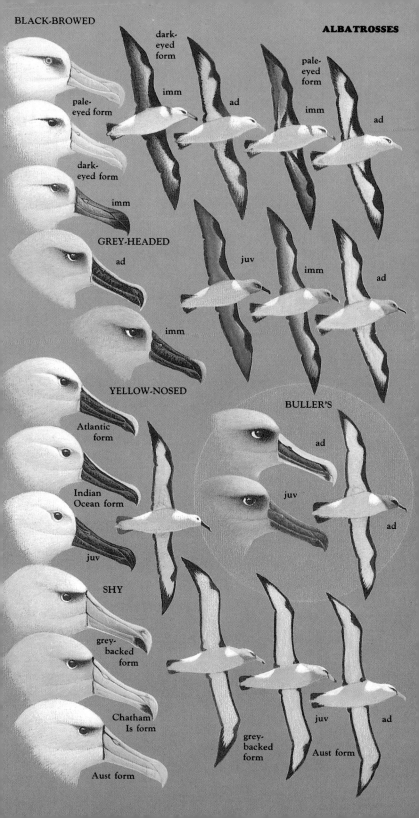

BLACK-BROWED

dark-
eyed
form

pale-
eyed
form

imm

ad

pale-
eyed form

imm

ad

dark-
eyed form

imm

GREY-HEADED

ad

juv

imm

ad

imm

YELLOW-NOSED

BULLER'S

Atlantic
form

ad

juv

Indian
Ocean form

ad

juv

SHY

grey-
backed
form

Chatham
Is form

grey-
backed
form

juv

ad

Aust form

Aust form

GREAT ALBATROSSES, SOOTY ALBATROSSES AND GIANT PETRELS

Great albatrosses: among world's largest birds, beautifully adapted for life at sea. Wandering has plumage dark brown to white depending on age; Royal lacks brown stage of plumage, less commonly sighted although it nests closer in NZ waters. **Sooty albatrosses:** graceful streamlined all-dark seabirds often difficult to tell apart at sea because of plumage wear; on beach colour of groove on lower mandible diagnostic. **Giant petrels:** Petrels are seabirds with nostrils fused into a tube on top of the bill, most exaggerated in the two giant petrels which are ungainly-looking scavengers, regularly found beached after storms. In flight flap more than albatrosses, often appear hump-backed in profile. Occasionally white form of Southern is sighted, unlike any other seabird in appearance.

Southern Giant Petrel   *Macronectes giganteus*   c. 90 cm

Large dark or all-white petrel with massive yellowish-horn bill tipped blue-green; most seen are dark evenly-coloured juveniles – adult and dark phase more mottled, with head, neck and upper breast ashy-grey. In flight, leading edge to underwing lighter than Northern; albatrosses are more elegant, and dark petrels (overleaf) are smaller, have less massive bills. *Range:* common visitor mainly June–Sep to southern waters, less common towards tropic.

Northern Giant Petrel   *Macronectes halli*   c. 90 cm

Large dark petrel with massive yellowish-horn bill tipped pinkish; most birds seen are dark juveniles – adults more mottled with whitish face, underparts more evenly grey than Southern. Rather ungainly in flight, flaps more often than albatrosses, settles often on water; leading edge to underwing darker than Southern. *Range:* uncommon visitor to southern waters north to tropic.

Sooty Albatross   *Phoebetria fusca*   c. 70 cm

Adult: all-dark albatross with bill black with pink groove, dark back may become paler with wear but not as pale as Light-mantled. *Imm:* bill all dark, hind neck and upper back pale but not lower back as in Light-mantled, shafts of primary and tail feathers dark (white in adult). Soars with elegant grace, unlike clumsy Giant Petrels. *Range:* uncommon visitor mainly June–Nov in southern waters north to Newcastle, NSW, and Perth, WA.

Light-mantled Sooty Albatross   *Phoebetria palpebrata*   c. 70 cm

Adult: dark albatross with back pale grey, bill black with blue or purple groove. *Imm:* bill all dark, back paler grey, more extensive than imm. Sooty; shafts of primary and tail feathers dark (white in adult). *Range:* very rare vagrant in south and south-eastern waters.

Royal Albatross   *Diomedea epomophora*   125 cm (ws <3·1 m)

Large albatross with yellowish bill having dark cutting edge and with black tip to tail only in young birds. Northern form (race *sanfordi*) has the upperwings black and body white at all ages (Wanderers with upperwings black have dark plumage on body), distinctive black edge to carpal area in underwing; southern form (race *epomophora*) similar in youngest birds (but lacking carpal mark), with upperwings becoming white with age from the leading edge back; old birds very similar to old Wandering, best told by bill, lack of black in tail (neither character obvious at distance). *Range:* rare in southern waters, mainly June–Nov, possibly more common well offshore.

Wandering Albatross   *Diomedea exulans*   135 cm (ws <3·25 m)

Large albatross with pink bill lacking black cutting edge and some black in sides of tail at all ages. Youngest birds are dark brown except for distinctive white face and underwings; body becomes white with age and upperwings whiten from centre outwards (Royal has body white at all ages, upperwings whiten from front to back and tail has black tip only in youngest birds). Old birds appear completely white with narrow black hind edge to wings, most have black outer tail feathers. *Range:* uncommon mostly July–Nov in southern waters north to tropic.

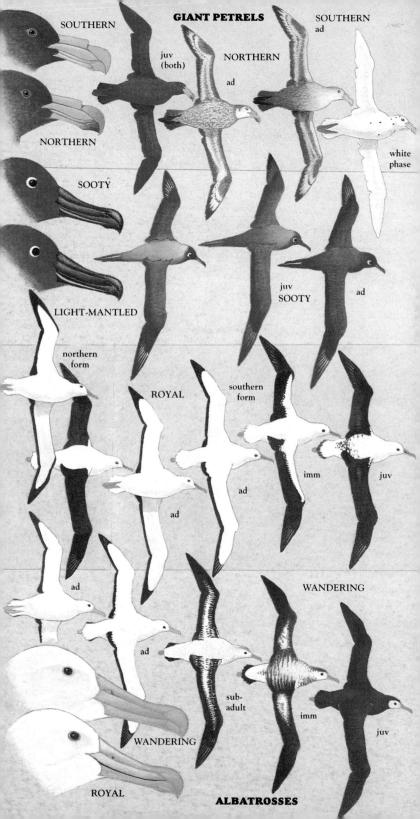

GIANT PETRELS

SOUTHERN

NORTHERN

juv
(both)

NORTHERN
ad

SOUTHERN
ad

white
phase

SOOTY

LIGHT-MANTLED

juv
SOOTY

ad

northern
form

ROYAL

southern
form

ad

imm

juv

ad

ad

WANDERING

ad

ad

sub-
adult

imm

juv

WANDERING

ROYAL

ALBATROSSES

FULMARS AND GREY PETREL

Petrels (including fulmars and shearwaters) are seabirds characterised by fusion of nostrils into tube on top of bill to improve breathing and salt excretion in marine environment. Bill is covered in plates of consistent design and has forcep-like down-drooping tip, though shape varies considerably. Long wings enable gliding for long periods in light to strong winds with little expenditure of energy; also swim and dive well. If unable to outrun storms, however, become exhausted and drift to beaches. Food mainly krill, fish and squid (prions take plankton with specialised bills). Nearly all species nest on islands, sometimes in great numbers; single egg laid in burrow, incubated for long period by one adult while other away fishing; chick fed at often-considerable intervals, and may be abandoned on fledging to learn to fish and to migrate for thousands of miles. Many species recorded only rarely in Aust. Apart from Cape Petrel, those on this page are seldom sighted; all nest on Antarctic continent and subantarctic islands.

Antarctic (Southern) Fulmar *Fulmarus glacialoides* 45 cm
Heavy-set gull-like fulmar with prominent windows in wings and dark hind edge to upperwings. Flies stiffly with alternate flapping and short glides on bowed wings, ignores ships, attracted to fishing boats. Battered beachdrift easily dismissed as Silver Gull. *Range:* normally rare in southern waters north to Ballina, NSW, and Perth, WA, but may periodically irrupt.

Antarctic Petrel *Thalassoica antarctica* 43 cm
Brown and white fulmar with broad white stripe in wing; plumage fades from dark chocolate to pale brown; typical fulmar flight, well above water. *Range:* very rare vagrant to southern waters.

Cape Petrel *Daption capense* 40 cm
Boldly marked fulmar, chequered black and white on upper surface; black plumage fades to dark brown between moults. Often swims in water in flocks around fishing boats. Dark-backed form (race *australe*) from islands near NZ has less chequering on back; chequered form (race *capense*) has more white on upperparts. *Range:* common visitor June–Dec to southern waters, less common towards tropic.

Snow Petrel *Pagodroma nivea* 32 cm
Unmistakable small to medium white fulmar with black bill, possibility of confusion with albinos of other seabirds should not be overlooked; fulmarine flight should distinguish it from albino prions and gadfly petrels. *Range:* extremely rare accidental to southern waters.

Grey Petrel *Adamastor cinereus* 48 cm
Large shearwaterlike petrel with upper surface grey or brown (depending on condition of plumage), pale below with dark underwing; bill distinctive in colour, green on sides with yellowish tip. Follows ships, flies with rapid wingbeats interspersed with long glides, plumps into water from some height when fishing. *Range:* rare vagrant in winter to south coast from south-west to Tas.

Barau's Petrel *Pterodroma baraui* 38 cm (ws c. 80 cm)
Medium-sized Indian Ocean gadfly petrel with white forehead, dark crown and lower back, brownish grey above with pale diagonal 'W' across upperwing, white below with dark diagonal bar in underwing. Not unlike a large Gould's Petrel (p 28) which, with other similar gadfly petrels, is unlikely in Indian Ocean. Little-known recent discovery, breeds on Reunion I., western Indian Ocean, may disperse south and east to about 20°S. Recorded in north-western Aust waters, possible anywhere off west coast.

BARAU'S PETREL

PETRELS

ANTARCTIC (SOUTHERN) FULMAR

ANTARCTIC PETREL

CAPE

SNOW

GREY

ALL-DARK PETRELS

Some petrels have black or blackish-brown plumage, easily confused with dark shearwaters (see p 32). **Black petrels:** three similar large black petrels wintering near Australia have pale yellowish bills, heavy bodies, wings appearing double-jointed, loose flapping flight interspersed with gliding, regularly follow ships. White-chinned is common offshore but rarely drifts to beaches. Westland and Black very rarely seen. **Gadfly petrels:** three smaller all-dark gadfly petrels rarely if ever follow ships. All have black bills. Magnificent fliers, with long swept-back swiftlike wings; Great-winged, with dashing wheeling arcs in strong winds, is most exciting bird to watch. Providence has white patches in wing and is not as dark – dark phases of Kermadec and Herald (p 26) are similar. Kerguelen smaller, paler and more often seen flying at some height with rapid wingbeats or 'hovering' well up on motionless wings. See also small Bulwer's Petrel (p 38).

White-chinned Petrel *Procellaria aequinoctialis* 50 cm

Large black petrel with pale yellow bill and variable white on chin, very extensive on rare 'Ring-eyed' form from south Atlantic, virtually none in some NZ birds. Great-winged Petrel smaller, may have pale face particularly to east and has dark bill, different flight. In hand, Westland has bill yellowish-horn with dark tip, 47–51 mm; Black smaller, bill yellowish-horn with less extensive dark tip, 39–44 mm; White-chinned yellow bill, 48–56 mm. *Range:* common away from southern shore, often following ships, occasionally beached after storms.

Black Petrel *Procellaria parkinsoni* 46 cm
Black petrel with pale yellowish, dark-tipped bill. Similar to larger White-chinned, some of which lack white chin, but distinguished by black tip to bill; not reliably separated from Westland at sea, perhaps good view may show more slender bill with less black at tip. *Range:* very rare vagrant, recorded as beachdrifts on south and east coast.

Westland Petrel *Procellaria westlandica* 51 cm
Black petrel with pale yellowish, dark-tipped bill. Similar to White-chinned, lacking white chin. Black is smaller, has slightly less black on bill-tip. Not reliably identified at sea from Black, but White-chinned specimens that lack pale chin told by lack of dark tip to bill. *Range:* very rare vagrant recorded as beachdrifts on south and east coast.

Kerguelen Petrel *Pterodroma brevirostris* 35 cm
Small rather stout dark-grey petrel with large head, small bill and whitish inner leading edge to underwing. Head often appears darker than body. Plumage has reflective quality and may look dark or pale according to light conditions. Has erratic zippy flight interspersed with glides, often flies higher than other petrels. Similar to dark-phase Soft-plumaged (p 26) which lacks whitish leading edge to wing, is browner, usually shows a darker 'breast band', flies more sedately. *Range:* uncommon visitor to southern waters, most likely June–Oct, has occasional irruptions.

Providence Petrel *Pterodroma solandri* 40 cm

Heavy all-dark petrel with whitish face and white patches on underwings. Head darker than belly, upperwing greyish with dark 'M' mark; similar dark Kermadec has white shafts to primaries and dark Herald has diagonal black stripe in underwing. Breeds on Lord Howe I in large numbers; extinct on Norfolk I. *Range:* uncommon visitor to central east coast, mainly from Apr–Oct, usually seen well offshore, less common farther north and south.

Great-winged Petrel *Pterodroma macroptera* 40 cm

Heavy all-dark petrel with long scythe-like wings, dark heavy bill and short wedged tail. Dashing flight with higher arcs in heavy weather distinguish it from shearwaters; in light winds carpals held well forward, more similar to some shearwaters; Black Petrel much heavier with slower flight. Australian form (race *macroptera*) has dark face, most likely to occur about south-west corner, and NZ form (race *gouldi*) with grey face, most likely to occur about south-east. *Voice:* 'ooi'; 'wik, wik, wik' heard at breeding islands. *Nesting:* in burrow in winter on islands of Recherche Archipelago west to Albany, WA; 1 white egg. *Range:* common in southern Australian waters, beachdrifts most likely Oct–Apr, sometimes becomes grounded well inland.

PETRELS

WESTLAND

WHITE-CHINNED

BLACK

KERGUELEN

PROVIDENCE

GREAT-WINGED

NZ form

Aust form

Great-winged 65 x 45

GADFLY PETRELS WITH MAINLY DARK UNDERWINGS

These are medium-sized petrels with some areas of white in plumage, but with underwings mainly dark, although a consistent feature is a pale patch at base of primaries. Some are very difficult to identify as they have plumage phases ranging from light to dark, but of these, Kermadec can be told in all plumages by white shafts to primaries in upperwing, visible in flight from considerable distance.

Tahiti Petrel   *Pterodroma rostrata*   38–40 cm
Solid-looking dark-brown gadfly petrel with dark head and breast sharply cut off from white belly and undertail coverts, pale bar in underwing. Most likely in north-eastern waters. Atlantic Petrel similar but has dark undertail coverts, lacks any pale marking on underwing, has southern range. *Range:* rare visitor from nearest breeding grounds in New Caledonia to north-east coast south to about Hasting R, NSW.

Atlantic Petrel   *Pterodroma incerta*   43 cm
Solid-looking dark-brown gadfly petrel with dark head and breast, white belly, and with no white bar in underwing, most likely in south-western waters. Follows ships. *Range:* extremely rare accidental from south Atlantic to south-west.

Soft-plumaged Petrel   *Pterodroma mollis*   32–37 cm
Grey-backed gadfly petrel with wings dark above and below, grey breast band and dark patch around eye. Likely to occur in large flocks unusual in gadfly petrels except during migration. Similar phases of Kermadec and Herald have more white in underwing and are browner above; Kermadec has white shafts to primaries. Rare dark phase rather like Kerguelen but browner, has dark leading edge to underwing and flight less wild with lower arcs. *Range:* regular visitor mainly June–Sep to southern waters, particularly south-west, but apparently not on east coast.

Herald Petrel   *Pterodroma arminjoniana*   35–40 cm
Variable gadfly petrel of north-east with light, dark and intermediate phases: all have underwing similar, dark with white patch on base of primaries and with dark shafts to primaries on upperwing. *Light phase* similar to (a) Soft-plumaged but browner, darker above, lacking distinct eye-patch, more extensive white in underwing, different range; (b) Kermadec intermediate phase has darkest part of underwing on trailing edge, white-shafts to primaries on upperwing. *Dark phase* similar to (a) Providence which lacks white basal leading edge to wing, has back greyer than head (uniform in Herald); (b) dark shearwaters have different shape, less-impetuous flight, lack white wing patches. *Intermediates* have variable amounts of dark plumage on underparts. *Nesting:* in burrows on Raine I off Cape York, Qld, extended breeding period; 1 white egg. *Range:* rare in waters of northern Barrier Reef, possibly offshore farther south.

Kermadec Petrel   *Pterodroma neglecta*   38 cm
Variable gadfly petrel with light, dark and intermediate phases: all have underwing similar, dark with white shafts to primaries in upperwing. Light phase similar to White-headed but without dark eye-patch and whitish tail, back darker. Intermediate and dark phases told from similar shearwaters and gadfly petrels by white shafts to primaries in upperwing. *Range:* rare vagrant to central east coast possibly from breeding areas of Lord Howe I or Kermadec I, NZ.

White-headed Petrel   *Pterodroma lessonii*   40 cm
Only petrel with white head and white tail and with dark underwings. Soft-plumaged has grey breast band and darker head, Grey has dark head and tail and undertail coverts. *Range:* rare to uncommon mainly June–Sep in southern waters; likely farther north in west than in east; beachdrifts mainly June.

GADFLY PETRELS

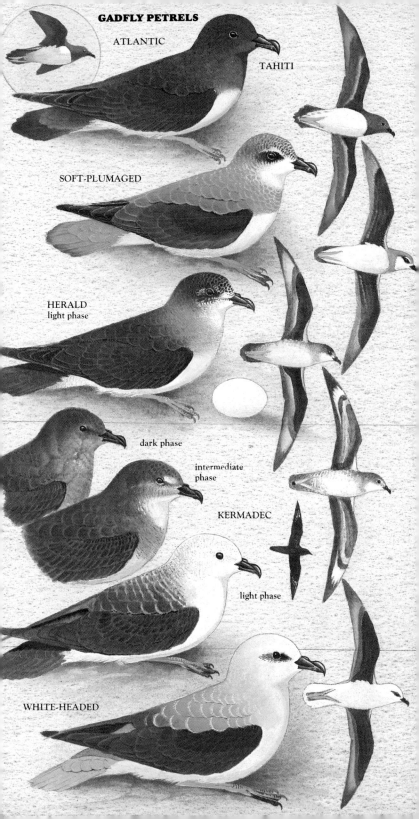

ATLANTIC

TAHITI

SOFT-PLUMAGED

HERALD
light phase

dark phase

intermediate
phase

KERMADEC

light phase

WHITE-HEADED

GADFLY PETRELS WITH LARGELY WHITE UNDERWINGS

Mainly small, fast-flying gadfly petrels with dark 'W' patterns on upperwings and white underwings with black diagonal stripe (much reduced in Cook's). Three small 'cookilaria' petrels (Cook's, Gould's and Black-winged) are similar; look for amount of black on head and underwing pattern. Similar species may occur as vagrants (Bonin, Pyecroft's and Stejneger's). Blue Petrel more like prions, but has unique white-tipped tail. White-necked, and Mottled, are larger; each has unique feature to aid identification.

White-necked Petrel   *Pterodroma externa*   43 cm (ws 95 cm)
Large gadfly petrel like much smaller Gould's but with distinctive white nape and less black on underwing. Flight less erratic than smaller gadflies, more majestic with higher arcs. *Range:* extremely rare vagrant off east coast. (Watch for Juan Fernandez Petrel *Pterodroma externa cervicalis* which is similar but has dark nape, looks like Buller's Shearwater (p 32) which lacks narrow diagonal bar on underwing.)

Mottled Petrel   *Pterodroma inexpectata*   33–35 cm (ws 75 cm)
Dark-bellied gadfly petrel with white throat and undertail coverts, and broad black diagonal stripe on underwing. Typical exciting gadfly flight with high bounding arcs. Some examples of smaller Collared form of Gould's have similar dark belly, confined to Coral Sea, while Mottled mainly sighted south of tropic, although it migrates annually to northern Pacific from breeding grounds in NZ. *Range:* very rare vagrant to south-east and Tas, beachdrifts most likely Nov–Apr, probably young birds back from northern migration but too young to breed.

Blue Petrel   *Halobaena caerulea*   28–30 cm (ws 60–66 cm)
Small dark-headed blue-grey petrel with unique white-tipped tail. Apart from tail, differs from prions (p 30) in darker head, more diffuse 'W' pattern on upperwings; differs from gadflies in white tail tip, paler upperparts, lack of black in underwing, less erratic flight, usually flying with stiff-winged 'flutter and glide' close to water in loose flocks, sometimes with prions. *Range:* uncommon visitor, mostly June–Oct, to south-eastern and southern waters particularly south-west, probably from Kerguelen I.

Cook's Petrel   *Pterodroma cooki*   26 cm (ws 65 cm)
Small gadfly petrel with grey head, back and tail, with black about eye and largely white underwing with narrow black diagonal stripe not often visible. Gould's has black head, thick black diagonal stripe on underwing; Black-winged has more black generally in underwing, more contrast between grey back and tail. Fast erratic 'gadfly' flight with towering arcs. *Range:* rare visitor Dec–Jan to mid east coast, probably from closest breeding island, Little Barrier I, NZ.

Black-winged Petrel   *Pterodroma nigripennis*   30 cm (ws 65–70 cm)
Small gadfly petrel with grey head, black about eye, and broad, black diagonal stripe on underwing. Gould's has black head; Cook's has largely white underwing, has more uniformly grey head, back and tail. Flight fast and erratic with high arcs. *Range:* apparently rare visitor to mid east coast; recently colonised Norfolk I, may be in the early stages of colonising southern Barrier Reef cays such as Heron I.

Gould's Petrel   *Pterodroma leucoptera*   30 cm (ws 70 cm)
Small gadfly petrel with black head, nape and semi-collar, and broad black diagonal stripe on underwing. Cook's and Black-winged have grey heads; much larger White-necked has white nape. Flight more sedate, closer to water than other gadflies, with occasional bursts of rapid banking and twisting. Collared form (race *brevipes*) nests in Fiji, likely to be seen off Barrier Reef, has pronounced collar; some have belly dark like larger Mottled Petrel. *Nesting:* Nov–Apr under rocks and vegetation on Cabbage Tree I, NSW, 1 white egg. *Range:* uncommon summer migrant Oct–May to central east coast, beachdrifts mostly Feb–Apr.

GADFLY PETRELS

WHITE-NECKED

MOTTLED

BLUE

COOK'S

BLACK-WINGED

GOULD'S

50 x 37

Gould's

PRIONS

Prions are small pale grey petrels with darker 'W' pattern across the upperwing and dark tip to the tail; fly swiftly and erratically, often in flocks, feeding on plankton while 'hydroplaning' across water. The various forms of prion are similar in appearance, but in hand, size and shape of bill are helpful. Features to look for are: (a) whether lamellae (hairlike fringes to help strain plankton from water) are visible when bill is closed, (b) length and width of bill, (c) whether edges of bill when viewed from above are bowed or straight. Different authorities regard prions as anything from one to six species; difficulty of drawing a line between some forms suggests that two species may be best compromise until detailed field studies settle the problem. For convenience, the old name of 'whalebird' is resurrected here to enable better-known names for various forms to be used for those who prefer to regard them as species. Blue Petrel (p 28) , somewhat similar in appearance to prions, is easily identified by its white-tipped tail.

Whalebird *Pachyptila vittata* 25–30 cm

Variable prion with head darker than back and narrow dark tip to tail. Some forms are doubtfully identifiable at sea and sometimes with difficulty in hand: Broad-billed form (race *vittata*) has lamellae visible when bill closed, bill blackish with sides strongly bowed, 30–38 mm long and 19–24 mm wide, forehead and eye-stripe darker than back; Medium-billed form (race *salvini*) similar but with less strongly bowed sides to bill, which averages smaller, 28–33 mm long, 13–19 mm wide; Antarctic forms (races *desolata, banksi, alta*) bill blue, lack visible lamellae when closed, edges straight; dark eye-stripe extends on to breast giving effect of half-collar, making it easiest prion to identify at sea; Thin-billed form (race *belcheri*) has small blue bill with straight edges and lacking visible lamellae; broad white eyebrow good field mark. *Range:* rare to common visitor to southern seas north to tropic depending on race, with most sightings in winter and early spring; beachdrifts most likely June–Aug; Antarctic commonest, followed by Medium-billed and Thin-billed, with Broad-billed rare.

Prion *Pachyptila turtur* 25 cm

Pale-headed prion with broad dark tip to tail. Common Australian form (usually called Fairy Prion, race *turtur*) has finer, less stout bill; and extremely rare Fulmar-like form from NZ (race *crassirostris*) has heavier bill, rather bluer back with more heavily marked 'W' on upper surface. *Nesting:* Nov–Feb in burrow on islands of western and central Bass Strait near Vic and Tas from Moncoeur I to Black Pyramid; 1 white egg. *Range:* Australian form common in south-eastern waters, less common further north and in west, where birds come from Marion I in Indian Ocean. Fulmar-like form breeds in NZ, extremely rare vagrant to south-east.

Note: Pink-footed Shearwater *Puffinus creatopus*, sighted at sea off Wollongong NSW in Mar 1986, is similar to pale phase Wedge-tailed Shearwater (see overleaf) but is larger and bulkier with a prominent pink bill. The underwing pattern is similar to the pale Wedge-tailed Shearwater (see p. 35) but the wings are broader and flight is heavier. It normally occurs in the eastern Pacific. Cory's Shearwater (*Calonectris diomedea*) from the Atlantic Ocean often winters in south-western Indian Ocean, straggling as far as New Zealand. It is like the Streaked Shearwater (overleaf and p. 35) but with a plain, diffuse, unstreaked cap. Similar to Grey Petrel but with largely white underwings and yellow rather than greenish bill.

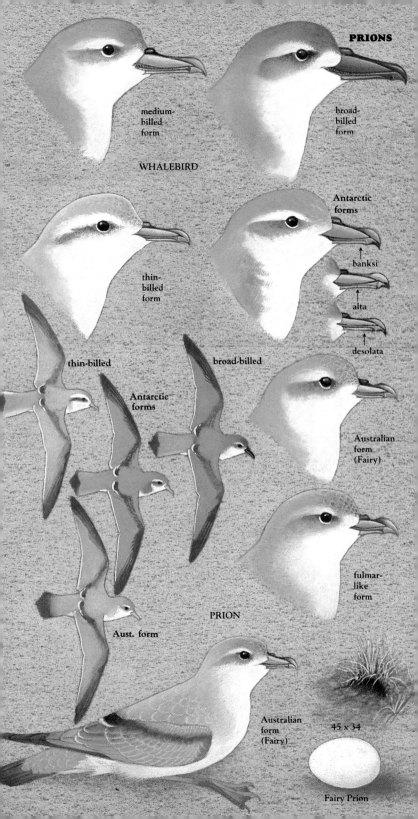

PRIONS

medium-billed form

broad-billed form

WHALEBIRD

thin-billed form

Antarctic forms

banksi

alta

desolata

thin-billed

broad-billed

Antarctic forms

Australian form (Fairy)

fulmar-like form

PRION

Aust. form

Australian form (Fairy)

45 x 34

Fairy Prion

32

LARGE SHEARWATERS

Shearwaters are distinguished from other petrels by more elongated appearance, finer bill with less obvious nostrils, and by flight, banking or flying into the wind to gain height then gliding down or across wind with one wing almost touching water to pick up speed for next bank. At other times flight direct with few flaps interspersed with glide. Often gather in large flocks at school of small fish or krill, swimming and diving, attracting other seabirds. When breeding large numbers congregate and may cover whole islands with burrows. Breeding islands noisy at night with constant cacophony of wailing cries but appear deserted by day. At sea usually silent. Each pair lays single egg. Most migrate after breeding, usually to north Pacific.

Wedge-tailed Shearwater *Puffinus pacificus* 41–46 cm (ws 1 m)

All-dark or white-bellied shearwater with dark grey bill, pink feet and wedge-shaped tail. Most are all dark, more slender than other dark shearwaters and petrels, and rather more buoyant in flight, gliding up into wind after leisurely flaps before banking down across wave; wedge-shaped tail usually folded in flight, only obvious while banking. Pale phase has white underparts and underwing coverts, occurs as breeding population Shark's Bay area, WA. *Voice:* mournful wailing 'ka-wooo-er'. *Nesting:* in burrow from Nov–May, on islands of west coast from Sable in Forrestier Group (between Roebourne and Port Hedland) to Carnac and on east coast between Raine and Montague; 1 white egg. *Range:* common off east and west coasts but rare on south coast.

Fleshy-footed Shearwater *Puffinus carneipes* 41–45 cm (ws 1 m)

All-dark heavy shearwater with pale horn black-tipped bill, pink feet and fanshaped tail; no other all-dark seabird in our area has pale feet *and* bill. Flies on stiff wings held straighter than Wedge-tailed with deeper flaps between glides. *Voice:* hoarse 'kuk-kuk-kuk – koooo-wah'. *Nesting:* in burrows on islands of south-west from Cape Leeuwin to Recherche; 1 white egg; breeding station closest to east coast is Lord Howe I. Migrates north in winter. *Range:* common in south-western and mid-eastern waters Sep–May, less common in remainder of southern and south-eastern waters.

Buller's Shearwater *Puffinus bulleri* 45 cm (ws 95–100 cm)

White-bellied shearwater with black cap and grey back, dark 'W' across wings. No other large petrel has this pattern; smaller petrels with this pattern are 'gadflies' with fast dashing erratic flight. *Range:* rare visitor to south-eastern waters from NZ, mainly Oct–June.

Streaked Shearwater *Calonectris leucomelas* 48 cm (ws 110 cm)

White-bellied shearwater with grey-brown upperparts streaked paler, particularly about head. Looks long and slender with large wings, rather pale-headed with white underwing coverts. *Range:* uncommon migrant from northern Pacific to north-east, more common off north coast.

Short-tailed Shearwater *Puffinus tenuirostris* 41–43 cm (ws 1 m)

All-dark shearwater with dark wing linings, dark slender bill and short rounded tail, often in great flocks. Sooty is similar, has longer bill, and most have paler wing coverts; Fleshy-footed has pale bill and feet; Wedge-tailed more slender appearance, larger wing area with bowed wings, wedged tail and pink feet; Great-winged Petrel solitary with dashing wheeling flight, heavier shorter bill with head tucked in wing. Other dark petrels have heavier pale bills or white windows in wing. *Voice:* wailing, crooning and sobbing. *Nesting:* Nov–early May in burrows in great number on islands of Bass Strait and off south-east coast; 1 white egg. Migrates northwards along east coast to northern Pacific. *Range:* common about south-east and Tas Sep–May; occasional 'wrecks' of beachdrifts on east coast, usually of migrant birds returning Oct–Dec.

Sooty Shearwater *Puffinus griseus* 40–46 cm (ws 95–105 cm)

Dark shearwater with whitish wing coverts, long thin dark bill, dark legs and short rounded tail, often in large flocks. Pale wing linings distinguish it from other dark shearwaters, cigar-shape with extended head and slim bill distinguish it from dark petrels. Some individuals have dark underwings. Beachdrifts separated from Short-tailed by bill length: Sooty 41–44 mm; Short-tailed 29–34 mm. *Voice:* alternately inhaled and exhaled 'koo-wah-koo-wah-koo-wah . . .'. *Nesting:* Nov–early May in burrows on island off NSW coast and southern Tas; 1 white egg; migrates after breeding into northern Pacific. *Range:* common about south-east and Tas Sep–May, less common farther west, rare WA.

SHEARWATERS

WEDGE-TAILED

pale phase

dark phase

FLESHY-FOOTED

STREAKED

BULLER'S

SOOTY

SHORT-TAILED

69 x 46
Fleshy-footed

Short-tailed 71 x 47   Sooty 77 x 48   Wedge-tailed 63 x 41

SMALL SHEARWATERS

Distinguished from small petrels by long thin bills, rather cigar-like shape, direct flight of several rapid flaps interspersed with short glide, habits of swimming and diving, often in flocks. Look for amount of white on undertail coverts and underwings and whether dark plumage on head extends below eye.

Manx Shearwater *Puffinus puffinus* 30–38 cm (ws 80–90 cm)
Small shearwater with black plumage on head extending below eye and white undertail coverts. Alternates fast wingbeats with twisting glides; Little has white face, more rapid wingbeats alternating with short flat glides close to water; Fluttering and Hutton's are browner above and darker on underwing; Audubon's has dark undertail coverts, broader dark edges to underwing. *Range:* extremely rare vagrant from Atlantic likely on southern beaches only after exceptional weather.

Hutton's Shearwater *Puffinus huttoni* 38 cm (ws 90 cm)
Small shearwater with brownish-black plumage on head extending well below eye and with largely dusky underwings. Broader-winged than Little with slower wingbeats, more banking during glides. Fluttering browner above, particularly late in year, with sharper demarcation between light and dark on side of neck, extends up east coast rather than towards Bight. *Range:* regular visitor from NZ to south-east and southern waters from southern Qld to about Perth, WA, most common near SA.

Fluttering Shearwater *Puffinus gavia* 31–36 cm (ws 75 cm)
Small shearwater with blackish-brown plumage on head extending below eye, white undertail coverts and white underwing coverts and grey axillaries. Upperparts fade to rusty brown by end of year. In flight, broader-winged than Little, with slower deeper wingbeats, often banking during glides. Manx and Little blacker above, both with white axillaries, latter with whiter face; Hutton's has darker underwing coverts and darker-headed appearance, does not fade to rusty brown with wear, commoner in more southerly waters. *Range:* regular visitor from NZ mostly immatures, most common Apr–Sep in Qld, July–Nov in NSW, with beachdrifts most likely Jan–Feb.

Audubon's Shearwater *Puffinus lherminieri* 30 cm (ws 70 cm)
Small shearwater with black plumage on head extending below eye and with dusky undertail coverts giving it long-tailed look. Rapid wingbeats (slower than Little but faster than Fluttering) interspersed with short glides. *Range:* recorded only rarely, but should occur in Coral Sea and outer Barrier Reef as it breeds in New Hebrides, and occasionally strays farther south.

Little Shearwater *Puffinus assimilis* 25–30 cm (ws 60–70 cm)
Small shearwater with eye clear of black plumage on head and with white undertail coverts. Four or five rapid fluttering wingbeats interspersed with short glides close to water with little banking in low winds (other small shearwaters bank during glides on longer, less-rounded wings). Back blacker than Fluttering and Hutton's, underwing cleaner. *Voice:* high cackle 'wah-ee-wah-ee-wah-oooo'. *Nesting:* June–Oct in burrows or under overhanging rocks on islands of south-west; 1 white egg. *Range:* common in southern waters particularly about south-west, less common eastward to southern Qld; beachdrifts most likely Mar–July and Nov.

SHEARWATERS

MANX

HUTTON'S

FLUTTERING

AUDUBON'S

LITTLE

52 x 36

Little

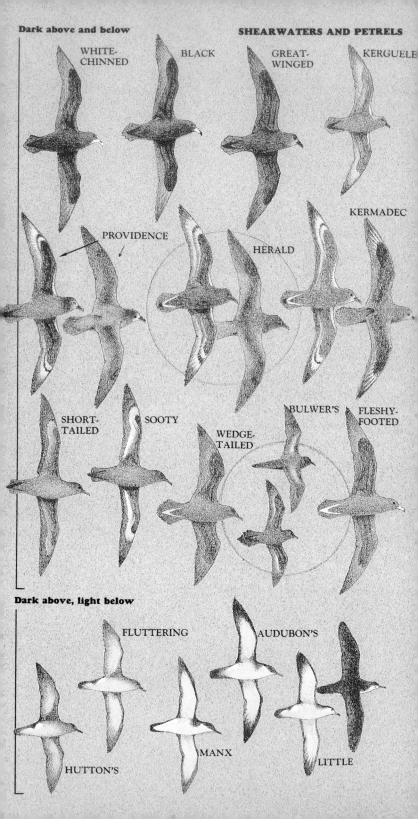

Dark above and below **SHEARWATERS AND PETRELS**

WHITE-CHINNED   BLACK   GREAT-WINGED   KERGUELE

PROVIDENCE   HERALD   KERMADEC

SHORT-TAILED   SOOTY   WEDGE-TAILED   BULWER'S   FLESHY-FOOTED

Dark above, light below

FLUTTERING   AUDUBON'S

HUTTON'S   MANX   LITTLE

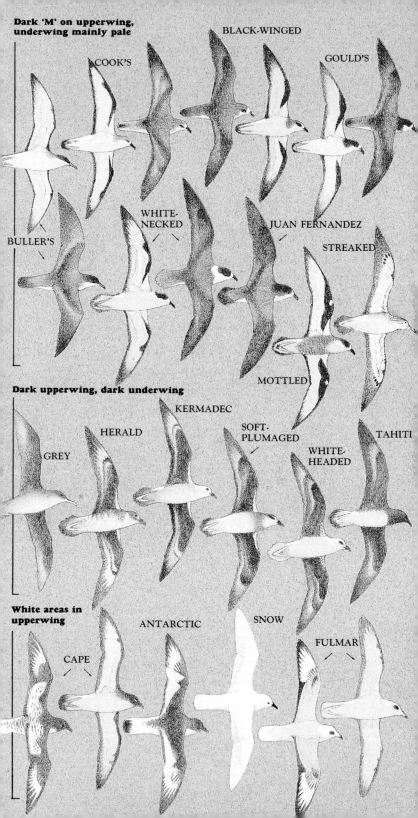

Dark 'M' on upperwing, underwing mainly pale

BLACK-WINGED

COOK'S

GOULD'S

BULLER'S

WHITE-NECKED

JUAN FERNANDEZ

STREAKED

MOTTLED

Dark upperwing, dark underwing

KERMADEC

HERALD

SOFT-PLUMAGED

TAHITI

GREY

WHITE-HEADED

White areas in upperwing

ANTARCTIC

SNOW

CAPE

FULMAR

DARK-BELLIED STORM PETRELS AND BULWER'S PETREL

Storm petrels: smallest of seabirds, often difficult to identify because of size, similarity of many forms and lack of opportunity for good views. Methods of flying and feeding important aids to identification. Species from northern hemisphere (Leach's, Matsudaira's) have longer, more pointed wings, forked tails (which are usually held closed except while banking) and shorter legs not protruding beyond tail; both are rarely sighted near Australia. Southern species have rounded wings, longer legs, more fluttering flight. Some are very common; Wilson's one of world's most abundant birds, feeds close to shore as well as far out like most others. Although looking so fragile, storm petrels seldom driven onshore after high winds like so many other seabirds, possibly avoid areas of low barometric pressure at sea. Nest in burrows on islands, coming ashore only at night; one egg is laid. **Bulwer's Petrel:** small black petrel rather like a large storm petrel, but with longer legs and tail, prionlike flight.

Wilson's Storm Petrel *Oceanites oceanicus* 15–19 cm (ws 38–42 cm)

Dark storm petrel with rounded wings, grey upperwing coverts, white rump extending to outer undertail coverts, squarish tail, long legs extending beyond tail in flight, with yellow webs (rarely visible at sea). Follows ships and boats, feeds in wake, pattering on water with wings held high; flight direct with fast wingbeats, swallowlike, often higher than other storm petrels. Leach's has dusky mid line dividing white rump (not always visible but making rump duller than Wilson's), forked tail, shorter legs, longer pointed wings with more darting erratic flight; dark phase White-bellied lacks yellow webs, lacks pale upperwing coverts and feeds differently.

Leach's Storm Petrel *Oceanodroma leucorhoa* 19–22 cm (ws 45–48 cm)
Dark storm petrel with pointed wings, grey upperwing coverts, white rump with dusky midline, short legs not extending beyond tail, forked tail (not always obvious). Longer winged and shorter legged than Wilson's with more darting erratic flight, slower deeper wingbeats; seldom follows ship; feeds with wings held just above horizontal while pattering on surface. Some examples lack white rump, look rather like Matsudaira's but smaller, narrower based wings, faster more darting flight, lack white base to primaries. *Range:* extremely rare accidental from N hemisphere, recorded once.

Matsudaira's Storm Petrel *Oceanodroma matsudairae* 24–25 cm (ws 56 cm)
Large dark storm petrel with pointed wings, pale brownish upperwing coverts, pale patch in outerwing caused by white shafts of primaries, and forked tail. Flies with slower wingbeats than other storm petrels, more direct leisurely progression; when feeding flutters above water with wings above horizontal; follows ships. (Watch for similar Swinhoe's Storm Petrel O. *monorhis* which has similar migration pattern – narrower wings at base, faster wingbeats, white shafts to primaries not visible at sea.) *Range:* migrates from Japan to Indian Ocean probably passing between Timor and Aust; probably common on passage off north-west coast.

White-bellied Storm Petrel *Fregetta grallaria* 20 cm (ws 45 cm)

Dark phase: Dark storm petrel with rounded wings, no greyish upperwing coverts but whitish patch on underwing coverts, variable dusky white rump. Flight wildly erratic with legs dangling, following contour of waves, often hits water with breast like bathing swallow, feeds swinging from side to side with wings horizontal and toes touching water; follows ships. *Range:* rare dark phase represents only small percentage of birds breeding on Lord Howe, likely to be dismissed as Wilson's unless observed closely; likely well offshore in south-eastern waters between Lord Howe and Tas (see overleaf for light phase).

Bulwer's Petrel *Bulweria bulwerii* 26–27 cm (ws 65–70 cm)

Small dark petrel with pale upperwing coverts and long wedged tail usually held closed but visible while banking. Flies close to water with rapid wingbeats alternating with twisting glides. Rarely follows ships. *Range:* mainly tropical, breeding in Pacific, migrating north of Aust into Indian Ocean, most likely off north-west coast on passage (Nov), but only record south of Cape Leeuwin, WA, Sep 1910, may have been accidental from Atlantic.

STORM PETRELS

WILSON'S

LEACH'S

MATSUDAIRA'S

WHITE-BELLIED
dark phase

BULWER'S

STORM PETRELS WITH WHITE BELLIES AND DIVING PETRELS

Storm petrels: these species have at least some white on their underparts. **Diving petrels:** small dumpy auklike petrels, most often seen swimming and diving, or taking short flights on rapidly whirring wings. They are greyish-black above shading to white below, with dusky underwings. It is not possible to distinguish the various species in the field, as diagnostic characters are shape of nostrils and pouch under bill.

Black-bellied Storm Petrel   *Fregatta tropica*   20 cm (ws 45 cm)

Dark-headed storm petrel with whitish chin, rounded wings and with rump, underwing coverts and sides of belly white; centre of belly black joining black breast and undertail coverts; legs long with toes projecting well beyond tail. Flight wildly erratic, hugging contour of waves, often hitting water with breast; feeds on horizontal wings swinging from side to side with toes touching water, White-bellied similar but chin black and back paler, legs shorter, projecting less beyond tail; in hand, toes are shorter and nails rounded; Grey-backed has paler back with no distinct white and more white on underwings. *Range:* rarely sighted near to shore, more common well offshore in southern waters, sightings north to Ballina, NSW. Breeding islands more southerly than White-bellied.

White-bellied Storm Petrel   *Fregetta grallaria*   20 cm (ws 45 cm)

Dark-headed storm petrel with black chin, rounded wings and white rump, underwing coverts and belly; legs shorter than Black-bellied, just projecting beyond tail in flight; feathers of back and upperwing coverts edged paler, giving greyish cast to upperparts. Small proportion of breeding birds on Lord Howe are all-dark (see p 38), and some show dark mottling on white belly. Flight similar to Black-bellied (see above); Black-bellied has darker upperparts, paler chin, dark line on centre of belly (difficult to sight at sea). Grey-backed has greyer back without distinct white rump and with more white on underwing, otherwise similar. *Range:* rarely sighted close to shore, more common offshore in south-eastern waters mainly between Lord Howe I and Tas.

Grey-backed Storm Petrel   *Oceanites nereis*   16–19 cm (ws 39 cm)

Dark-headed grey storm petrel with rounded wings, white belly and underwing; short square tail grey with broad black tip, legs black with dark webs, extending beyond tail in flight. Flight rapid and direct close to water, difficult to sight between waves; feeds skipping from side to side with wings horizontal; associates with White-faced which lacks dark head, larger with browner back. *Range:* reasonably common in south-eastern waters well offshore mainly Mar–Sep, rarely beachdrifts.

White-faced Storm Petrel   *Pelagodroma marina*   20 cm (ws 42 cm)

Greyish-brown storm petrel with rounded wings, white face and underparts; rump grey, tail square, legs long with yellow webs to toes projecting beyond tail in flight. Flight swift and erratic, often with jerky wingbeats; feeds while swinging from side to side, touching down between glides, legs not dangled except when pattering in strong winds. Sometimes associates with Grey-backed, which is greyer, has dark head and breast. *Voice:* scratchy 'tiu-tiu-tiu- . . . '. *Nesting:* in burrows Oct–Feb on islands of southern coast from Abrolhos in west to Broughton in east; 1 white egg. Migrates northwards after breeding. *Range:* common in southern waters May–Feb, probably seen less often than Wilson's as it does not follow ships and feeds further offshore; beachdrifts mainly Oct–Feb.

Common Diving Petrel   *Pelecanoides urinatrix*   20–25 cm (ws 35 cm)

Diving petrel with septal process at back of nostril opening and with area under bill, from which pouch unfolds, rounded. Distinguished in hand only from Georgian, which has septal process in middle of nostril opening and pointed pouch area. *Voice:* soft 'kooo-waka'. *Nesting:* in burrow mainly July–Jan on islands in Bass Strait and southern Tas; 1 white egg. *Range:* common in Bass Strait, rare elsewhere in south-east north to Stradbroke I, Qld; only record in Western Australia is race *exsul* (Kerguelen Diving Petrel) from subantarctic islands, which is larger, having broader bill with bowed edges and grey breast band.

Georgian Diving Petrel   *Pelecanoides georgicus*   18–21 cm (ws 30–33 cm)

Diving petrel with septal process in middle of nostril opening and with area under bill, from which pouch unfolds, pointed. A very rare accidental in Aust waters, recorded once, Bullambi Beach, NSW.

STORM PETRELS

BLACK-BELLIED

WHITE-BELLIED

GREY-BACKED

WHITE-FACED

DIVING PETRELS

COMMON

GEORGIAN

Common 38 x 24   White-faced 36 x 27

GANNETS AND BOOBIES

Large unmistakable seabirds, cigar-shaped with long narrow wings, long tails and pointed bills. Dive headlong into sea for food; nostrils not visible externally; feet have all four toes joined by webbing. Most nest on ground, in colonies on islands, but Red-footed builds in mangroves and bushes. Immatures generally are mottled brown.

Australian Gannet   *Morus serrator*   85–95 cm (ws 150–180 cm)

White gannet with golden head; black primaries, secondaries and central tail feathers; outer tail feathers white. Cape Gannet similar, has wholly dark tail; both have black line on chin, much longer in Cape. *Juv:* dark brown on head, neck and upperparts with white spots and speckles, underparts paler with brown speckles diminishing towards abdomen. *Nesting:* Oct–early May on Lawrence Rocks, Cat, Eddystone, Pedra Branca and Black Pyramid; 1 white egg. *Range:* common in southern waters from Shark Bay, WA, to Townsville, Qld, many from large colonies in NZ.

Cape Gannet   *Morus capensis*   85–95 cm (ws 150–180 cm)

White gannet with golden head, black primaries, secondaries and tail feathers; no white feathers in tail. Very similar to Australian Gannet, but has longer black line on chin and all-dark tail. Presumably immatures not identifiable in field. *Range:* very rare accidental from South Africa.

Masked Booby   *Sula dactylatra*   80–90 cm (ws 150 cm)

Heavy, mainly white gannet with white head, black primaries. secondaries and tail feathers; no white feathers in tail. Has more black in wing than Australian, lacks black line on chin, has pale grey feet and lacks golden head, with more extensive black mask on face. *Juv:* has body and underwings white, with head, tail and most of upperwings mottled dark brown. *Nesting:* throughout year on islands of north- west and north-east coasts from Bedout in west to Swain Reefs in east and on cays in Coral Sea off continental shelf, from Diana Bank to Cato I, preferring sites near deep water; 2 white eggs. *Range:* northern Australian waters, rarely extends into southern waters but occasionally may wander as far as SA, presumably in freak weather conditions.

Brown Booby   *Sula leucogaster*   65–75 cm (ws 130–150 cm)

Dark brown gannet with white or grey belly. *Adult:* rich brown with white belly and underwing coverts; *male* has blue naked skin on face, *female* yellow-green. *Juv:* duller brown plumage, and underparts are grey. *Nesting:* breeds throughout year on islands in northern waters from Bedout I in west to Lady Fairfax in east and on many cays in Coral Sea. *Range:* common in northern waters; occurs south of breeding grounds more often than other tropical gannets, nevertheless rare south of tropic.

Red-footed Booby   *Sula sula*   65–75 cm (ws 90–100 cm)

Lightly-built gannet of variable plumage, but red feet are diagnostic. *Light phase:* white with golden sheen on head, black primaries and secondaries and white tail; *dark phase:* uniform fawn-brown; *intermediates:* varying amounts of brown, but all have red feet. *Juv:* brown, slightly paler below but with dark breast band and pinkish grey feet. *Nesting:* June–Jan on islands of northern Barrier Reef and Coral Sea, where there are trees or bushes to hold nests; may also breed off north-west coast. *Range:* uncommon resident in northern waters, rarely south of tropic, usually confined to breeding areas.

GANNETS

juv

ad

CAPE

AUSTRALIAN

BOOBIES

♀ ♂

juv

MASKED

BROWN

♂ ♀

juv

dark form

intermediate form

light form

RED-FOOTED

78 x 48	76 x 48	64 x 46	61 x 42	61 x 41
Australian	Cape	Masked	Brown	Red-footed

GANNETS AND BOOBIES

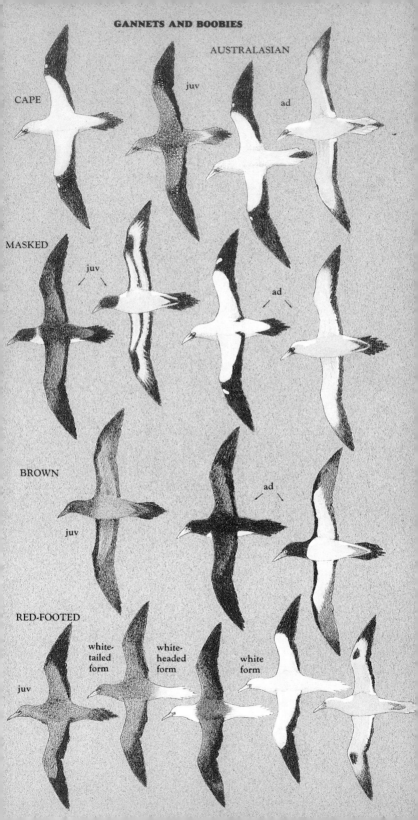

AUSTRALASIAN

CAPE

juv

ad

MASKED

juv

ad

BROWN

juv

ad

RED-FOOTED

juv

white-tailed form

white-headed form

white form

FRIGATEBIRDS

LESSER

juv   imm   ♀   ♂

GREATER

CHRISTMAS

TROPICBIRDS

RED-
TAILED
ad

WHITE-
TAILED

ad   juv

RED-
TAILED
juv

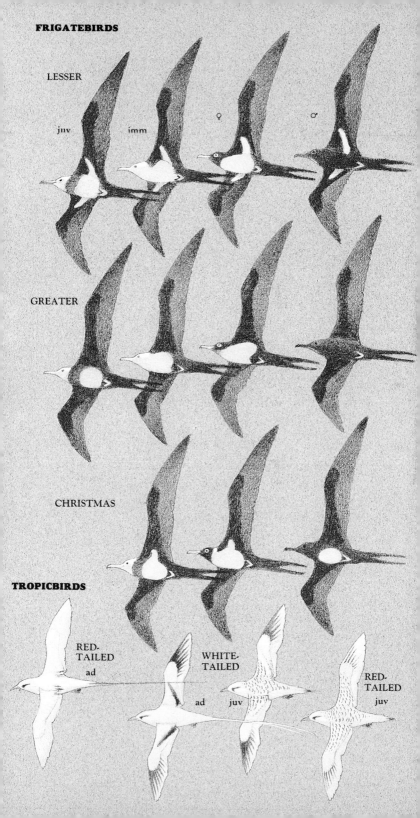

FRIGATEBIRDS AND TROPICBIRDS

Frigatebirds: large, long-winged, fork-tailed seabirds with weak feet, mostly observed in small groups soaring high on motionless wings. Magnificent fliers, sometimes feed by chasing gannets and terns and forcing them to disgorge their catches; more often catch own fish, particularly flying fishes. The two common species here are differentiated at all ages by axillary feathers, white in Lesser, black in Greater. When courting, males inflate red sac on throat. Nest of sticks and seaweed placed on ground, in mangrove or on bushes on coral cays; young birds have pale reddish to buff heads. **Tropicbirds:** mainly white birds, ternlike in many ways but adults are easily distinguished by very long tail streamers. Mainly tropical, but Red-tailed nests as far south as Cape Naturaliste in south-west and occasionally ventures around Bight. Usually fly well above water, dive ternlike for fish, may feed at night as well. Illustrations to much larger scale than Frigatebirds.

Lesser Frigatebird   *Fregata ariel*   70–80 cm (ws 175–190 cm)

Small frigatebird with white axillaries. *Male:* black with white patch on axillaries, upperwing coverts dark. *Female:* black with narrow collar, breast and axillaries white and with brownish bar on upperwing coverts. *Juv:* head varying from reddish to pale fawn, breast band dark brown and axillaries white. *Imm:* breast white without dark breast band, belly black, head black in male, brownish in female; white area progressively decreases in male. *Nesting:* on islands of northern coast from Bedout I in west to Swain Reef in east; 2 limy white eggs. *Range:* common around breeding islands, uncommon elsewhere around coast north of tropic, rare vagrant south.

Greater Frigatebird   *Fregata minor*   85–100 cm (ws 205–230 cm)

Large frigatebird with black axillaries in all plumages. *Male:* black with throat and breast white, chin grey, upperwing coverts pale brown. *Juv:* head white to russet, breast band blackish-brown, belly white. *Imm:* male becomes progressively blacker underneath, female at first loses breast band, then lower breast and belly become progressively darker. *Nesting:* cup-shaped platform in mangrove or on shrub on cays of outer Barrier Reef and Coral Sea; 2 smooth white eggs. *Range:* uncommon to common about coast north of tropic, rare vagrant farther south.

Christmas Island Frigatebird   *Fregata andrewsi*   90–100 cm (ws 205–230 cm)
Large frigatebird with belly white in all plumages. *Male:* black with belly white, axillaries black, upperwing coverts paler. *Female:* breast, belly and axillaries white, throat black, narrow white collar. *Juv:* similar to juv. Lesser, probably indistinguishable except by size. *Imm:* loses breast band, becomes progressively blacker on throat (female) and on breast and axillaries (male), very difficult to tell from smaller Lesser except as they approach adulthood. *Range:* very rare vagrant from Christmas I, records most reliably based on adult.

Red-tailed Tropicbird   *Phaethon rubricauda*   30–35 cm + tail

Only white or pinkish seabird with long red tail. If tailfeathers broken, distinguished from terns by white upperparts, shorter broader wings, black markings on flank, flight less buoyant, rarely follows ships. *Imm:* lacks long tail, heavily barred above with largely white wings and with stout black bill turning to red with age; White-tailed has yellow bill and prominent dark edges to outer primaries. *Nesting:* under overhanging rocks on offshore islands of south-west and north-east and on mainland at Busselton and Cape Naturaliste, WA. *Range:* uncommon from Cape Naturaliste, around north coast to cays of Barrier Reef and Coral Sea, rare farther south on east coast, very rare on south coast.

White-tailed Tropicbird   *Phaethon lepturus*   33–40 cm + tail
Only mainly white or golden-apricot seabird in our area with long white tail. Easily told from Red-tailed, even without tail feathers and yellow bill, by black pattern on upperwing and outer primaries. Beautiful race *fulvus* (Golden Tropicbird) from Christmas I is rich golden-apricot, recorded once in north-western waters. *Juv:* heavily barred on back, black edging to primaries, yellow bill. Follows ships. *Range:* rare but regular visitor to north-eastern waters, mostly immatures, very rare in north-west; cyclones may drive individuals farther south.

FRIGATEBIRDS

LESSER

64 x 44

Lesser

♂ ♀ juv

GREATER

68 x 48

Greater

♂ ♀ juv

CHRISTMAS ISLAND

♂ ♀

TROPICBIRDS

WHITE-TAILED

golden form

ad

juv

RED-TAILED

ad

juv

66 x 48

Red-tailed

CORMORANTS AND DARTER

Black or pied waterbirds with long necks, long tails and short legs with all four toes joined by web. (Shags are strictly marine cormorants.) Swim low in water and dive for fish and yabbies; after fishing sit in sun with wings spread. Nests usually built in trees over water but Pied in west and Black-faced build on ground.

Little Pied Cormorant   *Phalacrocorax melanoleucos*   50–55 cm

Small pied cormorant with yellow or brownish yellow bill. Some have rusty stains on underparts. *Adult:* only pied cormorant without black patch on flank, bill yellow. *Juv:* black patch on flank, bill brown with yellow base. Often solitary but also in flocks, often with Little Black; when flying does not form V-shaped flocks. *Voice:* croaking 'tuk tuk tuk'; cooing 'oo-oo' at nest. *Nesting:* bulky stick nest decorated with dead leaves, in tree over water; colonial, often with Little Black; 2–5 bluish eggs; chick has pinkish crown. *Range:* common resident or nomad in fresh water and estuaries throughout mainland and Tas; rarely in unsheltered marine situation.

Little Black Cormorant   *Phalacrocorax sulcirostris*   60–65 cm

Small black cormorant with black face. Great has yellowish face and is much larger. Often associates with Little Pied, sometimes swims in large flocks, rounding up shoals of fish. Flies in V formation with quick wingbeats. *Voice:* guttural croaking, whistling and ticking at nest site (male): ticking and croaking in fishing flocks. *Nesting:* bulky stick nest in tree over water; colonial, often with Little Pied; 3–4 pale green eggs; chick has dark grey crown. *Range:* same as Little Pied.

Australian Darter   *Anhinga melanogaster*   90 cm

Large black or greyish-brown darter with long thin neck and pale streaks on wing. *Male:* black with white stripe on neck and rusty patch on throat. *Female:* greyish-brown above, white or buff below, white stripe on neck edged black. *Juv:* dark brown above, pale brown below, stripe on neck less obvious. Usually solitary; each male has favourite calling perch. *Voice:* clicking rattle like ratchet on fishing reel. *Nesting:* large stick nest in tree over water; 3–5 pale green chalky eggs. *Range:* patchily common throughout mainland on fresh and salt water; offshore islands.

Great Cormorant   *Phalacrocorax carbo*   80–85 cm

Large black cormorant with yellowish face. *Adult breeding:* white patch on side of head and on flank, white plumes on neck lost after egg-laying. *Juv:* black with dark eye. *Imm:* back brown, neck and underparts mottled greyer, with some white. *Voice:* at nest braying and rattling (male), hoarse huffing (female). *Nesting:* bulky stick nest in tree over water; 3–4 pale blue-green chalky eggs. *Range:* common resident or nomad on estuaries, rivers, lakes, sheltered coasts, throughout southern and eastern mainland and Tas; less common north.

Pied Cormorant   *Phalacrocorax varius*   70–75 cm

Large pied cormorant with orange and yellow face. *Juv:* back browner, face dull yellow. Little Pied is different shape, much smaller, has yellow bill. Often occurs in large flocks particularly about islands off west coast. *Voice:* loud guttural grunting mostly heard when landing. *Nesting:* bulky stick nest on ground on island or in tree over water; colonial; 2–5 pale greenish eggs, often stained. *Range:* common on coast, islands and estuaries in west; on coast, islands, estuaries, lakes and rivers in east; found on Bass Strait islands but not Tas.

Black-faced Shag   *Leucocarbo fuscescens*   65 cm

Large pied cormorant with black face. *Adult breeding:* white plumes on neck, lost after egg-laying. *Juv:* back brown, pale edging to wing coverts brown; head and neck streaked brown, belly white. Pied has yellow and orange face, Little is smaller, has yellow bill. Occurs in flocks mainly about offshore islands along south coast. *Nesting:* bulky stick nest on ground on island; colonial; 2 pale green eggs. *Range:* locally common resident on islands on Recherche Archipelago, WA, and south-eastern coast from southern NSW and Kangaroo I, SA, to Tas.

CORMORANTS

LITTLE
BLACK

ad
LITTLE
PIED

imm   stained

GREAT
ad

AUSTRALIAN

♂

♀

imm

imm
BLACK-
FACED

PIED

ad

48 x 32	48 x 32	56 x 34		60 x 38	58 x 37
Little Pied	Little Black	Australian	Great 61 x 39	Pied	Black-faced

LARGE WATERBIRDS (PELICAN, STORKS, CRANES AND GREAT-BILLED HERON)

Pelicans: large waterbirds with enormous bill and pouch. Only one local species. Feeds often in swimming flocks, by dipping head under water to trap fish in large pouch. May round up shoal fish. Breeds in colonies on ground on islets in inland lakes and on offshore islands, mainly cays. After heavy rain inland may breed in enormous numbers. **Storks;** large stately birds with long bill and legs, frequent swamps, particularly with large areas of shallow water, but may occur on mangrove mudflats. Solitary while breeding, nest in crown of tall tree particularly if top has broken off. Young are brownish grey. **Cranes:** large graceful birds with grey plumage and red naked skin on head. Sociable birds, often seen 'dancing' (sudden outbursts of flapping, bowing, leaping and prancing on ground). Feed on plains, where they dig for tubers, or grasslands, where they hunt insects. Nests are built on ground, usually among reeds or sedges.

Australian Pelican
Pelecanus conspicillatus
150–180 cm (ws 250–260 cm)

Only pelican in Australia. *Breeding:* bill much brighter in colour, lost soon after egg-laying. Magnificent flier, capable of moving long distances in short time. *Nesting:* scrape in ground lined with grass, sticks and large feathers; 2–3 white eggs. *Range:* may turn up anywhere in Aust where conditions are suitable, most abundant in estuaries, but common on most waterways; more common in eastern Tas than western.

Great-billed Heron
Ardea sumatrana 100–110 cm

Large grey-brown tropical heron with massive bill. Unlikely to be confused with any bird except perhaps imm. Jabiru, which flies with neck extended, whereas Great-billed tucks neck in. Favours margins of tropical rivers, estuaries and mangroves. *Nesting:* large stick nest in mangroves, usually overhanging river. Appears to be shrinking from civilisation, so if you find one nesting, have a quick look and leave it undisturbed; 2 pale green or pale blue eggs. *Range:* rare on coastal rivers and mangroves from Derby, WA, to Rockhampton, Qld.

Black-necked Stork (Jabiru)
Xenorhynchus asiaticus
112–115 cm (ws 200 cm)

Large pied stork with iridescent neck. Male has brown eye; female has yellow eye. *Imm:* brown with buff-white areas where adult is white; may be confused with Great-billed Heron, but flies with neck extended. *Nesting:* large stick nest on tree top; 2–4 white eggs. *Range:* uncommon nomad over most of range from Broome, WA, to about Newcastle, NSW, a rare vagrant farther south. (Watch for White Stork *Ciconia ciconia* from northern hemisphere, a large white stork with black wing quills, red legs and with bill black (eastern race *boyciana*) or red (western race *ciconia*), recorded once doubtfully from Cardwell, Qld, 1889; likely anywhere along northern coastal swamps.)

Brolga
Grus rubicundus 100–125 cm (ws 200 cm)

Large crane with heavy dewlap and red adornment confined to head. Sarus has less dewlap and red extends down neck; it is darker grey and slightly larger. *Nesting:* large heap of grass and sticks, usually in shallow water, sometimes well away from water; 2 spotted and blotched white eggs. *Range:* most of northern and eastern Aust. In north it is very common, congregating in marshlands in dry season, but has become much less common in south.

Sarus Crane
Grus antigone 112–115 cm (ws 240 cm)

Large crane with little dewlap and with red adornments extending down neck. Brolga is much more common and widespread, has heavy dewlap and lacks red on neck. *Nesting:* large heap of grass and rushes in shallow water; 2 greenish or pinkish eggs, sometimes spotted and blotched. *Range:* uncommon, most often seen in small flocks of up to 100, often on artificial wetland and farms. Confined to scattered localities in northern Aust, from Kununurra, WA, to Townsville, Qld.

PELICANS

AUSTRALIAN

HERONS

GREAT-BILLED

STORKS

BLACK-NECKED

imm

ad

WHITE

CRANES

BROLGA

SARUS

90 x 59	69 x 47	72 x 53	92 x 61	102 x 65
Pelican	Great-billed	Black-necked	Brolga	Sarus

EGRETS

White herons that develop fine plumes in breeding season. Reef has a dark slate-grey phase, and Cattle develops buff plumage in breeding season. Feed by wading in shallow water, spearing fish; Cattle feeds with livestock, catching insects, toadlets, etc, disturbed by grazing horses and cattle. Some prefer fresh water, others also exploit estuarine mudflats, and Reef is confined to coasts, reefs and islands. Scanty cup-shaped nests of sticks built in trees and mangroves, usually in large colonies. In breeding season colours of bill and legs change, making Great and Intermediate easier to identify; in non-breeding season these two can be distinguished by size, stance and extent of gape below eye. Most fly with slow wingbeats, Cattle faster, often flying in V-shaped flocks particularly in late evening.

Cattle Egret   *Ardeola ibis*   48–53 cm

Small sociable egret with heavy jowl, yellow bill and legs, frequents livestock. *Breeding:* white, with buff coarse plumes on head and neck; bill turns red with yellow tip, and legs become reddish. *Imm:* white with dark bill and legs at first, but bill changing to yellow within three months of fledging; immature Great and Intermediate have yellow bills at all stages; Little has black bill with yellow face. *Nesting:* colonies in trees, usually over water, often with other species; 3–6 pale blue eggs. *Range:* becoming more common and expanding range in western, northern and eastern mainland, winter visitor Tas.

Eastern Reef Egret   *Egretta sacra*   60–65 cm

White or slate-grey egret found on coasts, reefs or islands. Dark phase unmistakable, much darker than White-faced, having white confined to throat (often hard to see). White phase shorter-legged, longer-billed and more hunched when fishing than other egrets. *Nesting:* stick platform Pandanus or Pisonia on islands, on ground if trees unavailable. Dark and light phases interbreed, offspring white or grey, no intermediates. *Range:* common in coastal areas and on islands around mainland, less likely on Vic and Tas coasts; dark phase more abundant in south.

Little Egret   *Egretta garzetta*   56 cm

Small white egret with black legs and black bill with yellow face. *Breeding:* similar with plumes on back and breast and two long plumes on head. At distance its high-stepping daintiness and jerky running after fish are useful characters. In flight has slower wingbeat than Cattle, faster than Plumed and Large, usually flies alone. *Nesting:* colonial in trees or mangroves over water; 3–5 pale blue-green eggs. *Range:* generally less common than other egrets in eastern and northern Australia, occasional sightings in WA, winter visitor Tas.

Intermediate Egret   *Egretta intermedia*   65 cm

White egret with gape of yellow bill not extending behind eye and with dark legs. *Breeding:* bill red with greenish face, legs reddish down to knees; plumes on back and breast. Great has larger gape extending behind eye, often stands motionless in water with neck in 'question mark' position; Intermediate tends to stand with neck straight. *Nesting:* colonial in trees in freshwater or mangrove swamps; 3–4 pale blue-green eggs. *Range:* more common than Little or Great in northern and eastern wetlands, rare visitor to north-west and Tas.

Great Egret   *Egretta alba*   75–90 cm

Large white egret with gape of yellow bill extending behind eye, and with dark legs. *Breeding:* similar, with black bill and greenish face, legs reddish down to knees; plumes on back only. Large size should aid identification when comparison available, otherwise long slender neck crooked into question mark is characteristic. *Nesting:* colonial in trees in swamps, sometimes apart from other egrets but often with cormorants and night herons. *Range:* great variety of wetlands, particularly in interior; only egret commonly observed in inland waters in WA; winter visitor to Tas.

EGRETS

CATTLE

juv

br

non-br

REEF

light phase

dark phase

LITTLE

br

non-br

INTERMEDIATE

br

non-br

br

non-br

GREAT

45 x 33   48 x 35   46 x 34   46 x 34   53 x 38

Cattle   Reef   Little   Intermediate   Great

HERONS

Herons are long-necked, long-legged waterbirds much like egrets but with coloured plumage. Adult of each of Australian species is distinctive and presents no problems in identification; some immatures may give trouble; immature Pied is sometimes mistaken for Pacific, and immature Night Heron is often thought to be a bittern, particularly as it is nocturnal in habits, roosting by day. Herons usually feed in shallow water, but White-faced and more occasionally Pacific feed on grasshoppers in pastures, and Pied visits rubbish dumps and meatworks. Nest in trees, building untidy cup of sticks; food is regurgitated for chicks.

White-faced Heron
Ardea novaehollandiae 65–70 cm

Grey heron with white face. *Juv:* white face much reduced. Unlikely to be mistaken for any other bird even when on coast and reefs where dark phase Eastern Reef Egret may occur, being much paler and more slender with fine bill and prominent white face. Probably commonest heron Australia-wide, occurring wherever there is suitable habitat. *Nesting:* solitary in treetops often well away from water; 3–5 pale green eggs. *Range:* common in or near wetlands.

Pacific Heron
Ardea pacifica 75–100 cm

Large glossy black heron with white neck spotted down centre, and with white spots on leading edge of wings, visible in flight. *Breeding:* similar with maroon plumes on back. *Juv:* duller with dirty white neck more heavily spotted down front. Bill shorter than other herons, useful character when in silhouette; similar immature Pied is much smaller, occurs in large numbers with adults, lacks black spots on neck, lacks white patches on wings in flight. Most often feeds in marshy shallow water, but also frequents pastures, rivers, dams and lakes, less inclined to salt water. *Nesting:* less solitary than White-faced, may nest in small loose groups of up to a dozen pairs but lone individual pairs also found in trees over water; 3–5 blue eggs. *Range:* not as common as White-faced but just as likely to occur anywhere in suitable habitat on mainland; in Tas only rarely encountered.

Grey Heron
Ardea cinerea 100 cm

Large grey heron with black on wing quills and white spots on leading edge of wing, whitish neck with black spots down centre and black crest, yellow bill and yellowish-brown legs. *Imm:* black crest reduced to black line behind eye, bill blacker. Behaves like Pacific Heron. (Watch for Purple Heron *A. purpurea*, smaller, darker and with rusty neck.) *Range:* extremely rare vagrant to northern Aust.

Pied Heron
Ardea picata 43–48 cm

Small pied heron with crown black (adult) or white (imm). *Juv:* sometimes confused with much larger Pacific, lacks black spotting on neck, lacks white spots on wings in flight. Neat-looking heron often in large number in tropical wetland, also scavenges at rubbish dumps and abattoirs. *Nesting:* colonies with egrets in mangroves along tropical rivers spreading into colonies on east coast; 2–4 blue-green eggs, darker than egrets'. *Range:* tropical wetlands from Drysdale R, WA, to Ayr, Qld, common in NT, rarer towards extremes.

Nankeen (Rufous) Night Heron
Nycticorax caledonicus 56–64 cm

Rufous heron with black crown sometimes bearing three white nuptial plumes. Mainly nocturnal, usually seen hunched in trees over water or flying about in small flock when disturbed; usually gives loud 'skeowk' when disturbed, also at night. *Juv:* often confused with bitterns, has regular pattern of 'tear drop' white spots on back and wings – bitterns have fine barring and black streaking, are yellowish-brown rather than rufous-brown, when surprised bitterns extend necks with bills pointed up, mostly in reeds except Black, rarely circle about like night herons when disturbed but drop back into reeds. *Nesting:* loose colonies often near egrets, in trees over water; 2–5 pale bluish-green eggs. *Range:* common near permanent water, being nocturnal not obvious unless flushed; limited in Tas.

HERONS

PACIFIC

juv

ad

WHITE-
FACED

GREY

juv

ad

PIED

ad

juv

ad

juv

NANKEEN
(Rufous)
NIGHT

juv

ad

53 x 38

48 x 35

Pied 41 x 30

53 x 37

Pacific

White-faced

Nankeen

BITTERNS AND MANGROVE HERON

Bitterns are skulking nocturnal herons usually found in reeds, but Black frequents riverside trees, bushes and mangroves. Mangrove Heron is often regarded as a bittern, but feeds according to tides, so often seen hunting in daylight. Most bitterns build a cup-shaped nest of reeds or rushes in reeds, but Black and Mangrove build stick nests in trees. Two small bitterns occur, one very rarely; they are easy to distinguish in adult plumage, but immatures are very similar, probably not separable in field, due to difficulty of good view; however, the cat may bring one in and they sometimes crash into windows at night, so can be examined more closely; bill length is deciding factor.

Mangrove (Striated) Heron   *Butorides striatus*   45–50 cm

Small black-capped mangrove heron with back blue-grey over most of range, reddish-brown in vicinity of Onslow, WA. *Breeding:* face and legs orange. *Juv:* streaked head and neck with back brown spotted buff; like juvenile Night Heron but very different in size, shape and habits; bill much longer and finer than Night Heron, much smaller and more slender. Skulks among mangrove roots and on mudflats, seldom upright, usually with body horizontal. Beautiful golden pattern to edges of wing feathers, visible at close range. *Nesting:* solitary in mangroves; 2–4 pale blue-green eggs. *Range:* common in mangroves from Shark Bay, WA, to southern NSW.

Little Bittern   *Ixobrychus minutus*   25–35 cm

Tiny bittern with dark back, rarely seen unless flushed from reeds. *Male:* black-crowned, black-backed bittern with yellow patch in black wing. *Female:* streaked brown with dark brown back. *Juv:* more heavily streaked on back, lacking dark cap, short bill (>45 mm); Yellow separated in hand only by longer bill (<50 mm). *Nesting:* solitary in reeds; 3–4 white eggs. *Range:* rare in northern, eastern and south-western Aust.

Yellow Bittern   *Ixobrychus sinensis*   30–35 cm

Tiny yellowish-brown bittern with brown back contrasting with black wing quills and with longer bill (>50 mm) than Little (>45 mm). *Male:* black-crowned, black-backed bittern with yellowish-brown patches in wings not as distinct from back as Little. *Female:* streaks brown with pale brown back. *Juv:* strongly streaked, probably not separable in field from shorter-billed juvenile Little apart from bill length. *Range:* extremely rare accidental, possible in summer in any tropical reedbeds.

Black Bittern   *Ixobrychus flavicollis*   54–66 cm

Sooty black or dark brown bittern with yellow plume on neck, usually flushed from trees. *Male:* black with yellow neck plumes. *Female:* dark brown, often with faint freckling on edges of back feathers, visible close, more yellowish plumes and streaking on neck. *Juv:* feathers on back pale edges. Other bitterns are lighter in colour. More likely to feed in open than other bitterns, more diurnal. *Nesting:* solitary in dense tree overhanging water or mangrove Sep–Jan; 3–5 (usually 3) pale blue-green eggs. *Range:* in wetlands and mangroves from south-west around northern coastal areas to southern NSW.

Brown Bittern   *Botaurus poiciloptilus*   66–76 cm

Large mottled brown bittern with black streak on side of head and neck; favours reeds, rarely emerges unless flushed, then usually flops back into reeds, not circling like Nankeen Night Heron (juvenile Night Heron lacks black streak, has spotted back, is rufous-brown not golden or yellowish-brown). *Nesting:* solitary in reeds; 3–6 olive eggs. *Range:* uncommon in reedbeds in south-west and south-east; in Tas mainly in eastern marshes.

BITTERS

MANGROVE
(STRIATED)

juv

♀ YELLOW
♂

♀ LITTLE
♂

juv

juv

♀ BLACK

♂

♂

AUSTRALASIAN

32 x 26

42 x 30   45 x 35

52 x 38

Little

Mangrove   Black

Australasian

SPOONBILLS AND IBIS

Spoonbills are white, long-legged, deepwater waders with long spatulate bills. Feed by wading slowly, swishing bills from side to side, relying on hair-trigger reflexes to snap up any aquatic creature touched, thus can fish productively in muddy water. Fly with neck and bill extended. Build large cup-shaped nest in trees, lay 3–4 white eggs. Small chicks similar to those of ibis at first.

Ibis: have downcurved bills, Straw-necked and Sacred often feed away from water, probing into marshy ground or under cowpats; Glossy probes small clumps of sedge, etc, in shallow water. Fly in flocks, often in V formation. Nest colonially on lignum, mangroves, bushes or trees in marshes.

Royal Spoonbill   Platalea regia   75–80 cm
White spoonbill with black bill. *Breeding:* long plumes on head. At close range, yellow spot over eye and red patch on forehead. *Nesting:* builds large stick nest in colonies in trees or on lignum over water; 2–4 spotted white eggs. *Range:* common throughout northern and eastern Aust wherever suitable water exists; not widespread in southern WA nor in Tas.

Yellow-billed Spoonbill   Platalea flavipes   76–90 cm
White spoonbill with yellow bill. *Breeding:* small stiff plumes on breast and small black feathers on wing. More likely on small dams than Royal, much less sociable. *Nesting:* often solitary, large stick nest in tree; 2–3 white eggs. *Range:* common throughout mainland on suitable waterways. Rare in Tas.

Straw-necked Ibis   Threskiornis spinicollis   58–76 cm
Pied ibis with long yellow plumes on neck. Black plumage of back and wings glossed with blue, purple and green iridescence. Naked skin on underwing yellow. *Juv:* lacks plumes and has little gloss on wings. Feeds in large flocks on pasture and marshes, often with Sacred, sometimes on carcasses. *Nesting:* builds large cup-shaped nest in dense colonies on lignum or in bushes and trees over water; 4–5 white eggs. *Range:* common throughout Aust near wetlands or on pastures. A great traveller, could turn up anywhere, yet much less often seen in Tas than White.

Sacred Ibis   Threskiornis aethiopica   68–75 cm
White ibis with black naked head and neck and with black plumes and feathers near tail. Sometimes shows scarlet naked skin on underwing in flight. At close range black head shows transverse red bars on nape. *Juv:* dusky feathered head; *Imm:* white feathered neck but has naked black skin on head, often dirty-looking. Usually associates with Straw-necked. *Nesting:* builds large cup-shaped nest in colonies on lignum or in mangroves, trees or bushes over water; 2–4 white eggs. *Range:* common over northern and eastern Aust; rare south of Kimberley in WA but may be colonising south-west; winter visitor to Tas, some years in large numbers.

Glossy Ibis   Plegadis falcinellus   48–61 cm
Small all-dark ibis with reddish-brown neck and dark iridescent body. *Juv:* dull brown with streaked neck. Usually seen in small flocks, often flying in V formation, looks like dark curlew in flight. Mainly in shallow water and mudflats, thrusts bill into semi-submerged clumps of grass for frogs and tadpoles. *Nesting:* in colonies with other birds, sometimes other ibis but also egrets, generally builds cup-shaped nest deeper into lignum or bushes than others; 3–6 dull green-blue eggs. *Range:* less common than other ibis, could occur on suitable water anywhere, but mainly confined to northern and eastern humid areas; rare in WA south of Kimberley and in Tas.

SPOONBILLS

ROYAL

br

non-br

YELLOW-BILLED

IBIS

STRAW-NECKED

juv

ad

SACRED

ad

juv

GLOSSY

ad

imm

65 x 45	68 x 45	67 x 46	66 x 45	51 x 34
Royal	Yellow-billed	Straw-necked	Sacred	Glossy

SWANS, GEESE AND SHELDUCKS

Ducks, geese and swans are aquatic fowl having long necks, with flat blunt beaks and short legs with webs between three front toes, enabling them to swim expertly but waddle rather clumsily on ground (except longer-legged 'geese'). They have compact waterproof feathers often with glossy patches in wings. Downy young can swim and dive within hours of hatching. A variety of feeding methods may aid identification: (a) *grazing* on pasture and grasses out of water; (b) *dabbling* and *upending*, feeding on aquatic vegetation in shallow water; (c) *filter feeding*, straining small animal or vegetable organisms from surface of water with specially adapted bill; (d) *diving* feeding on submerged vegetation in deep water. Nests are usually built in reeds, but some species use hollow logs, sometimes distant from water and often quite high in trees; chicks jump to the ground.

Burdekin Duck (Radjah Shelduck)   *Tadorna radjah*   48–60 cm
Large white-headed duck with pink bill. Much smaller Cotton Pygmy Goose has black cap, black bill and spends most of time swimming. Usually in small flocks, occasionally up to 200, prefers to stand on bank, on log or in shallow water. Feeds in shallow water mainly on small aquatic creatures but also water weeds. *Nesting:* tree hollow with small amount of down; 6–12+ cream-white eggs. *Range:* becoming rare at extremes of range which extends from Derby, WA, to Mackay, Qld.

Australian Shelduck   *Tadorna tadornoides*   55–72 cm
Large black-headed duck with chestnut breast. *Male:* white ring around neck, breast more buff. *Female:* white ring around eye and base of bill. Feeds by grazing or upending. In flight, prominent white shoulders and underwing. *Voice:* loud hollow 'ang-ownk'. *Nesting:* tree hollow lined with grass and large amount of down, often very high and far from water; also rabbit burrows; 8–10+ cream-white eggs. *Range:* common around wetlands and dams in southern mainland and Tas, vagrant elsewhere.

Cape Barren Goose   *Cereopsis novaehollandiae*   75–100 cm
Large grey goose on or near south coast and islands. *Goslings:* dark grey-brown, with underparts, stripe down back and eyebrow pale grey. Usually in small flocks on offshore islands and coast but visits pasture farther inland in larger numbers (up to 250). *Nesting:* heap of grass and sticks with down-lined depression, 3–6 grey eggs. *Range:* limited numbers (probably 5000–8000) based on islands of Recherche and Furneaux Group but also along coast of Vic, SA, Kangaroo I, and northern Tas (introduced Maria I).

Pied Goose   *Anseranas semipalmata*   70–90 cm
Large pied goose of tropical wetlands. *Gosling:* dark grey with rusty head and neck and white underparts. Usually in large flocks in shallow water where it grazes, wading on abnormally long legs and semi-webbed toes; also grazes in pasture and crops. *Nesting:* large floating nest of aquatic vegetation in emerging rushes or grasses in water. *Range:* common in subcoastal wetlands from Derby, WA, to Mackay, Qld; has disappeared from former range in south-east, but vagrants still occur in south.

Black Swan   *Cygnus atratus*   120–130 cm
Large black swan with white wingtips. *Juv:* grey, darkening quickly as it grows. Often in large flocks, may nest in big colonies on islands in lakes. *Voice:* evocative musical bugling; soft crooning. *Nesting:* large heap of aquatic vegetation on ground crowned by depression lined with down; 4–10+ greenish white eggs. *Range:* common in wetter areas of mainland and Tas, less common inland and north.

Mute Swan   *Cygnus olor*   130–160 cm
Large white swan with orange-red bill and black knob on forehead. *Juv:* grey, quickly turning white with growth. Introduced to ornamental parks and gardens, established as feral bird in south-west on Avon R at Northam and on Blackwood R at Bridgetown and in northern Tas.

SHELDUCKS

AUSTRALIAN

♂

♀

BURDEKIN
(RADJAH)

PIED
(MAGPIE)

GEESE

CAPE BARREN

SWANS

MUTE

BLACK

59 x 42	68 x 50	83 x 56	72 x 53	104 x 67	115 x 75
Burdekin	Australian	Cape Barren	Pied	Black	Mute

WHISTLING-DUCKS, FRECKLED DUCK, PERCHING DUCKS and POCHARDS

Whistling-ducks: grazing ducks with elongated flank feathers and loud whistling calls, noisy wingbeats making whistling sound in flight. **Freckled Duck:** unusual primitive waterfowl with ski-jump bill, unadorned freckled dark grey plumage, subdued displays, feeds by filtering, upending and dabbling. **Perching Ducks:** have maned heads, short stubby bills, long legs, feed by grazing. **Pochards or Diving Ducks:** swim low in water, have streamlined shape and usually have large patches of white in wing, fly swiftly, dive expertly, walk awkwardly.

Plumed Whistling-duck   *Dendrocygna eytoni*   41–61 cm
Light-crowned whistling-duck with long flank plumes, upright stance, pinkish bill and legs. Flies with head lower than back and legs trailing in characteristic posture; when disturbed often circles to considerable height until intruder goes away. Feeds at night, often flies some distance from water to graze on short grasses. *Voice:* high-pitched rather nasal whistle, often heard overhead at night. *Range:* common on or near dams, waterholes and lagoons in northern and eastern Aust, rare to uncommon vagrant in south-east, accidental south-west.

Wandering Whistling-duck   *Dendrocygna arcuata*   55–61 cm
Dark-crowned whistling-duck with short flank plumes, horizontal stance, dark bill and legs. Flies like Plumed Whistling-duck but looks darker, more chestnut, has dark legs. Feeds in water on aquatic weeds gathered on surface or underwater. *Voice:* high-pitched nasal whistle. *Range:* common on wetlands, mainly deep lagoons, in northern Aust, rare vagrant farther south.

Freckled Duck   *Stictonetta naevosa*   48–58 cm
Dark-looking duck with 'ski-jump' bill and large 'peaked' head. *Male:* bill red at base while breeding, plumage blackish-brown freckled with white or buff. *Female:* all-grey bill, plumage paler. In flight, the only all-dark duck without prominent white patches in wings or on rump; large-headed, short-necked, sharp-winged, mostly dark with paler belly and wing linings. *Voice:* seldom heard, rather swanlike, fluting 'see-you'; grunting roar. *Range:* rare, mainly in freshwater swamps and lakes in south-western and south-eastern Aust; vagrant Tas.

Wood Duck   *Chenonetta jubata*   48 cm
Greyish duck with dark maned head and rather goose-like stance. *Male:* head dark brown with prominent mane, black undertail coverts. *Female:* head paler brown with pale eyebrow stripe and less obvious mane; white undertail coverts. *Flight:* only duck with white secondaries and dark primaries except much smaller white-cheeked Green Pygmy Goose. *Voice:* goose-like nasal 'now'. *Nesting:* hole in tree lined with down, 7–10+ white eggs. *Range:* common on wetlands throughout mainland, particularly lakes with standing dead timber, and earth dams, also occasionally on estuaries; visitor to Tas, sometimes breeds.

Hardhead   *Aythya australis*   41–54 cm
Streamlined brown duck with dark head, white tip to bill and white patch under tail. *Male:* eye white. *Female:* paler brown, eye dark. *Flight:* dark duck with a lot of white in wings and a pale belly; prominent white window extending whole length of upperwing, most of underwing white. Usually seen in flight or swimming. *Voice:* usually silent; wheezing whistle, soft quack. *Nesting:* usually well-hidden neat cup of reeds and other waterside vegetation near water; 9–12 creamy white eggs. *Range:* common to uncommon on wetlands in southern mainland, uncommon farther north; non-breeding visitor Tas.

Red-crested Pochard   *Netta rufina*   56 cm
Male: unmistakable with reddish head, bill and legs; black neck, breast, centre of belly and undertail coverts; broad white stripe in wings; in eclipse similar to female but with red bill. *Female:* similar to female Hardhead but with prominent white cheek, giving appearance of black cap. *Voice:* male, harsh wheezing; female, harsh 'churr'. *Range:* very rare vagrant, one record, Darwin, 1982.

DUCKS

PLUMED

WANDERING

♂

♀

FRECKLED

WOOD
(MANED)

♂

♀

HARDHEAD

♂

♀

RED-CRESTED
POCHARD

♂

♀

50 x 37	48 x 36	63 x 47		
Plumed	Wandering	Freckled	Wood 57 x 42	Hardhead 57 x 42

DABBLING DUCKS

Dabbling ducks take insects, seeds and floating vegetation from on or just below the surface, also up-end in shallow water to dredge food from mud and may graze near water. Shovelers have fine hairs (lamellae) along edges of bill to help strain insects and seeds from water and are unlikely to graze on land. The Mallard is introduced – it interbreeds with the Pacific Black. Some observers speculate that hybridisation may lessen the Black's ability to survive drought, so believe shooters should be encouraged to bag the Mallard.

Pacific Black Duck *Anas superciliosa* 47–61 cm

Large dark duck with two dark lines on buff face, green or purple speculum and white underwing lining. Possibly commonest duck, darker than Teal, which have white in upperwing, and darker than female Mallard, which has orange feet and whitish tail, blue speculum with narrow white edges. *Voice:* male, soft three-noted quack; female, loud descending six-noted quack; 'crank' when flushed. *Nesting:* down-lined tree hollow or grass cup in grass or reeds; 8–10 whitish or pale greenish eggs. *Range:* common resident or nomad on wetlands throughout Aust, most common in south-east and south-west.

Mallard *Anas platyrhynchos* 52–68 cm

Male: yellow-billed duck with dark green head, narrow white collar, purplish-brown breast, grey body, black undertail and rump with two upturned feathers. *Female:* paler brown than Pacific Black with less obvious lines on face, and blue rather than green/purple speculum, orange legs. *Eclipse male:* like female but darker, bill yellow. *Voice:* male, 'queek'; female, familiar barnyard 'quack'. *Range:* introduced, spreading into wetlands in south-east and Tas, interbreeding with Pacific Black; localised about Perth in west.

Northern Shoveler *Anas clypeata* 45–53 cm

Male: long-billed duck with bottle-green head, white breast and chestnut belly. *Female:* long-billed mottled brown duck with blue shoulder, difficult to tell from female Australasian. *Eclipse male:* similar to female but with orange feet, difficult to tell from eclipse Australasian. *Voice:* male, 'took took'; female, double quack. *Range:* very rare vagrant or accidental to wetlands mainly fresh water; likely anywhere in suitable habitat.

Australasian Shoveler *Anas rhynchotis* 45–53 cm

Long-billed duck with blue shoulders and chestnut or brown belly. *Male:* grey-green head marked with white crescent, breast and belly chestnut, white patch on flank (Chestnut Teal smaller, has no white face mark, no blue shoulder). *Female:* mottled brown with blue shoulders, yellowish brown feet. *Male eclipse:* similar with orange feet. Usually swims well out with low hunched profile, beak half submerged; loud whirr of wings in flight, looks faster and more erratic than other ducks. *Voice:* male, 'took took'; female, double quack; chatter in flight. *Nesting:* cup of grass lined with down, in grass near water or in hollow; 8–10+ white eggs. *Range:* uncommon nomad in wetlands mainly southern mainland and Tas, irregular nomad elsewhere.

DUCKS

PACIFIC BLACK

HYBRID MALLARD–BLACK

MALLARD

♀

♂

NORTHERN SHOVELER

♂

♀

AUSTRALASIAN SHOVELER

♂

♀

58 x 41

57 x 41

54 x 37

Black

Mallard

Australasian

TEAL

Teal are small dabbling ducks often occurring in flocks. They are surface feeders, taking mainly seeds and insects as well as up-ending in shallow water to reach the bottom. Favourite nesting site is hollow limb of dead tree often well away from water. Chestnut is concentrated in southern wetlands, Grey is great nomad likely to turn up anywhere suitable conditions occur, may fly a hundred or more kilometres overnight. Northern hemisphere teal are migratory – Garganey may visit Australia more commonly than presently reported, Baikal has possibly been sighted once but is less likely to occur than Common (or Green-winged) and Indonesian Grey.

Australian Grey Teal
Anas gracilis 37–47 cm

Small brown duck with pale throat, small white patch in upperwing and underwing. Similar to female Chestnut, which is brown with dark throat and has higher forehead; female Garganey has blue-grey shoulders; Pacific Black Duck is darker, has lines on face and white on underwing only. *Voice*: male, 'pip'; female, vigorous chuckling quacks 15 syllables. *Nesting*: down-lined hollow in tree, often some distance from water; 7–10+ white eggs. *Range*: common nomad in wetlands throughout mainland and Tas. (Watch for Indonesian Grey Teal *A. gibberifrons* which is similar but has domed forehead, giving it a profile like female Chestnut Teal but even more pronounced; possible in Kimberley and Top End.)

Chestnut Teal
Anas castanea 38–48 cm

Male: small dark duck with bottle-green head and speculum, chestnut breast and belly, white patch on flank (male Australasian Shoveler, similar in pattern, has white line on face, much larger bill). *Female*: similar to Grey but darker, lacking pale throat. *Voice*: male, 'pip' similar to male Grey; female, rapid cackle, higher and shorter than female Grey. *Nesting*: down-lined hollow in tree, grass, reeds or under bush; 7–10 cream eggs. *Range*: common nomad in wetlands in Tas, less common south-east and south-west mainland, uncommon farther north.

Garganey
Anas querquedula 38 cm

Small short-billed duck with blue-grey shoulders and pale belly. *Male*: dark head with prominent drooping white line from eye to hind neck. *Female and eclipse male*: mottled brown with white stripes on face, above and below eye, more greyish shoulders. Grey Teal and female Chestnut Teal lack lines on face and blue-grey shoulders. *Voice*: male, unusual cackle, unlikely to be heard while in Aust; female, low quack. *Range*: rare to uncommon summer visitor to wetlands, likely anywhere.

Baikal Teal
Anas formosa 40 cm

Male in breeding plumage unmistakable with yellow, white, green and black pattern on head. Female like other female teal but has white spot at base of bill, dark eye-stripe interrupted in front of eye, rusty tinge to plumage, and in flight has buff to pale rufous bar in upperwing (white in other teal). *Range*: very rare vagrant sighted once near Darwin.

Northern Pintail
Anas acuta 20–26 cm

Slender-necked duck with pointed tail, pale underwing and belly, narrow white bar in upperwing. *Male*: chestnut head, white breast, barred flanks, long pointed tail. *Eclipse male*: like female, but greyer above with buff edges to feathers. *Female*: brown with buff edges to feathers, underwing pale with freckled coverts, short pointed tail. Very rare vagrant from northern hemisphere, possible on any wetlands, Sept–Apr.

NORTHERN PINTAIL   eclipse ♂   ♀   ♂   ♀

TEAL

AUSTRALIAN GREY

CHESTNUT

♀

♂

GARGANEY

♂

♀

BAIKAL

♂

♀

50 x 36   52 x 37

Australian Grey   Chestnut

DUCKS (PYGMY-GEESE, PINK-EARED AND STIFF-TAILS)

Pygmy-geese: small dainty ducks with stubby goose-like bills, usually seen swimming among waterlilies. In flight prominent white windows in wings, except female Cotton. Feed mostly on surface on aquatic vegetation, but also dive Coot-like. Nest in hollow or hidden in grass near water. **Pink-eared Duck:** aberrant dabbling duck with large bill adapted for filter feeding by fleshy flaps towards tip. Flies with head up, bill pointed 45° down, prominent white rump. **Stiff-tails:** bulky diving ducks with stiff pointed tails often raised in display. Prefer to dive to escape, rather than fly. Usually seen well out in large stretches of water, Musk sometimes in open sea. Nest is a well-hidden cup in reeds, etc; also use old Coot's nests.

Green Pygmy-goose   *Nettapus pulchellus*   30–36 cm
Small stubby-billed duck with prominent white cheek. *Male:* head, neck and back glossy green, cheek white, flanks grey with dark bars. *Female:* faint white eyebrow, neck, breast and flanks grey barred darker. *Flight:* secondaries white, primaries black. *Voice:* musical whistling 'peeyou', 'peewhir'. *Nesting:* in tree hollow over water or in vegetation near water; 8–12 white eggs. *Range:* locally common on waterlily lagoons, dams and swamps in northern Aust from Broome, WA, to Rockhampton, Qld; rare vagrant farther south.

Cotton Pygmy-goose   *Nettapus coromandelianus*   33–38 cm
Small duck with stubby black bill, mainly white head and neck. *Male:* neat black cap well above eye, black band on breast, faintly freckled flanks, glossy green back; in flight, broad white window in whole length of hindwing. *Female:* black stripe through eye, prominent white eyebrow, brownish bars on breast; in flight, no white window in wing (female Green has faint white eyebrow, heavily barred breast and flanks and white window in wing). *Voice:* male, loud cackle; female, soft quack. *Nesting:* unlined hollow in tree, 6–15 white eggs. *Range:* locally common on lily-covered lagoons, dams and ponds in north-east Aust, rare vagrant farther south to north-eastern NSW.

Pink-eared Duck   *Malacorhynchus membranaceus*   39–42 cm
Unmistakable zebra-flanked duck with long bill ending in leathery flap; black patch around eye, small pink ear spot. *Flight:* 'head up, beak down', narrow white hind edge to wing, white rump. Often feeds in V formation, swimming with head half submerged. *Voice:* 'chirrup'. *Nesting:* heap of down on log stump or bush over water; 5–8 white eggs. *Range:* common nomad on wetlands throughout Aust; mainly north-east Tas.

Blue-billed Duck   *Oxyura australis*   35–44 cm
Dark diving duck with broad blue or blue-grey bill. *Male:* chestnut with black head, bill blue; cocks his tail in display. (Hardhead somewhat similar, has prominent white patch on undertail.) *Female:* dark brown, finely freckled lighter brown; bill grey-blue. *Flight:* tail-heavy, laboured, long pattering takeoff. *Voice:* male, low-pitched rattle; female, soft 'quack'. *Nesting:* cup of reeds, often domed, on trampled reed platform; 5–6 pale green eggs. *Range:* uncommon on deep vegetated swamps in south-west, south, south-east and Tas.

Musk Duck   *Biziura lobata*   47–72 cm
Large low-slung freckled diving duck with large head and wedge-shaped bill. *Male:* large lobe under bill; in display, lobe inflated, tail cocked, kicks out jets of water. *Female:* lobe absent or rudimentary. Seldom flies by day, usually seen well out from shore in small groups or singly. *Voice:* male, loud whistle, deep 'plonk'. *Nesting:* flimsy cup in reeds, etc, sometimes with canopy; 1–3 pale green eggs, often resting in water, become stained. *Range:* common resident in deep permanent swamps in south-west and south-east, nomadic elsewhere on swamps, estuaries, occasionally open sea.

ANATIDAE

DUCKS

GREEN

♂

♀

COTTON

♂

♀

PINK-EARED

BLUE-BILLED

♂

♀

MUSK

♂

♀

44 x 32	47 x 35	49 x 36	66 x 48	79 x 54
Green	Cotton	Pink-eared	Blue-billed	Musk

No white in Upperwing

PLUMED

WANDERING

Little white in Upperwing

PINK-EARED

NORTHERN SHOVELER

AUST SHOVELER

Much white in Upperwing

RED-CRESTED

HARDHEAD

COTTON ♂

GREEN ♂

♀

BURDEKIN

WATERFOWL

BLUE-BILLED

BLACK

FRECKLED

MALLARD ♂

GREY

MALLARD ♀   GARGANEY

CHESTNUT

COTTON
♀

WOOD
♂

♀

AUST
SHELDUCK

♂

PIED
GOOSE

GOSHAWKS

Small to large long-tailed broad-winged hawks with powerful talons and red or yellow eyes. Nervous-looking, often flick tail, generally sit hidden in tree, make sudden dash at small birds, or fly rapidly through trees close to ground hoping to startle out prey. Often soar, wings generally flat, may be upswept slightly in strong thermal. Loud hawk-alarm calls of honeyeaters reveal presence. Male considerably smaller than female. Build large stick nest in tree, lay 2–4 white or faintly marked eggs. Young generally more heavily marked than adults except primitive Red which has a juvenile-like adult plumage.

Collared Sparrowhawk   *Accipiter cirrhocephalus*   ♂ 30 cm ♀ 38 cm
Barred brown hawk with long square tail. *Adult:* grey- brown above with rufous collar, barred pale rufous below. *Juv:* dark brown above, heavily streaked brown on breast, heavily barred on flanks. Similar to Brown Goshawk but smaller with longer toes, square tail diagnostic; looks shorter-tailed, longer-winged in flight. *Voice:* rapid chatter, higher-pitched than Brown; soft 'seet'; slow 'swee-swee-swee-swee' with downward inflection; coughing 'bckt'; 'swee-swit' in display. *Nesting:* small stick nest in tree often quite low, 2–4 blotched white eggs. *Range:* uncommon in woodlands and riverine vegetation throughout Aust.

Brown Goshawk   *Accipiter fasciatus*   ♂ 42 cm ♀ 50 cm
Barred brown hawk with long rounded tail. *Adult:* greyish-brown above, finely barred pale rufous below, complete rufous collar. *Imm stages:* 1st year (juv): dark brown above, heavily streaked brown on breast and broadly barred abdomen; two forms, one dark brown with broader bars, other more reddish-brown, narrower bars. 2nd year: slaty brown above, narrower pale rufous bars on breast and abdomen, no chestnut on collar. 3rd year: bars finer, rufous collar incomplete. Secretive, generally more common than suspected. Tropical form (race *didimus*) smaller, longer toes, greyer above, more apricot bars below. *Voice:* rapid chatter 'ki-ki- ki . . . '; slower 'swee-swee . . . '; slow 'swee-it swee-it . . . '. *Nesting:* large stick nest in tall tree; 2–4 white eggs sometimes spotted. *Range:* common resident or migrant in woodlands, forest and riversides throughout Aust; southern form overlaps tropical form in north, may have regular migration.

Grey Goshawk   *Accipiter novaehollandiae*   ♂ 35 cm ♀ 54 cm
Grey or white goshawk with powerful yellow legs and red eyes, usually in forests. (Grey Falcon smaller legs, brown eyes, dark wingtips, dark 'moustache', different shape, usually on wooded plains.) *Adult grey phase:* finely barred on upper breast, tail and wings unbarred. *Juv grey phase:* heavily barred on breast, tail and wings barred. *Voice:* slow ascending piping 'swee-swee-swee . . . '; loud 'swee-swit'. *Nesting:* bulky stick nest in forest tree; 2–3 white eggs. *Range:* uncommon in forest in eastern and northern Aust, from Tas to Broome, WA. White phase most common in Tas and Kimberley, WA. Grey phase most common in centre of range.

Red Goshawk   *Erythrotriorchis radiatus*   ♂ 45 cm ♀ 58 cm
Large reddish goshawk with enormous yellow legs. Upperparts attractively marked with black 'arrowheads', underparts striped black, thighs unmarked; wings and tail prominently barred. Among rarest Aust birds. Many records probably reddish juveniles of other raptores – Spotted and Swamp Harriers, Square-tailed and Black-breasted Kites, Little Eagle. Massive yellow legs, black striped breast and heavily barred wings and tail are best features to watch for. Little Eagle has feathered legs, short faintly barred tail; harriers have long lanky legs, weightless flight with upswept wings; Square-tailed Kite has small legs, much longer wings; Black-breasted has short unbarred tail, long unbarred 'windowed' wings. *Nesting:* large stick nest in tall tree; 2–3 blue-white eggs sometimes blotched. *Range:* rare in forests and woodlands in north-eastern and northern Aust, occasionally wanders farther.

GOSHAWKS

COLLARED

juv

ad

BROWN

imm

ad

tropical form

juv

GREY

RED

38 x 30

47 x 37

56 x 46

48 x 40

Collared

Brown

Red

Grey

CRESTED HAWK AND HARRIERS

Crested hawks (Bazas): small greyish hawks with neat crest, large yellow eyes, compactly feathered lores (bare in most hawks) and barred abdomens. Feed mainly on tree frogs and phasmids, taken among outer leaves and on fruit such as native figs. Fly in small flocks, on long pinched-in wings, often engage in tumbling aerobatics. Loud call 'ee-chew' reveals presence. Small nest like pigeon's placed in leafy upright fork towards top of tree. **Harriers:** large lightly-built hawks with long, broad wings, upswept in flight, long tails and long slender legs. Usually seen floating low over reeds, grasslands or crops, dropping down to take grasshoppers, rodents, reptiles and birds. Most nest on ground in reeds or long grass, but Spotted builds in trees.

Crested Hawk (Pacific Baza) *Aviceda subcristata* ♂38cm ♀43cm
Bar-bellied hawk with small crest. *Adult:* mainly grey with brown wings and pale rufous bars. *Juv:* mainly brown with reddish-brown bars. Usually in small flocks seen either in trees periodically taking prey from outer leaves or flying with slow wingbeats, inspiring few alarm calls from honeyeaters. *Voice:* loud 'ee-chew'. *Nesting:* small cup in upright leafy fork; 2–3 white eggs. *Range:* uncommon to scarce resident in forest and woodland in eastern and northern Aust.

Swamp Harrier *Circus approximans* ♂50cm ♀58cm
Large harrier with white rump and faintly barred tail. *Male:* brown with grey wings and tail. *Female:* brown with pale patch in underwing, streaked darker. *Juv:* dark chocolate brown. *Flight:* wings in deep V, white rump, tail unbarred or faintly barred; juvenile soars more than adult, is all dark except for pale rump and patch in outer underwing. *Voice:* loud 'kee-oo' in display; low 'kok-kok-kok'; high-pitched 'psee-uh'. *Nesting:* flat cup in reeds or crops; 3–6 white eggs. *Range:* common resident, nomad or migrant in grasslands, reedbeds and crops in east, south-east and south-west mainland; common migrant to Tas, rare nomad elsewhere.

Spotted Harrier *Circus assimilis* ♂53cm ♀60cm
Large harrier with heavily barred tail. *Adult:* blue-grey with rufous face, shoulder and abdomen, profusely spotted white. *Imm:* dark grey-brown above, heavily mottled brown and buff below. *Juv:* pale reddish-brown streaked darker – pale rump, often mistaken for Swamp (and Red Goshawk). *Flight:* barred wings in steep V, heavily barred tail. *Voice:* high-pitched chipping. *Nesting:* flat stick cup in tree; 2–4 white eggs. *Range:* common to uncommon in grasslands, crops, open mulga and mallee woodland throughout Aust, mainly in drier areas.

Papuan Harrier *Circus spilinotus spilothorax* ♂45cm ♀55cm
Male: an unmistakable pied harrier with black head, breast, back and wingtips; white belly and shoulders; remainder of wings and tail grey; underwings white with black tips. *Female:* similar to female Swamp but with heavily barred tail, more barring in wings. *Juv:* dark chocolate brown, like juv Swamp but with large white patch on hind-neck (some juv Swamp have obscure white freckles on hind-neck). *Range:* extremely rare vagrant from New Guinea, likely over any grassland or marsh in Cape York or Top End.

BAZAS

HARRIERS

juv

ad

CRESTED
(PACIFIC)

ad

imm'

SPOTTED

juv

juv

SWAMP

♀

♂

juv

♀

PAPUAN

♂

42 x 35

55 x 40

55 x 40

Crested

Spotted

Swamp

LARGE KITES

Kites are long-winged, weak-footed raptores mostly seen soaring. Although they take live prey such as lizards, grasshoppers and fish, much of their food is carrion (except aberrant Square-tailed Kite, which is a treetop harrier preying mainly on honeyeaters, particularly nestlings; Black-breasted is also a nest robber). Nests are large and bulky, placed high in tall trees, usually very obvious, often used for many years. Young differ from adults in plumage, giving rise to identification problems – watch for angle of wings in flight, length and shape of tail, underwing patterns, habitat and behaviour.

Black Kite   *Milvus migrans*   52 cm

Dark kite with forked tail, often twisted in flight. *Adult:* dark brown. *Juv:* brown with head and upperparts streaked and spotted buff, similar to juvenile Brahminy. Usually seen in large flocks near slaughter yards, rubbish dumps, cattle properties, soaring with flat wings; in flight, spread tail looks square, but Square-tailed Kite never flocks, has redder plumage and indistinct bullseye in upswept wings. *Voice:* weak whinnying. *Range:* common in northern Aust, less common farther south, occasional irruptions of large numbers.

Square-tailed Kite   *Lophoictinia isura*   50–55 cm

Long-winged kite with square tail and upswept wings in flight. *Adult:* reddish-brown with white face and dark streaks on breast. *Imm:* reddish-brown with paler wing coverts. Usually seen soaring through treetops in open woodland; obscure bullseye in wing; when sitting, legs barely visible, wings longer than tail. *Voice:* hoarse 'kee-ep'. *Nesting:* large stick cup on horizontal branch; 2–3 blotched white eggs. *Range:* resident or nomad in open woodland throughout Aust; each resident pair has enormous territory, +100 km².

Black-breasted Kite   *Hamirostra melanosternon*   ♂ 55 cm ♀ 60 cm

Large short-tailed kite with prominent white bullseye in wings. *Adult:* similar to juvenile Wedge-tail in colour; head, breast and back black; nape and shoulders rufous; tail pale buff. *Imm:* straw-coloured with darker centres to feathers on back. *Juv:* rusty red with darker centres to back feathers (often mistaken for Red Goshawk). *Voice:* short hoarse yelp. *Nesting:* large stick cup; 2 blotched white eggs. *Range:* common to rare resident or nomad in open woodland and plains throughout interior of Aust, most common in north-west.

Whistling Kite   *Haliastur sphenurus*   53 cm

Dingy-looking kite with long pale tail and distinctive underwing pattern. *Adult:* head, breast and tail pale buff, wings dark brown with paler shoulders. *Juv:* reddish-brown with prominent spots on wings. Soars with slightly bowed wings, often uttering loud whistle. *Voice:* loud vigorous whistle 'pee-aar-wh-wh-wh-wh-wh-wh'. *Nesting:* large bulky stick cup in upright fork of tall tree; 2–3 blue-white eggs sometimes blotched. *Range:* common to uncommon resident or nomad throughout Aust.

Brahminy Kite   *Haliastur indus*   ♂ 45 cm ♀ 51 cm

Short-tailed broad-winged kite usually seen along beaches and mangroves; occasionally on rivers. *Adults:* unmistakable deep chestnut kite with head and breast white. *Imm:* brown with paler head and breast; underwing pattern similar to Little Eagle but more diffuse. *Juv:* dark brown with head and breast streaked buff. *Voice:* quavering 'pee-aaar'. *Nesting:* large stick cup in mangrove; 2–3 blotched whitish eggs. *Range:* common around coastal northern Aust from Carnarvon, WA, to Hastings R, NSW.

White-eyed Buzzard   *Butastur teesa*   50 cm

Unusual hawk with falcon-shaped wings, Baza-like pattern and kite-like behaviour; grey-brown above with white throat divided by black streak, dark breast, barred abdomen and unbarred tail; (may have dark subterminal bar); eyes white. Recorded once at Lithgow, NSW, in 1889, possibly an aviary escapee; if not, more likely to be similar Grey-faced Buzzard B. *indicus* which breeds in northern Asia, migrating south as far as New Guinea, and has pale yellow eyes, tail with 3–5 dark bars and more heavily barred underwings.

KITES

SQUARE-TAILED

BLACK

ad

juv

ad

juv

BLACK-BREASTED

juv

imm

ad

WHISTLING

BRAHMINY

juv

ad

juv

ad

62 x 50

50 x 42

53 x 38

Black-breasted

Black

Square-tailed

Whistling 57 x 44

Brahminy 50 x 40

EAGLES AND OSPREY

Eagles are large raptores with massive legs and long, broad wings. True eagles have feathered legs; sea-eagles have bare tarsi. Males are usually much smaller than females, have spectacular diving displays. Calls are very loud. Nests are large stick structures in trees or on ground. Eaglets are clad in white down. **Osprey** is a large fish predator probably not related to raptores. Plunges into water for fish. Builds large nest on ground or in tree; young are clad in dark brown down, have heavily spotted plumage on leaving nest.

Osprey   *Pandion haliaetus*   ♂ 55 cm   ♀ 63 cm

A large fishing hawk with dark brown back and tail, white head and underparts with black band extending through eye to brownish 'necklace'. *Juv*: similar but with buff-white tips to feathers of back and upperwings, heavier band on breast. *Flight*: long wings with pronounced kink, underwing white with barred flight feathers, and black carpal patch; often hovers over water, dives headlong after fish (Sea-Eagle scoops fish from surface without getting wet). *Voice*: plaintive 'chip-chip'. *Nesting*: massive pile of sticks in tree or on ground; 2–3 heavily blotched white eggs. *Range*: common around northern coasts and islands, uncommon to rare on south-eastern and southern coasts. Also rivers, large dams; may be vanishing from southern range.

Little Eagle   *Hieraaetus morphnoides*   ♂ 48 cm   ♀ 55 cm

A small eagle with short square tail. Two phases occur: Common *light phase*: pale brown above, darker on wings except pale shoulder patch; pale buff below with distinctive underwing pattern; faint bars in tail; *Juv*: similar but head and breast rufous. Rarer *dark phase*: dark brown all over with paler shoulder patch, underwing dark with obscure bullseye; *Juv*: similar but dark reddish brown. Soars on flat wings, flies with slow wingbeats. Often confused with Whistling Kite (longer paler tail, bowed wings), immature Brahminy Kite (marine habitat, broader rounder wings), Black-breasted Kite (more prominent bullseye in wing). *Voice*: loud rapid 'wh-whee-whit', middle note highest; penetrating 'kuk-kuk-kuk . . .'. *Nesting*: large stick nest in fork or on horizontal limb; 2 blotched white eggs. *Range*: uncommon in woodland and along tree-lined watercourses throughout Aust but not Tas. (Watch for Short-toed Eagle *Circaetus gallicus*, 63–69 cm, recorded doubtfully near Derby, WA, 1961, a rather variable harrier-like 'eagle' with dark rounded head, narrow nipped-in wings, barred abdomen and shortish tail with three bars. *Flight*: slow deep flaps, soars on flat faintly barred wings with dark tips, often hovers. Some birds are very pale below, lacking dark hood, but retain dark wingtips and three tail bars. Ranges widely in northern hemisphere and migrates south as far as Timor.)

White-breasted Sea-Eagle   *Haliaeetus leucogaster*   ♂ 76 cm   ♀ 84 cm

Large sea-eagle with short pale tail. *Adult*: grey above with white head, breast and abdomen; tail pale grey with white tip. *Juv*: speckled slaty brown with paler face, large white bullseye in wing. *Imm*: second-year birds lack pale edges to dark feathers, head and breast pale. Soars with wings upswept (more than Wedge-tail) like huge butterfly. *Voice*: loud goose-like 'ang-ank'. *Nesting*: enormous pile of sticks in tree or on ground (on island); 2 white eggs. *Range*: common resident or nomad along coast and on rivers, lakes and dams.

Wedge-tailed Eagle   *Aquila audax*   ♂ 90 cm   ♀ 1 m

A large brown or black eagle with long wedge-shaped tail. *Adult*: black with rufous or blond nape and shoulder patch and pale undertail coverts. *Juv*: dark brown above with black throat, breast and abdomen, feathers tipped pale at first, and with pale rufous nape and shoulder patch; immatures vary considerably but generally darken progressively for 5–7 years before adult plumage. Soars to great heights on upswept wings with faint pale base to primaries. *Voice*: loud, seldom-heard 'coo-weee-el'; yelping 'seet-you'; soft 'soo-wee-ya'. *Nesting*: enormous stick nest in large tree; 1–3 blotched white eggs. *Range*: common resident or nomad in most habitats throughout mainland and Tas; rare to uncommon in settled areas. (On Cape York and Torres Strait islands watch for Gurney's Eagle *Aquila gurneyi* 29–34 cm (ws c. 2 m), a large all-dark eagle with square to rounded tail. Broad-winged blackish adult and mottled brown juvenile soar on flat wings unlike Wedge-tailed and juvenile White-breasted Sea-eagle. Recorded on Australian islands near New Guinea, unlikely farther south but not impossible.)

OSPREY

juv

ad

60 x 44

Osprey

EAGLES

LITTLE

juv

dark phase

light phase

WHITE-BREASTED

WEDGE-TAILED

ad

juv

56 x 46

Little

70 x 53

White-breasted

73 x 58

Wedge-tailed

FALCONS AND HOVERING KITES

Falcons are small but powerful raptors with long pointed wings and notched bills. Some fly swiftly, easily outpacing other birds, but several species are relatively slow and frequently hover when searching for food. Most eat birds, but mice, grasshoppers and reptiles are staple diet of others. Calls are basically similar, loud rapid hectoring 'kik-kik-kik . . .'. Males much smaller than females, perform incredible aerobatics during courtship. Do not build own nests but use cliff ledges, tree hollows and appropriated nests of other raptors or corvids. Young are generally darker and more heavily patterned than adults and wander considerably after fledging. **Hovering kites:** two small, mainly white and grey hawks with black shoulders, feeding on rodents; easily distinguished by underwing pattern. One widespread, often seen on telegraph poles, dead trees or hovering; other nocturnal, confined to interior where plague rats occur, except during periodic irruptions.

Letter-winged Kite   *Elanus scriptus*   ♂ 33 cm ♀ 38 cm

Small nocturnal grey and white kite with black shoulder and black line like letter 'W' on underwing. *Male:* head white. *Female:* head grey. *Juv:* head, breast and back tinged rufous. Sleeps during day in eucalypts along dry watercourses, often in large flocks. *Voice:* plaintive 'tew' or 'pitsyou', higher than Black-shouldered; harsh 'skew . . .'; loud 'skow-skow-skow' from juv. *Nesting:* large stick nest in eucalypt; loose colonies >30 nests; 3–5 heavily blotched white or buff eggs. *Range:* locally common in channel country of Qld, NT and SA and Barkly Tableland, NT. Occasional irruptions all over mainland after rat plagues die out.

Black-shouldered Kite   *Elanus notatus*   36 cm

Small grey and white kite with black shoulders and small black patch on underwing. *Juv:* pale rufous tinge to head, breast and back. Perches conspicuously along roadsides and paddocks, hovers more heavily than Kestrel, usually with tail depressed and legs dangling. *Voice:* loud 'chip', screeching hiss. *Nesting:* stick cup in upright leafy fork; 3–4 blotched white eggs. *Range:* common to uncommon nomad or resident in open woodland, grasslands and paddocks throughout mainland; vagrant Tas.

Black Falcon   *Falco subniger*   ♂ 45 cm ♀ 55 cm

Dark slaty-brown falcon with dark underwing. *Adult:* up to about 10 years old similar to juvenile but with slightly shorter tail and wings and with paler throat and hindcheeks; after 10 years, throat and breast become speckled with white, some very old birds have extensive white bib; wing and tail feathers may acquire pale bars but never as many as Brown Falcon. *Juv:* darker slaty-brown with faint bars in primaries and outer tail feathers. *Flight:* long-winged, long-tailed, outer tail feathers shorter than others, soars with flat wings and spread tail. Lazy-looking when flapping but can fly very fast with flickering wingbeats. Dark phase and juv. Brown Falcons are darker brown but have pale buff edges to wing feathers, sit much higher on perch with longer legs, larger head; in flight always show more barring in wings and tail. *Voice:* whining 'kee-ar'; loud rapid 'kek-kek-kek . . .'. *Nesting:* appropriates old stick nest of corvid or raptore; 2–4 blotched pink eggs. *Range:* uncommon to rare nomad on tree-scattered plains or along tree-lined watercourses throughout drier areas, increasing in agricultural clearings; rare in Tas.

Brown Falcon   *Falco berigora*   ♂ 45 cm ♀ 55 cm

A very variable, long-legged falcon from pale sandy brown to almost black, but always with dark thighs, pale flight feathers on underwing. Upswept wings in flight. Three basic forms occur: *red phase* with reddish plumage; *brown phase* with brownish plumage; and *dark phase*, which remains all dark-brown for life. *Juv:* most are dark-breasted, usually with yellowish-buff throat and incomplete collar and buff undertail coverts; some completely dark, looking like dark phase; some red phase juveniles are white-breasted. *Second year:* breast speckled in brown phase; many birds in northern Aust remain speckle-breasted for life. *Third year:* brown phase may remain speckle-breasted but less speckled than second year; some males assume adult form. In *fourth year* most birds are adult but many continue to become paler in subsequent years. Distinctive *adult forms* are: (a) dark phase, (b) speckle-breasted, (c) stripe-breasted, (d) white-breasted, (e) grey-headed, (f) sandy-breasted, (g) red-breasted;. Usually seen perched on dead trees or telegraph poles. *Flight:* slow overarm wingbeats, soars with upswept wings, can fly fast; hovers, usually at considerable height, more clumsily than Kestrel. *Voice:* noisy 'yeah-cook', 'yeah-cook-uk-uk'; 'tek'; juvenile, rasping 'keea-keea . . .'. *Nesting:* appropriates old stick nest of corvid or raptore; 3–5 blotched pinkish-buff eggs. *Range:* common resident or nomad in open woodland, tree-scattered plains and paddocks throughout mainland and Tas.

KITES

LETTER-WINGED

BLACK-SHOULDERED

juv

ad

juv

ad

44 x 32

Letter-winged

BLACK

old
ad

FALCONS

BROWN

dark

juv

red-breasted

grey-headed

speckle-breasted

stripe-breasted

white-breasted

40 x 30

Black-shouldered

54 x 40

Black

50 x 38

Brown

FALCONS

Kestrel is common small slow falcon, hunts either from convenient perch, eg telegraph pole or dead tree, or by hovering skilfully. Feeds mostly on grasshoppers, mice and small lizards. Other falcons shown here are magnificent aerialists, among most exciting of all birds to watch. Although each is distinctive, can be difficult to differentiate without experience, especially in juvenile plumages. Hobby is more slender than others, appears to fly faster but is in fact slower than Grey and Peregrine.

Nankeen (Australian) Kestrel
Falco cenchroides ♂ 31 cm ♀ 35 cm

A small hovering falcon with pale rufous upperparts and whitish underparts. *Male:* head grey, tail grey with black subterminal band. *Female:* head pale rufous, tail paler rufous with numerous dark bars; above pale rufous with black spot in each feather. *Juv:* more brownish rufous above with more extensive dark centres to feathers. *Flight:* rapid flickering wingbeats, soars on flat wings with tail spread; hovers frequently with tail fanned, shallower faster wingbeats than Black-shouldered Kite. Usually seen on telegraph poles or dead trees. *Voice:* rapid 'ki-ki-ki . . .'; slow 'tek-tek-tek . . .'; indrawn 'ee-agh'. *Nesting:* in hollow dead limb of eucalypt or hollow in rock face, but in central Aust, often in old stick nest of corvid or raptore; 3–5 blotched pale buff eggs. *Range:* common resident or nomad (migratory in north) in open woodland, plains and paddocks, coastal cliffs and dunes, towns and city ledges throughout Aust; most widely recorded Aust bird.

Australian Hobby (Little Falcon)
Falco longipennis ♂ 30 cm ♀ 35 cm

A small falcon with dark head, restricted black cheeks and pale half collar. Several forms occur. *Dark form* of humid areas: *adult:* dark blue-grey above, paler on rump; cap and cheeks black, buff collar, throat off-white; breast and abdomen dark brown streaked on breast and obscurely barred on flanks; *juv:* dark brown above with paler edges to feathers, below similar to adult but darker. *Light form* of drier areas: *adult:* pale blue-grey above, darker on shoulders, head and cheeks; pale buff collar; white throat; breast and abdomen vary from pale brown to pale dusty rufous, almost off-white in some individuals, commonly mistaken for Grey Falcons; *juv:* pale brown above with much brown in dark cap and cheeks; lighter rufous brown below. Very fast little falcon, seen best hunting dragonflies in late evening; often chases large birds in play. Peregrine is larger, bulkier with more extensive white or pale buff on breast, more black on cheeks. *Voice:* rapid excited 'ki-ki-ki . . .'. *Nesting:* appropriates old stick nest of crow, large raptore; 3–4 heavily blotched eggs, generally later than other raptores. *Range:* uncommon resident or nomad in open woodland, forest and tree-scattered plains throughout Aust; restricted range in Tas.

Grey Falcon
Falco hypoleucos ♂ 34 cm ♀ 43 cm

A grey falcon with wispy black on cheek. *Adult:* pale blue-grey above with centres of feathers darker, paler almost white below with fine dark streaks. *Juv:* dark blue-grey above, more extensive dark cap and cheeks, underparts white with dark spot in each feather; looks rather like a pallid Peregrine; juveniles in Kimberley are darker than in central Aust. *Flight:* pale-looking with dark wingtips; wingbeats slower, deeper than Peregrine, more rapid in pursuit of prey; stoops vertically after birds, also takes reptiles on ground. *Voice:* loud slow 'kek-kek-kek . . .'. *Nesting:* appropriates stick nest of raptore or corvid; 2–4 heavily blotched buff eggs. *Range:* rare resident or nomad on tree-scattered plains or along desert watercourses, mainly in interior.

Peregrine Falcon
Falco peregrinus ♂ 38 cm ♀ 48 cm

A large falcon with dark head and extensive dark cheeks. *Adult:* dark blue-grey above, paler on rump; cap and cheeks black; underparts pale grey and buff, profusely barred with narrow black bands; breast usually white or faintly buff. *Juv:* dark brown above, with darker cap and cheeks, underparts pale fawn with dark brown streaks; desert birds are paler above, some almost ginger-looking with dark cap less obvious. South-western form (race *submelanogenys*) much smaller, adults paler grey above, bars below reduced in old birds, often confused with Little Falcon; birds of north-east coastal areas darker buff on breast, sometimes rufous in fresh plumage, similar to New Guinea race *ernesti*. *Voice:* hoarse indrawn 'ee-agh . . . '; loud slow 'kek-kek-kek . . .'; soft 'ke-kik' from male. *Nesting:* 3–4 heavily blotched eggs laid in scrape on cliff ledge or in old nest of raptore or crow; sometimes tree hollow. *Range:* uncommon resident or nomad in wide range of habitats mainly near cliffs, throughout mainland and Tas; some migration across Bass Strait.

FALCONS

AUSTRALIAN HOBBY

juv

light form

dark form

GREY

NANKEEN KESTREL

juv

ad

♀

♂

38 x 30

Nankeen

south-western form

PEREGRINE

buff-breasted form

coastal juv

inland juv

ad

46 x 34

50 x 38

50 x 40

Australian

Grey

Peregrine

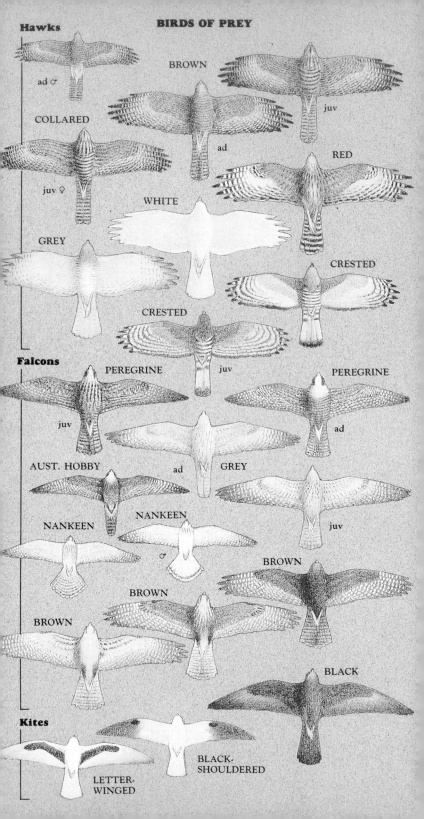

BIRDS OF PREY

Hawks

BROWN

ad ♂

COLLARED

juv

RED

juv ♀

WHITE

GREY

ad

CRESTED

CRESTED

juv

Falcons

PEREGRINE

PEREGRINE

juv

ad

AUST. HOBBY

ad   GREY

juv

NANKEEN

NANKEEN

♂

BROWN

BROWN

BROWN

BROWN

BLACK

Kites

LETTER-
WINGED

BLACK-
SHOULDERED

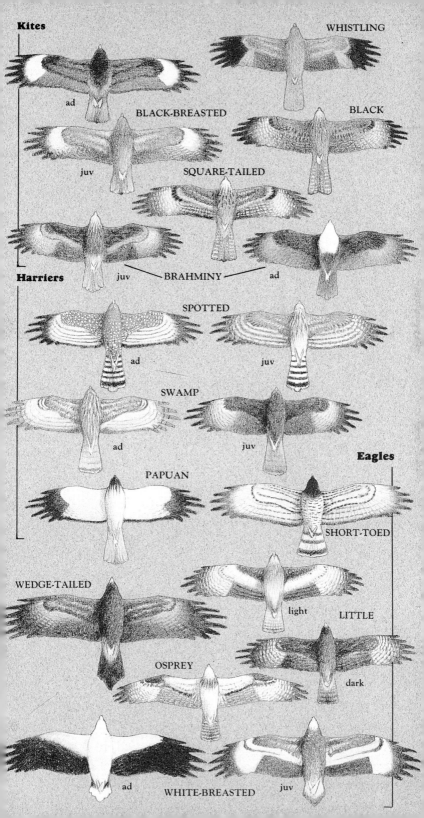

Kites

WHISTLING

ad

BLACK-BREASTED

BLACK

juv

SQUARE-TAILED

Harriers

juv — BRAHMINY — ad

SPOTTED

ad

juv

SWAMP

ad

juv

PAPUAN

Eagles

SHORT-TOED

WEDGE-TAILED

light

LITTLE

OSPREY

dark

ad

WHITE-BREASTED

juv

MOUND-BUILDERS, THICK-KNEES AND BUSTARDS

These are large long-legged ground birds. **Mound-builders, or megapodes:** have stout legs with powerful claws for scratching up earth and vegetation into large incubators where the eggs are laid; temperature is controlled by opening or closing mound. **Thick-knees:** have long legs without noticeable thick knees; nocturnal feeding on invertebrates and small vertebrates, usually lie hidden during day; loud wailing calls at night reveal presence. **Bustards:** heaviest of all flying birds; walk majestically with bill carried superciliously at 45°.

Malleefowl *Leipoa ocellata* 60 cm

Large megapode with eye-like patterns on back and black streak on breast. Moves slowly through mallee woodland, singly or in pairs, freezes if disturbed, walks away or flies clumsily if pressed. Males are localised near mounds, females wander more. *Voice:* male, loud three-noted booming; female, high-pitched crow. *Nesting:* large earthen mound, 4 m across, 75 cm high, filled with leaves and twigs, tended daily during breeding season Sep–Apr; up to 30 eggs, pale pink at first, brown later. Chick on hatching digs out of mound, looks after itself. *Range:* locally common, generally rare in mallee woodland in southern mainland.

Australian Brush-turkey *Alectura lathami* 70 cm

Medium-sized black megapode with red and yellow wattles, found in coastal north-east. Birds on northern Cape York have bluish-white wattles (race *purpureicollis*); birds of southern Cape York have orange legs. When disturbed, flies into tall tree, looks like strange vulture. *Voice:* hoarse grunt. *Nesting:* large mound of leaves and earth, 4 m across, 1–2 m high, up to 15 white eggs, Sep–Mar. Chick has silky chestnut and brown down, cares for itself. *Range:* common in upland rainforest and wet sclerophyll forest, uncommon in lowland forest, becoming rare in brigalow scrub; from Cape York to Gosford, NSW.

Orange-footed Scrubfowl *Megapodius reinwardt* 40 cm

A small plain megapode with prominent crest, found in rainforest, monsoon forest, riverine forest and mangroves. Runs quickly when disturbed or flies into tree; generally more common in lowland forest than Brush-turkey in north-east. *Voice:* loud double crow. *Nesting:* huge mound of leaves and soil up to 12 m across and 3–5 m high; many pinkish-white eggs, quickly becoming stained. *Range:* common resident from Kunmunya, WA, to about Ingham, Qld; disappearing population between Mackay and Yeppoon, Qld.

Beach Thick-knee *Burhinus neglectus* 55 cm

A large-billed thick-knee found on northern beaches and reefs. Usually rests near driftwood on beach during day, begins feeding late afternoon on beaches or when tide is out on reefs. *Voice:* wailing 'weelow', higher and hoarser than Bush. *Nesting:* 1–2 blotched white eggs laid on sand. *Range:* uncommon resident on beaches, islands, reefs and mangrove mudflats around Aust, south to Point Cloates, WA, and Tweed R, NSW, rare farther south to Shoalhaven R, NSW.

Bush Thick-knee *Burhinus magnirostris* 55 cm

A small-billed thick-knee with streaked neck, found in any habitat with ground litter, from rainforest to open woodland and paddocks. Disappearing from settled areas in south. *Voice:* loud wailing 'weeloo'; 'will-aroo, will-aroo . . .'. *Nesting:* 1–2 blotched pale umber eggs on ground. *Range:* common to rare resident or nomad throughout Aust, less common in south and interior, occurs in large groups on northern pastures, eg Atherton Tableland, possibly regular migrant.

Australian Bustard *Ardeotis australis* 75–150 cm (ws 1·5–2 m)

Large stately ground bird with long neck, long legs and heavy body. Usually seen singly or in small loose groups on plains, grasslands or open woodland; easily approached in vehicle but very shy when approached on foot, either squatting to hide or flying off with slow, powerful wingbeats, seldom higher than 30 m. Male has throat pouch extended in display. *Voice:* hollow grunting or croaking; in display, male roars like distant lion. *Nesting:* 1–2 buff to greenish-buff eggs laid on bare ground. *Range:* once a common nomad in open country throughout Aust, now generally rare except in isolated areas.

MEGAPODIIDAE, BURHINIDAE, OTIDIDAE

MOUND-BUILDERS

eastern form

northern form
♀

northern form
♂

♂

northern form

AUSTRALIAN
BRUSH-TURKEY

MALLEE FOWL

ORANGE-FOOTED

THICK-KNEES

BEACH

BUSH

BUSTARDS

AUSTRALIAN

58 x 39
Bush

65 x 45
Beach

78 x 55
Bustard

92 x 61
Mallee Fowl

92 x 64
Brush-turkey

90 x 52
Orange-footed

BUTTON-QUAIL AND TRUE QUAIL

The **button-quail** on this page are closely related, differing mainly in size of bill and extent of plumage patterns. Inhabit open forest, favouring hill slopes and ridges where grasses tend to be sparse. When flushed fly rapidly, zig-zagging through trees. Calls similar, low booming like bronzewing pigeons. **True quail** are bulkier and smaller-headed than button-quail, have a hind toe. Keep hidden in grasses, pastures and stubble, revealed by characteristic calls. Usually run when disturbed, but if startled burst from cover in loud clattering flight. Feed on small seeds and occasional invertebrates. Nest well-hidden, lined scrape in tangled vegetation; female incubates eggs and cares for often large number of young (7–14), may be assisted by male.

Painted Button-quail *Turnix varia* 17–20 cm
Large button-quail with reddish shoulder patch, back heavily mottled with black and fine bill. *Male:* smaller, less brightly coloured. *Imm:* lacks prominent reddish shoulder patch, has barred breast. Pale island form (race *scintillans*) found on Houtman Abrolhos, near Geraldton, WA. Inhabits open forest, particularly stony ridges. When flushed usually flies some distance to cover. *Nesting:* lined scrape under grass clump or fallen branch; 4 spotted whitish eggs. *Range:* uncommon in open forest in south-west; and Tas, south-east and eastern mainland north to about Cairns, Qld.

Chestnut-backed Button-quail *Turnix castanota* 15–20 cm
Large button-quail with reddish shoulder, back sparsely mottled with black, streaked breast and heavy bill. *Male:* smaller, less brightly coloured. Buff-breasted form (race *olivei*) inhabits Cape York, Qld; grey-breasted form (race *castanota*) inhabits Arnhem Land, NT, and Kimberley, WA. More southern Painted has finer bill, more heavily mottled with black above. *Nesting:* lined scrape under bush or grass clump; 4 freckled whitish eggs. *Range:* uncommon in far north.

Brown Quail *Coturnix australis* 18 cm
Large quail with variable plumage, but always with finely barred underparts. *Male:* more finely barred than female; two main variations: (a) back feathers with white shaft, grey centre and rufous edges heavily blotched with black, (b) back feathers with more extensive grey centre, finely vermiculated with black, giving silvery appearance, more common from Tas to southern Qld. *Female:* more coarsely barred below, more heavily marked black above. Female King also barred below but darker, smaller and has brownish face. Jumps high when flushed, flies short distance. Runs with neck stretched giraffe-like. *Voice:* sad ascending whistle 'ph-phweeep'. *Nesting:* lined scrape in long grass; 4–7+ spotted buff eggs. *Range:* common in rank long grass near water and stubble in northern, eastern and south-western Aust; also in Tas, where it is called Swamp Quail.

Stubble Quail *Coturnix pectoralis* 18 cm
Only quail with streaked flanks. *Male:* pale buff to chestnut face, black patch on chest. *Female:* whitish throat, streaked breast and flanks. Usually found in stubble and short grasses. Flies rapidly when flushed, usually alights at considerable distance; appears greyish-brown in flight, paler than Brown Quail. *Voice:* sharp 'pippy-wheat'. *Nesting:* lined scrape in long grasses, 7–14 smudged buffish eggs. *Range:* south-western, south-eastern Aust to central Qld and Tas; after good seasons may occur outside usual range.

King Quail *Coturnix chinensis* 13–14 cm
Male: small dark quail with blue-grey breast, chestnut belly, brown back, and black and white pattern on throat. *Female:* small dark brown with barred underparts and brownish face. Flies rapidly with loud whirr when flushed, often in 'covey' of several birds. Brown is much larger and paler than female, has finer bars on underparts and greyish rather than brown face. *Voice:* high-pitched three-noted descending whistle. *Nesting:* lined scrape in long grass; finely spotted pale brown eggs. *Range:* uncommon to rare in moister grasslands of northern and eastern mainland and Tas.

PHASIANIDAE, TURNICIDAE

BUTTON-QUAIL

CHESTNUT-BACKED

PAINTED

♀

grey-breasted form

♂

buff-breasted form

♂

TRUE QUAIL

♂

♂

BROWN

♀

♀

STUBBLE

KING

♀

♂

♂

25 x 19	33 x 23	28 x 22	33 x 24	25 x 19	27 x 20
King	Stubble	Brown	Brown (Tas)	Chestnut-backed	Painted

BUTTON-QUAIL AND PLAINS WANDERER

Button-quail: look like true quail (p 88) but lack hind toes, have larger head, polyandrous behaviour with male smaller than female. Upper plumage cryptic like true quail but more attractively patterned. Some are common, but Black-breasted and Red-breasted becoming rare. Most feed on seeds, Little favours cranesbill, Black-breasted wild tobacco, fine-billed Red-backed more insectivorous. Leave circular feeding scrapes on ground. Call is low repetitive booming 'oom', rather like Bronzewing Pigeon. Female more brightly coloured than male, initiates courtship. Male builds nest scrape lined with grasses under bush or clump of grass, incubates usually four eggs and cares for young. **Plains Wanderer:** unusual, rare quail-like bird probably better regarded as aberrant wader, has hind toe, polyandrous habits, upright stance. Once common in south-east, now rare, restricted in range. Nesting behaviour is like button-quail.

Red-backed Button-quail *Turnix maculosa* 11–15 cm

Small buff-breasted quail with red nape, prominent black spots on buff wings, buff undertail coverts, fine yellow bill, yellow legs and eyes. *Male:* less brightly coloured with red nape small or absent, sides of neck heavily spotted, breast pale buff. *Female:* large red nape patch, face and breast buffish in eastern Qld and NSW (race *melanota*), face and breast rufous in east Kimberley and NT (race *pseutes*). When disturbed flies fast in pairs or covey, alights in cover at considerable distance. *Nesting:* lined scrape under low bush or green clump; 2–4 eggs pale yellowish-grey, heavily spotted darker. *Range:* uncommon to rare in swampy grasslands and pasture in northern and eastern Aust.

Red-chested Button-quail *Turnix pyrrhothorax* 13–15 cm

Small buff-breasted quail with pale buff undertail coverts, stout bluish-grey bill, pink legs and white eyes. *Male:* less brightly coloured, pale buff breast, black and white bars on side of neck. *Female:* rich buff breast, faint white bars on side of neck. When flushed jumps high in pairs or covey, seldom flies far, shows buff flanks. *Nesting:* lined scrape under grass clump or in stubble; 4 spotted buff-white eggs. *Range:* uncommon in lightly timbered woodland with grassy understorey or pasture and stubble in northern and eastern Aust.

Little Button-quail *Turnix velox* 13–15 cm

Small quail with whitish or pale burnt sienna breast, white undertail coverts, stout bluish-grey bill, pinkish legs and white eye. *Male:* breast whitish only faintly tinged buff, sides of neck heavily scalloped white, brown and black. *Female:* breast pale burnt sienna, with faint white and black scallops in some examples, probably younger females. When flushed, jumps in pairs or covey with twittering giggle, flies fast and low, showing white flanks, seldom far. *Nesting:* lined scrape in grass clump; 3–4 spotted buffish eggs. *Range:* common to uncommon in grasslands and grassy woodlands throughout Aust.

Black-breasted Quail *Turnix melanogaster* 15–18 cm

Large quail with profusely barred breast, black or whitish. *Male:* brownish with breast whitish with black and chestnut barring. *Female:* head and breast black with white spots and bars, mainly on sides of breast. Normally runs when disturbed; watch for circular scrape among leaves on forest floor; 2–4 boldly blotched buff eggs. *Range:* rare in drier rainforest, vine scrub, lantana, usually about edges infested with wild tobacco, in central eastern Aust from Shoalwater Bay, Qld, to north-eastern NSW, rare vagrant (or aviary escapee?) farther south.

Plains Wanderer *Pedionomus torquatus* 16–17·5 cm

Small long-legged quail-like plover with nipped-in neck and finely scalloped plumage. *Male:* less brightly coloured, breast more coarsely scalloped white and dark grey. *Female:* black collar heavily spotted white, narrow breast band bright chestnut. When disturbed usually runs, if pressed flies rather rail-like with legs dangling. *Nesting:* lined depression under bush or grass tuft, 3–4 pointed blotched buff eggs. *Range:* now rare over most of former south-eastern range in natural grasslands, well-grown pasture and crops.

TURNICIDAE, PEDIONOMIDAE

RED-BACKED

♂

♀

BUTTON-QUAIL

RED-
CHESTED

♂

♀

LITTLE

♀

♀
imm

♂

BLACK-
BREASTED

♂

♀

PLAINS WANDERER

♀

♂

30 x 23	28 x 21	23 x 18	23 x 18	22 x 17
Plains Wanderer	Black-breasted	Red-chested	Little	Red-backed

SMALL RAILS AND CRAKES

Crakes are small secretive birds of reedbeds with short bills and laterally compressed bodies; large crakes are called gallinules; **rails** look similar but have longer bills. Flick tails constantly and fly with dangling legs. Also swim well, looking like small moorhens. If undisturbed, crakes emerge from reeds to feed along water's edge or on lily pads; Lewin's Rail more secretive, rarely seen. Food mainly aquatic invertebrates and frogs, some plants and seeds. Voices loud with creaky tones. Nest is well-hidden cup made from grass and reeds with surrounding vegetation pulled overhead in canopy, often with 'walkway' of flattened grass or reeds; chicks clad in black down, leave nest quickly.

Lewin's Rail   *Rallus pectoralis*   21–23 cm
Small rail with long pink bill, chestnut neck patch and barred belly, undertail and wings. *Imm:* streaked darker above, lacks chestnut nape. Rarely leaves reeds; flies clumsily with legs dangling. Crakes are slightly smaller, have shorter bills, less extensive barring on sides; Buff-banded Rail (overleaf) larger, shorter stouter bill, buff patch on chest, rufous barred wings. *Voice:* 'jik-jik . . . '; whistling grunt. *Nesting:* 4–6 spotted pale umber eggs. *Range:* rare, possibly disappearing, south-west, south-east and Tas.

Spotless Crake   *Porzana tabuensis*   18–19 cm
Small all-dark crake with barred undertail and reddish legs and eyes. *Imm:* less sooty with white throat, dark eyes and legs. *Voice:* varied high-speed rattling, chittering and bubbling. *Nesting:* 4–6 mottled umber eggs. *Range:* uncommon to rare nomad or migrant in south-west and south-east north to Rockhampton, Qld; more common in Tas and north-east (Atherton Tableland) Oct–May.

Australian Crake   *Porzana fluminea*   19–21 cm
Small crake with short red and green bill, dark grey breast, olive back spotted with white, barred belly and white undertail coverts. *Imm:* browner above, head and breast streaked darker and spotted white. Often emerges from cover. Water Rail has longer bill, rufous neck and more barring on sides and undertail; Marsh Crake has green bill, paler underparts, barred undertail. *Voice:* 'krr-ek'; buzzing 'bzzzt'. *Nesting:* 4–8 eggs. *Range:* uncommon to rare in reedbeds, possibly decreasing, south-west, south-east, and Tas; and Kimberley, WA.

Baillon's Crake   *Porzana pusilla*   15–16 cm
Very small crake with dull green bill, pale grey breast, olive back streaked with black and white, barred belly and undertail. Lewin's Rail has longer bill, more profusely barred; Australian Crake is darker with red and green bill (appearing almost yellow in some lights) and white undertail. *Voice:* 'krek-krek'; 'churrr'. *Nesting:* 4–8 eggs. *Range:* common in reedbeds and other subaquatic vegetation over much of Aust and Tas where suitable conditions occur.

Red-legged Crake   *Rallina fasciata*   19–24 cm
Small red-legged crake with reddish breast, brown upperparts sparsely spotted with pale buff or white, heavily barred belly and undertail, and barred flight feathers. Very rare vagrant recorded once in Kimberley, WA. Similar crakes likely to turn up are: Band-bellied Crake *Porzana paykulli* narrower bars on belly, spots on upperparts confined to wing coverts, no barring on flight feathers; Slaty-legged Crake *P. eurizonoides* lacks spots on upperparts, belly more finely barred, legs slaty grey, bars in flight feathers visible only when wing opened, longer bill; Ruddy-breasted Crake *P. fusca* lacks spots on upperparts and flight feathers, less barring on belly.

White-browed Crake   *Poliolimnas cinereus*   18–19 cm
Small tropical crake with dark cap, two white stripes on face and no bars on pale underparts. *Juv:* similar but with brown cap and buff sides to breast. Ventures from cover more than other crakes, often on lily pads, inhabits long grasses and mangroves near water as well as reeds. *Voice:* high-pitched reedy piping, usually at sunrise and sunset. *Nesting:* 4–6 spotted and blotched pale green eggs. *Range:* uncommon in swamps and mangroves in north from Kimberley, WA, to Townsville, Qld.

CRAKES

RAIL

LEWIN'S

ad

juv

SPOTLESS

juv

ad

AUSTRALIAN

BAILLON'S

RED-LEGGED

juv

ad

WHITE-
BROWED

35 x 26	30 x 23	28 x 20	31 x 23	29 x 22
Lewin's	Spotless	Baillon's	Australian	White-browed

RAILS, LARGE CRAKES AND JACANAS

These rails are found in tangled vegetation and long grasses rather than in reeds, not always close to water. Buff-banded extends to offshore islands; Red-necked is most often seen in rainforest along streams. **Jacanas:** have abnormally long toes to spread weight while walking on waterlily leaves and other floating vegetation. Fly clumsily with long toes trailing, generally move between lagoons at night. Nest is small floating pile of vegetation, highly-glossed eggs often half-submerged.

Buff-banded Rail   *Rallus philippensis*   31 cm

Short-billed rail with heavily barred underparts broken by buff patch on breast. Lewin's (p 92) is smaller, has longer bill, less extensive barring on underparts. *Voice:* rapid 'tuk-e-te-ka'; throaty 'krek'; 'swit-swit'. *Nesting:* well-hidden grass cup in dense vegetation; 5–8 blotched eggs. *Range:* patchily distributed in swamps and rank grasses near water and on islands in south-western, northern, eastern, south-eastern mainland and Tas.

Red-necked Crake   *Rallina tricolor*   30 cm

A large dark-brown crake with head and breast reddish-chestnut. Faint buff bars on flanks, more prominent bars in wing showing when it (rarely) flies. *Voice:* hollow 'coot-coot-coot . . .' while walking; grunting 'gurk-gurk . . . '. *Nesting:* cup of grass or leaves among tree roots; 3–6 spotted and blotched cream eggs. *Range:* locally common in rainforest in north-eastern Qld from Cape York to about Ingham.

Bush-hen   *Gallinula olivacea*   26 cm

A small grey-breasted gallinule with belly and undertail coverts buff; bill bright green with red base and olive-green legs. *Voice:* extended call of 10–15 loud notes, harsh and slow at first tailing away more quickly; single 'peet'. *Nesting:* grass cup in tall grass or tangled vegetation, 4–7 freckled pinkish eggs. *Range:* common resident in north, summer visitor in south, in thick marshy vegetation, in north-east from Weipa to Brisbane, Qld.

Corncrake   *Crex crex*   27 cm
Large buff-coloured crake with chestnut shoulders, inhabiting long grass and thick undergrowth. Loud rasping call at night like thumbnail on teeth of comb. Very rare vagrant from northern hemisphere, normally migrating to southern Africa and Arabia; recorded near Sydney, NSW, 1893, and on a ship off Jurien Bay, WA, 1944.

Comb-crested Jacana   *Jacana gallinacea*   20–24 cm

A dark-shouldered jacana with red or orange chicken-like comb on forehead. *Adult:* dark bronze back and wings, crown black extending to broad breastband. *Juv:* reddish-brown above, whitish-buff below. Active on floating vegetation on lagoons and ponds. *Voice:* thin piping. *Nesting:* low pile of floating vegetation; 3–4 beautiful glossy brown eggs covered with black lines. *Range:* common to uncommon resident or nomad on lagoons and ponds with floating vegetation in northern and eastern Aust.

Pheasant-tailed Jacana   *Hydrophasianus chirurgus*   25–30 cm + tail
A white-shouldered jacana without crest or comb on forehead. Most likely to be in non-breeding plumage while in Aust, superficially similar to Comb-crested but easily told by whitish shoulders. Breeding plumage is distinctive with long drooping tail and yellow hindneck. Very rare vagrant from south-east Asia, which normally migrates to Indonesia, recorded once at Paraburdoo, WA.

CRAKES AND RAILS

BUFF-
BANDED

RED-
NECKED

BUSH-
HEN

CORNCRAKE

JACANAS

COMB-
CRESTED

juv

ad

PHEASANT-
TAILED

non-
br

br

36 x 26	35 x 26	40 x 29	30 x 23
Red-necked	Buff-banded	Bush-hen	Comb-crested

LARGE RAILS AND GALLINULES

Rails have long bills; the single species here is largest of all rails; lives among mangroves on northern coast. **Gallinules** are large crakes with short bills and coloured frontal shield on forehead; much of their food is taken by grazing, usually near water or on emerging aquatic weeds; Coot often dives for food, has flaps on toes to aid swimming; Swamphen often pulls up reeds and grasses to feed on succulent bases and roots. Nests are large, usually well hidden, Coot less so. Chicks are covered in black down, some have coloured down on head.

Eurasian Coot   *Fulica atra*   32–42 cm
A black gallinule with frontal shield and beak white. *Juv:* greyish-brown, paler below, bill greyish. Less dependent on cover than most rails, often feeds in open water, jumping slightly to gain impetus for dive, or grazing on shoreline. *Voice:* loud 'kwok'; hoarse whinnying screech. *Nesting:* large cup of waterweeds and grass; 5–7 spotted umber eggs. *Range:* common resident or nomad on wetlands, throughout mainland and Tas.

Dusky Moorhen   *Gallinula tenebrosa*   35–42 cm
A dusky brown gallinule with frontal shield red, thin bill with yellow tip and with white patches on each side of undertail. *Juv:* greyish-brown, paler below, with dusky greenish bill. Usually seen swimming not far from cover; often perches in groups in half-submerged logs. *Voice:* loud sharp 'krek'. *Nesting:* large well-hidden cup of grass or reeds among reeds, usually on a solid base; 7–10 blotched buff eggs. *Range:* common to uncommon resident or nomad on wetlands in south-west and eastern Aust.

Purple Swamphen   *Porphyrio porphyrio*   45–52 cm
Large gallinule with frontal shield and stout bill red; breast blue (south-western race *bellus*) or purple (widespread race *melanotus*). Often grazes well away from water, but makes for cover if disturbed. Builds resting platforms of trampled flags, roosts in trees overhanging water. *Voice:* loud screeches. *Nesting:* large well-hidden cup of grass and reeds; 3–5 blotched buff eggs. *Range:* common resident or nomad in wetlands in south-west, east and Tas; less common in northern Aust.

Black-tailed Native-hen   *Gallinula ventralis*   30–37 cm
Bantam-like gallinule with small bright green frontal shield, bill green above, red below; white spot on flank; black undertail; bright red legs. Large groups arrive overnight, stay awhile then vanish, perhaps for years. *Voice:* sharp 'kek'. *Nesting:* well-hidden flat grass cup, often in loose colonies; 5–7 blotched green eggs. *Range:* locally abundant during irruptions, highly nomadic in wetlands throughout Aust, vagrant Tas.

Tasmanian Native-hen   *Gallinula mortierii*   45 cm
Large gallinule with small frontal shield and bill green; brown above and grey below with large white patch on flank. Usually singly or in small groups often seen grazing in open, runs swiftly to cover when disturbed, unable to fly; builds roosting nests in tussocks. *Voice:* rasping 'see-saw'. *Nesting:* trampled hollow in grass tussock; 6–9 blotched and spotted umber eggs. *Range:* common resident in grasslands and wetlands with suitable cover in Tas.

Chestnut Rail   *Eulabeornis castaneoventris*   43–52 cm
Large chestnut-bellied rail with long yellow-tipped green bill and powerful yellow legs. Very shy, one of the most difficult birds to view satisfactorily, keeping out of sight in mangrove thickets. *Voice:* loud screech; loud hollow drumming. *Nesting:* large flat stick cup in mangrove; 3–4 spotted pinkish eggs. *Range:* uncommon to common in thick mangroves, mainly in estuaries, in northern Aust, from about Derby, WA, to about Karumba, Qld.

GALLINULES

DUSKY

ad

juv

COOT

ad

juv

SWAMPHEN

eastern form

western form

BLACK-TAILED

CHESTNUT

RAILS

TASMANIAN

53 x 36	52 x 35	45 x 30	53 x 36	60 x 40	55 x 35
Dusky	Coot	Black-tailed	Swamphen	Tasmanian	Chestnut

LARGE PLOVERS

Long-legged wading birds with plump bodies, short necks and short bills with 'nail' at the tip causing a characteristic subterminal thickening. Some large plovers have facial wattles and are called lapwings because of their method of flight. The two species occurring in Australia live and breed here. Other larger plovers are migratory, two common and one rarely sighted, distinguished by its white axillaries and underwing coverts. Lesser Golden is sometimes regarded as two species, Pacific Golden *P. fulva* and American Golden *P. dominica*. For fascinating discussion, see Peter G. Connors, *The Auk*, 100: 607–620. Smaller plovers on following pages are referred to as sand-plovers, plovers or dotterels, somewhat arbitrarily.

Lesser Golden Plover   *Pluvialis dominica*   23–28 cm

Large migratory plover with slender bill, clear white eyebrow, buff-grey 'armpit' visible in flight, dark rump, faint wingbar formed by white primary shafts. *Breeding:* above speckled black and gold; face and underparts, including undertail coverts, black. *Non-breeding:* head, neck, breast, back and wings grey-brown with pale yellow speckles, belly white. *Juv:* similar to adult but rather paler and with obscure bars on underparts. Easily told from Grey by more slender build, gold spotting on upperparts and greyish armpit. Two forms may occur, regarded by some as different species: most are Pacific form (race *fulva*) with complete white flanks in breeding, more mottling on breast and belly in non-breeding, and more golden spangled in juv; and occasional birds are American form (race *dominica*), with incomplete white flanks in breeding, less mottling on breast in non-breeding, and greyer in juv. (Watch for Golden Plover *P. apricaria*, which may come as vagrant; in non-breeding plumage, stouter build, less clear eyebrow, larger in body size but shorter legs and bill, slightly more boldly speckled above, pure white 'armpits' and slightly more obvious white wingbar in flight with white on primary shafts extending to webs.) *Voice:* musical 'tuill' or 'fiu'; 'too-weet' or 'too-lee-e'; 'queedle' or 'deedleek'. *Range:* common migrant from Arctic to coastal and subcoastal wetlands, fields and inland swamps around mainland and Tas.

Grey Plover   *Pluvialis squatarola*   25–30 cm

Large migratory plover with stout bill, black axillaries (or 'armpit', visible only in flight), white wingbar and white rump. *Breeding:* above speckled black and white, face and underparts (except lower belly and undertail coverts) black. *Non-breeding:* above grey-brown with white edges to feathers, underparts white with some mottling on breast. *Juv:* above grey-brown spotted with buff and gold; underparts white with mottling and streaking on crown, face, breast and flanks. Rather similar to Golden, but bulkier with stouter bill and showing diagnostic black 'armpit' when flushed. *Voice:* plaintive 'pee-co-ee' or 'pee-er-ee'. *Range:* uncommon migrant from arctic to mudflats, salt marshes and estuaries around mainland and Tas.

Banded Lapwing   *Vanellus tricolor*   25–28 cm

Large resident plover with small red facial wattles and broad white band in upperwing. *Juv:* wattles small, back feathers edged paler. *Voice:* sad 'er-chill-char . . .'. *Nesting:* scrape on ground lined with twigs, stones or animal droppings, 4 blotched olive eggs. *Range:* common resident or nomad in open spaces throughout mainland and Tas, less common in north.

Masked Lapwing   *Vanellus miles*   35–38 cm

Large resident plover with yellow facial wattles and no white in upperwing. Two forms occur: eastern form (race *novaehollandiae*) with extensive black patch on side of neck and northern form (race *miles*) with sides of neck white; intergrades with varying amount of black on neck occur where two forms meet. *Juv:* similar to adult, but with smaller wattles and pale edges to back feathers. *Voice:* staccato 'keer-kik-ki-ki . . .'; 'tek'; trilling 'krrr' on landing. *Nesting:* scrape on ground sparsely lined with twigs, stones or animal droppings; 4 blotched olive eggs. *Range:* common resident or nomad in open spaces throughout Tas, eastern and northern Aust, but some wander extensively, so individuals could turn up anywhere.

PLOVERS

LESSER GOLDEN

Pacific form

br

juv

non-br

American form

br

GOLDEN

non-br

br

GREY

juv

br

non-br

LAPWINGS

MASKED

eastern form

northern form

BANDED

42 x 32

43 x 32

Banded

Masked

SMALL RESIDENT PLOVERS

These small plovers reside throughout the year and breed in Aust. Unlike most other wading birds they do not have a plain non-breeding plumage, so present no difficulties in identification. Three are dependent on expanses of water, but Inland Dotterel is very much a dry-country bird, living in often arid stony country. All nest on ground, laying in a scrape, usually near a stone or piece of wood; Inland covers eggs with sand when it leaves nest.

Red-kneed Dotterel   *Erythrogonys cinctus* 17–19 cm

Medium-sized plover with black cap and breastband, white throat and chestnut flanks, usually seen feeding in shallow water on fresh, brackish or moderately saline lakes and swamps throughout Aust. *Juv:* clean-cut, with cap, back and wings brown, lacks breastband and chestnut flanks but has typically front-heavy appearance of adult. *Flight:* broad white trailing edge to wing, centre of rump and tail dark, sides white. *Voice:* liquid 'chet-chet' or 'wit-wit'; trilling 'prrrp prrrp'. *Nesting:* scrape in ground, usually on island in lake, well lined with twigs; 4 streaked and blotched pale umber eggs. *Range:* patchily common on lakes, swamps and ponds throughout Aust.

Hooded Dotterel (Plover)   *Charadrius cucullatus* 19–21 cm

Medium-sized plover with black head, white nape and black shoulder patch, usually seen on sandy beaches in Tas, southern and south-eastern mainland and on saltlakes in southern WA. *Juv:* very pale looking with sandy brown cap, earpatch and shoulder, faintly scalloped sandy brown back and pale yellowish legs. *Flight:* bold white wing stripe, dark centre to rump. *Voice:* short piping 'peet-peet . . .'; barking 'fow-fow'. *Nesting:* shallow scrape in sand; 2–3 blotched buff eggs. *Range:* uncommon to locally common on southern sandy beaches or seasonal visitor to western saltlakes.

Black-fronted Dotterel (Plover) *Charadrius melanops* 16–18 cm

Small plover with black forehead and breastband, chestnut patch on shoulder, bright red bill with black tip, usually seen on edges of fresh lakes, ponds and dams. *Juv:* prominent white eyebrow, no breastband or chestnut shoulder patch, but with reddish wing coverts. When disturbed often flies a short distance, runs swiftly on twinkling legs or stands with back to observer relying on camouflage. *Flight:* slow hesitant wingbeats, prominent white stripe in wing, dark centre to rump. *Voice:* high pitched metallic 'tink-tink'; rapid 'tik-ik-ik-ik . . .'. *Nesting:* small scrape in sand or among stones, sometimes well away from water; 2–3 finely freckled pale umber eggs. *Range:* common on edges of lakes, ponds and dams, less frequent on salt water, throughout mainland and Tas.

Inland Dotterel   *Peltohyas australis* 19–23 cm

Medium-sized sandy-coloured plover with narrow black Y-shaped band on breast and vertical black line through eye, usually seen well away from water on stony ground, mainly in interior. *Juv:* lacks black line through eye and on breast. When disturbed runs, stopping frequently, bobbing head with back to intruder. *Flight:* buff patch in upperwing; underwing coverts pale buff, undertail white, contrasting with chestnut belly. *Voice:* metallic 'quoik', 'krrrooot'. *Nesting:* scrape in sand, often scratches soil over eggs when leaving nest; 2–3 heavily blotched buff or olive eggs. *Range:* uncommon nomad in interior of Aust, mainly on stony ground in arid areas but also paddocks, isolated unsealed roads.

DOTTERELS

juv

ad

HOODED

juv

ad

RED-KNEED

juv

ad

BLACK-FRONTED

juv

ad

INLAND

39 x 27	30 x 22	29 x 20	37 x 27
Hooded	Red-kneed	Black-fronted	Inland

SAND-PLOVERS

These small to medium-sized plovers have reddish breasts in breeding plumage, but for most of their time in Australia are rather plain grey-brown above and white below with grey-brown sides to breast. The two sand-plovers are like larger versions of the common resident Red-capped Dotterel (overleaf) and are mostly seen on beaches or mudflats – some Mongolians look big and may be mistaken for Large, but lack the longer bill. Oriental Plover is more often seen in small flocks on drier ground, such as dry floodplains, airfields, paddocks, etc, and is rather difficult to approach. Caspian is very similar but smaller, very rarely recorded but may be more common than suspected. Illustrated are a juvenile Oriental and an adult Caspian in worn plumage, but apart from size each could equally represent the other species. Both have less hunched stance than other plovers, and cross wings over tails like pratincoles.

Caspian Plover
Charadrius asiaticus 22–25 cm

Rare vagrant plover very similar to Oriental Plover but with white underwing in flight, most likely to occur on dry ground. *Adult breeding:* broad chestnut breastband with narrow black line below; black streak behind eye in Sep or Mar–Apr. *Non-breeding:* broad grey-brown breastband, feathers on back with buff edges in spring, wearing away towards autumn. *Range:* very rare summer vagrant from central Asia, most likely on dry plains of north and north-west.

Oriental Plover
Charadrius veredus 22–25 cm

Medium-sized long-legged plover with broad breastband, grey underwing in flight, elegant head-up stance, usually seen in small flocks on dry ground sometimes with pratincoles and Lesser Golden Plovers which are more mottled above. *Adult breeding:* broad chestnut breastband with broad black band below (female duller), head and face largely white with diffuse brown crown. *Non-breeding:* broad grey-brown breastband, feathers on back edged buff in spring, wearing away towards autumn. *Juv:* broader, brighter buff edges to back and wing feathers. Flocks could be examined for rare Caspian, which is smaller, has white underwing linings but is very difficult to differentiate except in breeding plumage. *Flight:* rump, tail and lower back grey-brown, faint narrow white stripe in wing, underwing brownish-grey; when disturbed generally flies off close to ground in fairly tight flock with assertive wingbeats, alights at short distance, but may zig-zag erratically away if pressed. *Voice:* piping 'klink' in flight; soft 'chip-chip-chip' or 'tick-tick-tick'. *Range:* common summer migrant from eastern Asia to plains, paddocks, airfields, saltmarshes and occasionally mudflats in northern Aust; less common in south.

Large (Greater) Sand-Plover
Charadius leschenaultii 20–23 cm

Medium-sized coastal plover with long bill and narrow breastband, legs greyish often with olive or yellowish tinge, variable in size, leg length and bill stoutness. *Adult breeding:* prominent black mask, broad white forehead, narrow orange-chestnut band on upper breast. *Non-breeding:* white eyebrow, breastband pale grey-brown, mainly restricted to sides of breast but usually joined at centre by narrow line. *Juv:* pale buff edges to feathers of back and wings, breastband restricted to sides of breast; very similar in plumage to juvenile Mongolian, best told by larger size and longer bill. *Flight:* narrow white wingbar, more obvious than Mongolian; dark centre to rump, with sides and outer tail feathers white. *Voice:* quiet 'treep'; 'treep-treep . . .'; 'trr-dit, trr-dit . . .' in flight downward inflected trill, shorter and slower than Mongolian. *Range:* common summer migrant on sandy beaches and coastal mudflats on north and west coasts, less common in east and south, uncommon Tas.

Mongolian (Lesser) Sand-Plover
Charadrius mongolus 19–20 cm

Small coastal plover with small bill and breastband narrow in non-breeding plumage or broad in breeding plumage, legs dark. *Adult breeding:* prominent black mask with white patch on forehead, broad chestnut breastband with colour extending to flanks (west Asian birds have small forehead spot and no black on breastband; east Asian birds have large forehead spot and narrow black line on breastband). *Non-breeding:* narrow white eyebrow, grey-brown breastband usually confined to sides of breast, but sometimes joining in centre. *Juv:* pale buff edges to feathers of back and wings, buff breastband restricted to sides of breast. *Flight:* narrow white wingbar, centre of rump dark, sides of rump and tail less white than Large. *Voice:* trilling 'drrrit'; quiet 'trik', 'trik-it'. *Range:* common summer migrant on sandy beaches and coastal mudflats on east and north coasts, less common in south-west and south, uncommon Tas.

PLOVERS

ORIENTAL

CASPIAN

br

non-br

juv

br

SAND-PLOVERS

LARGE

ad

juv

non-br

MONGOLIAN

East
Asian
form

juv

non-br

West
Asian
form

Caspian

Oriental

Large

Mongolian

RINGED PLOVERS

Small plovers with a prominent dark breastband in breeding plumage. In non-breeding plumage the collar is lighter in colour. Red-capped has incomplete breastband, breeds here. Double-banded is migrant from New Zealand with large number of juveniles mostly in winter; Ringed is rare summer migrant from northern hemisphere; Little Ringed may also be from northern hemisphere, but could be from South-East Asia or New Guinea – may yet be found breeding along rivers of western Cape York.

Double-banded Plover (Dotterel) *Charadrius bicinctus* 17·5–19 cm

Small migratory plover from NZ, either with two bands on breast (breeding Aug–Mar) or with single breastband, indistinct eyebrow, green-grey legs and buff-brown upperparts (non-breeding and juv Feb–Aug). *Adult breeding:* unmistakable with black band on upper breast and chestnut band on lower breast. *Non-breeding:* combination of indistinct eyebrow, no prominent white patch on forehead and buff tinge to upperparts distinguishes from all similar small plovers. *Imm:* similar to non-breeding but with two sandy-brown breastbands. *Juv:* similar to non-breeding but with single incomplete breastband and with pale edges to feathers of upperparts – arrives Feb–Mar after all other juvenile plovers have moulted. *Flight:* narrow wingbar, dark centre to rump and tail. *Voice:* high-pitched 'pit' or 'peet'; lower pitched 'chip-chip'; 'tink' in flight. *Range:* common migrant from NZ, most abundant Feb–Aug on mudflats, beaches, marshes, lakes, dams, paddocks and open fields in south-east and Tas, wandering to Cairns, Qld, and southern WA.

Ringed Plover   *Charadrius hiaticula*   18–19 cm

Smallish plump migratory plover with white ring around neck, single complete breastband, white wingbar, orange eye-ring and yellow legs. *Adult breeding:* yellow bill with black tip, black breastband, black band over forehead meeting brown crown. *Adult non-breeding:* dark bill with yellowish base, white forehead, brown or black breastband. *Juv:* black bill with yellow base to lower mandible, white forehead extending as far as eyebrow, wings and breastband edged buff with fine black subterminal line. *Flight:* narrow but prominent white wingbar, dark centre to rump, white edges to tail. *Voice:* loud musical 'too-lee', with emphasis on second syllable. *Range:* rare vagrant from northern hemisphere, likely on any sandy or muddy shores around Aust.

Little Ringed Plover   *Charadrius dubius*   15 cm

Small slender migratory plover with white ring around neck, single complete breastband, no white wingbar, yellow eye-ring and flesh-coloured to orange legs. *Adult breeding:* black bill with yellow base to lower mandible, white forehead, black band over forehead not extending behind eye, black areas separated from brown crown by white band – narrow in Asiatic form (race *curonicus*) to broad in New Guinea form (race *dubius*). *Adult non-breeding:* black areas replaced by brown. *Juv:* black bill, pale buff diffuse forehead, buff edges to back, wings and sometimes incomplete breastband, looks paler and more buff than juvenile Ringed. *Flight:* only faintest of wingbars, dark rump, subterminal band in white-edged tail. *Voice:* whistling 'pee-u', with emphasis on first syllable. *Range:* extremely rare vagrant, could be from northern Asia (race *curonicus*), South-East Asia (race *jerdoni*) or New Guinea (race *dubius*).

Red-capped Plover (Dotterel)   *Charadrius ruficapillus*   15 cm

Small resident plover with incomplete breastband, incomplete white ring about neck, broad white forehead not extending behind eye, and pale sandy-brown upperparts. *Male:* rufous cap and breastband. *Female:* brown cap and breastband, sometimes tinged rufous. *Juv:* very pale with pale buff edges to feathers of crown and upperparts. *Flight:* narrow white wingbar, dark centre to rump, white sides to tail. *Voice:* sharp 'wit', sometimes extending into rapid trill; 'prrrt' in flight; piping 'weet'. *Range:* common resident or nomadic plover mainly associated with salt or brackish water throughout Aust.

Kentish Plover   *Charadrius alexandrinus*   15–17 cm

Similar to Red-capped Plover but in all plumages has white collar; breeding male has heavier black patch on sides on neck, not extending to hind-neck. Malaysian Plover *Charadrius peronii* (shown opposite) differs in that (a) black breast-band extends to hind-neck (b) female has rufous cap and breast band. *Range:* Very rare vagrant from northern hemisphere or Indonesia, possible on any northern salt or brackish wetlands.

PLOVERS

DOUBLE-BANDED

imm

br

non-br

juv

RINGED

non-br

br

juv

Asiatic form

New Guinea form

br

LITTLE RINGED

non-br

juv

RED-CAPPED

br

non-br

juv

MALAYSIAN

29 x 21

Red-capped

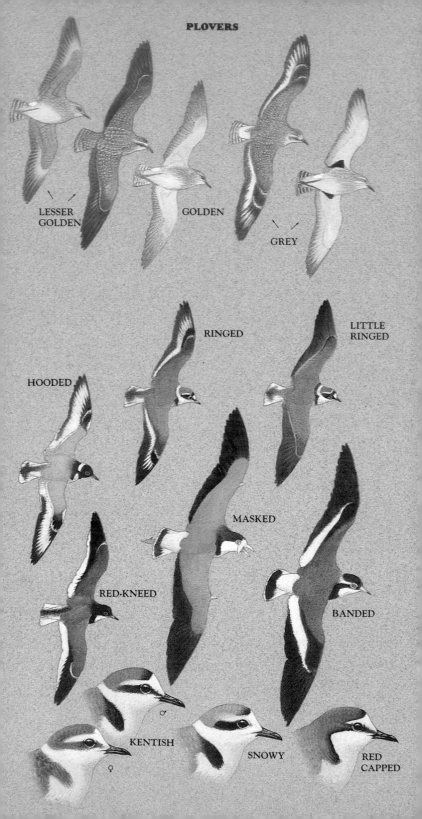

PLOVERS

LESSER
GOLDEN

GOLDEN

GREY

HOODED

RINGED

LITTLE
RINGED

MASKED

RED-KNEED

BANDED

KENTISH

♂

♀

SNOWY

RED
CAPPED

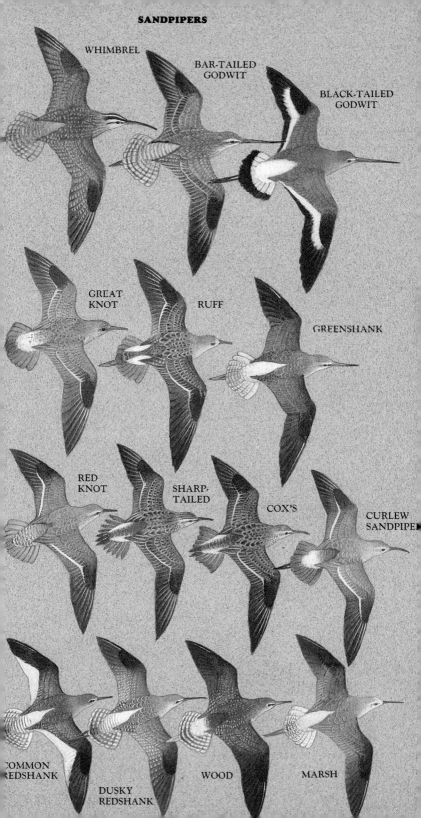

SANDPIPERS

WHIMBREL

BAR-TAILED
GODWIT

BLACK-TAILED
GODWIT

GREAT
KNOT

RUFF

GREENSHANK

RED
KNOT

SHARP-
TAILED

COX'S

CURLEW
SANDPIPE

COMMON
REDSHANK

DUSKY
REDSHANK

WOOD

MARSH

STINTS

Smallest sandpipers characterised by short bills, shortish legs and dumpy bodies, usually seen working energetically and methodically picking over mudflats and along lake margins. Red-necked is commonest migratory wader in Aust, easily distinguished from other common shorebirds by its small size. However, the rarer, equally small stints are very similar and are notoriously difficult to differentiate, so they may easily be overlooked: the best way to find them is to look critically at all stints. Long-toed is easiest to find, being darker and with yellowish legs; Western has longer legs; smaller Little is hardest, requires subjective assessment of finer bill; Temminck's has yellowish or greenish legs, and tail extends beyond wingtip. A brilliant paper by P.J. Grant and Lars Jonsson, in *British Birds*, 77: 293–315, is worth obtaining; it includes all stints likely to occur in Aust.

Red-necked Stint   *Calidris ruficollis*   15 cm

Tiny active sandpiper with black legs, short stout bill, slightly bulbous at tip, commonest small wader. *Adult breeding:* throat chestnut merging with narrow black-speckled gorget on upper breast. *Non-breeding:* above grey-brown, below white with greyish unstreaked smudge on side of breast. *Juv:* least brightly coloured of juvenile stints, lack white 'V' on back of juv Little. *Flight:* broad wingbar, dark centre to rump and tail, grey-brown sides to tail. *Voice:* weak 'tewp'; trilling 'krrrt'; 'pit' harsher than Little. *Range:* common migrant from northern Asia (Aug–May) to coastal mudflats and estuaries, often seen inland on salt and fresh wetlands particularly on migration Aug–Oct, throughout Aust; many overwinter.

Little Stint   *Calidris minuta*   14 cm

In non-breeding plumage very difficult to tell from Red-necked, but bill tends to be finer, particularly towards tip. *Adult breeding:* brighter, more orange-chestnut than Red-necked, throat white, broad gorget on upper breast of buff or grey finely speckled black; distinct white edges to mantle, forming prominent 'V' when viewed from behind. *Juv:* similar to breeding but usually darker above nape pale and streaked buff patch on sides of breast; distinct 'V' pattern on back from white edges to mantle and scapulars. *Voice:* short 'pit'. *Range:* rare vagrant from northern Eurasia, likely wherever Red-necked occurs.

Long-toed Stint   *Calidris subminuta*   15 cm

Tiny sandpiper with yellow legs and long central toe (20–23 mm), looking like small Sharp-tailed Sandpiper. In flight, toes extend beyond tailtip. Other stints so far recorded in Aust have black legs (but watch for (a) Least Sandpiper *Calidris minutella* from America, which is very similar, has slightly thicker bill and in hand has shorter central toe (16–19 mm), toes just extend beyond tail in flight; (b) Temminck's Stint *Calidris temmincki* which is plainer dark grey than brown in all plumages with white sides to tail visible in flight, dark patch on side of breast like Common Sandpiper, and tail extending beyond wingtip). *Juv:* similar to much larger Sharp-tailed, with yellowish legs but lacking bright buff breast. *Voice:* soft 'chrrup-chrrup-chrrup'; 'seepseep . . .'. *Flight:* dull wingbar, less conspicuous than Red-necked, and less pale at side of tail, looks darker in flight, erratic towering flight when flushed. *Range:* scarce regular visitor from northern hemisphere (Sep–Apr) to freshwater and brackish lakes and swamps, only rarely on coast, most common in west.

Western Sandpiper   *Calidris mauri*   16 cm

Tiny black-legged sandpiper with bill slightly longer than other stints, often looking downcurved at tip; in hand shows partial webbing between toes (Semipalmated Sandpiper *Calidris pusilla* from America very similar, but straighter bill averages shorter with noticeably bulbous tip). *Breeding:* rusty cheeks and crown, breast white strongly streaked and spotted black, with spots extending along flanks. *Non-breeding:* pale grey sides to breast with fine dark streaks, virtually identical to non-breeding Red-necked but lacks white 'V' on back, has rufous edges to scapulars; cap and ear coverts rufous, pronounced white eyebrow. *Voice* trilling 'chrreep'; 'chiet'. *Range:* rare vagrant from eastern Siberia and Alaska normally wintering in South America but odd birds may occur with Red-necked (Aug–Apr).

STINTS

RED-NECKED

juv

non-br

br

non-br

LITTLE

juv

non-br

br

LONG-TOED

juv

non-br

br

non-br

WESTERN

juv

br

non-br

TEMMINCK'S

SMALL SANDPIPERS

These are larger than stints but smaller than the common Sharp-tailed and Curlew Sandpipers. None is really common, and the two lower species, White-rumped and Baird's, are only rarely sighted – they differ from all other small to medium sandpipers in the length of the wings, which at rest project well beyond the tailtip. The Sanderling is well-named, being virtually restricted to sandy beaches and coastal lagoons; it lacks a hind toe. The Broad-billed has a noticeably drooping bill and, in breeding and juvenile plumages, a snipe-like pattern on the head; it is an active feeder on mudflats, regular visitor in north, uncommon in south, absent some years.

Sanderling *Calidris alba* 20 cm

Small to medium sandpiper with black shoulder patch and prominent white wingbar, usually found on sandy beaches. *Adult breeding:* head and breast rufous with black and buff speckles, back feathers rufous with black centres. *Adult non-breeding:* pale grey above, white below – palest sandpiper. *Juv:* back black with attractive pattern of white spots, below white with pale buff breastband that fades with wear. *Flight:* broad white wingbar, dark centre to rump, white sides to tail. *Voice:* high-pitched liquid 'twick-twick' in flight, quiet twittering when feeding in groups. *Range:* uncommon to locally common migrant from Siberia mostly to sandy beaches and coastal sand lagoons around Aust, restricted in Tas; some overwinter.

Broad-billed Sandpiper *Limicola falcinellus* 18 cm

Small active sandpiper, larger than Red-necked Stint with much longer downcurved bill, dark shoulder patch. *Adult breeding:* snipe-like double stripe on head, feathers of back black with grey tips and chestnut edges, when worn looking black, dark streaks on breast extending along flanks. *Adult non-breeding:* grey-brown above, white below with faint streaks on breast. *Juv:* snipe-like pattern on head, feathers of back and rump black with buff and white edges, breast buff with pale brown streaks. *Flight:* white wingbar, dark centre to rump, dark shoulders. *Voice:* trilling 'chree-eep' in flight less musical than Curlew Sandpiper; 'drrt-drrt . . .'. *Range:* locally common migrant from Eurasia to mudflats, saltmarsh estuaries and occasionally fresh swamps in northern and north-eastern Aust, scarce elsewhere.

Baird's Sandpiper *Calidris bairdii* 18 cm

Small sandpiper with buff tone to plumage, particularly on breast, wings at rest extending well beyond tail, fine bill pointed at tip, dark rump with little white at sides. *Adult breeding:* crown and breast buff with brown streaks, back strongly scalloped black and buff. *Adult non-breeding:* grey-brown above, crown and breast pale buff streaked darker. *Juv:* similar to breeding but buff breast brighter although less extensive, back with paler buff edges to feathers. *Flight:* short narrow wingbar, dark rump. *Voice:* trilling 'kreep' or 'kree-eep' not unlike Curlew Sandpiper. *Range:* rare vagrant from North America, most likely on coastal or subcoastal wetlands of eastern mainland or Tas.

White-rumped Sandpiper *Calidris fuscicollis* 19 cm

Small to medium sandpiper with white band on rump, black legs and wings at rest extending well beyond tail (only other sandpiper with white rump and black legs is larger Curlew Sandpiper, which has downcurved bill). *Adult breeding:* reddish crown, feathers of back black with chestnut, buff and white edges, breast pale buff with streaks extending to flanks (similar to Western but with shorter, straight bill and lacking chestnut ear coverts.) *Adult non-breeding:* grey-brown above, grey breast with darker streaks extending to flanks. *Juv:* similar to non-breeding but with distinct pale edges to feathers of back and wings. *Flight:* faint short wingbar, white band across rump. *Voice:* distinctive high-pitched thin 'jeet' or 'jee-et'. *Range:* very rare vagrant from North America, most likely on swamps, lakes, flooded pastures or brackish coastal wetlands throughout southern Aust.

SANDPIPERS

SANDERLING

non-br

juv

br

BROAD-BILLED

non-br

juv

br

BAIRD'S

non-br

juv

br

WHITE-RUMPED

non-br

juv

br

SANDPIPERS

Medium-sized waders with bills solid at base, finer at tip, slightly decurved but looking more or less straight in the field except through good optics. In flight, all show dark centre to rump and narrow white wingbar. Sharp-tailed is among commonest waders, once learned is useful reference point for less-common species. Its numbers tend to obscure presence of others which are not sighted often; Cox's in particular being a newly-described species should inspire field observers to extend knowledge by recording observations. Juvenile, for instance, is unknown. To distinguish species, look for colour of legs, amount of speckling on breast, presence or absence and shape of eyebrow.

Sharp-tailed Sandpiper   *Calidris acuminata*   18–23 cm

Medium-sized sandpiper with boldly mottled upperparts, rufous crown, green legs, greenish-yellow base to bill and finely streaked breast. *Adult breeding:* breast heavily streaked, extending to boomerang-shaped chevrons on belly; edges of feathers on upperparts chestnut. *Non-breeding:* crown pale rufous, breast buff-grey finely streaked darker, underparts white with grey mottling on flanks, dark streaks on undertail coverts, feathers of upperparts grey with whitish fringes. *Juv:* crown bright rufous, breast bright buff with faint streaking, back feathers black with chestnut and white edges. *Flight:* narrow white wingbar, dark centre to rump. *Voice:* sharp 'whit-it' or 'tree-trit' in flight; when flushed, metallic 'pliep'. *Range:* common migrant from Siberia to coastal, subcoastal and inland wetlands throughout Aust.

Pectoral Sandpiper   *Calidris melanotos*   18–24 cm

Medium-sized sandpiper similar to Short-tailed but with heavily streaked breast abruptly demarcated from pure white belly and flanks, lower and darker in centre than at sides; bill longer with yellowish base, off-white eyebrow widest in front of eye, fading away behind (Sharp-tailed broader behind eye, giving different angle to 'cap'). *Breeding:* pale buff heavily streaked breast, feathers of back and scapulars dark brown edged chestnut. *Non-breeding:* whitish or faintly buff heavily streaked breast, edges to scapulars brown (looks darker above than Sharp-tailed). *Juv:* more brightly marked on back, wing coverts darker. *Flight:* narrow white wingbar, darker centre to rump with white oval patches on sides. *Voice:* loud chirruping 'chrrt' or 'kreek' like Budgerigar; 'trrt-trrt' usually heard only in flight. *Range:* scarce regular migrant from Siberian and American Arctic, mainly Nov–Mar, to fresh and salt marshes, flooded pastures, saltworks and sewerage farms, possible on any suitable wetlands.

Cox's Sandpiper   *Calidris paramelanotos*   18–23 cm

Very rare, recently discovered sandpiper similar to Pectoral but with bill longer, darker, slightly decurved with yellow (in any) confined to base of lower mandible, legs dark grey to yellow-olive (usually look dark in field); eyebrow less distinct with faint streaking; upperparts more mottled than Curlew Sandpiper but less than Sharp-tailed. *Flight:* in non-breeding rump like Sharp-tailed, but in breeding lacking solid dark centre; narrow pale wingbar, more obvious than Sharp-tailed and Pectoral, less than Curlew Sandpiper and Dunlin. *Adult breeding:* rusty ear coverts; feathers on upperparts dark centred with grey, buff and chestnut fringes, rusty edges to scapulars and tertials; breast rusty toned with black streaks, belly white with dark barring on flanks and dark streaks on lateral undertail coverts; rump mainly white with dark grey-brown and buff barring concentrated in centre. *Non-breeding:* greyer overall, darker and more mottled above than Curlew Sandpiper, darker and more streaked on breast. *Voice:* similar to Pectoral but quieter. *Range:* rare migrant from northern hemisphere to south-eastern wetlands; specimens from SA at Price Saltfields, upper Gulf St Vincent and Mosquito Point, Lake Alexandrina; sight records at Altona, Werribee, Laverton and Geelong, Vic, mainly on saltworks.

Ruff   *Philomachus pugnax*   ♂ 30 cm ♀ 23–27 cm

Medium-sized (female) or large (male) sandpiper with upright stance, pale face, no obvious eyebrow, unstreaked breast and underparts, legs variable in colour (red, orange, yellow, green or brown), larger, longer-legged than Sharp-tailed. *Male breeding (Ruff):* large ruff of feathers around neck – assumed on northern hemisphere breeding grounds, unlikely in Aust. *Female breeding (Reeve):* speckled plumage with dark feathers mixed with grey-brown plumage, tertials strongly barred black and chestnut. *Non-breeding:* grey-brown above, some males have broad white collar; face and underparts white, pale brown base to bill. *Juv:* brighter, with buff breast and pale buff edges to back feathers, tertials unbarred with buff edges; occasional juveniles much brighter than shown. *Flight:* narrow white wingbar, narrow dark centre to rump, with white oval patch on each side; looks lazy in flight, often glides. *Voice:* low 'tu-whit'. *Range:* rare migrant from northern Eurasia to marshes, saltponds and sewerage farms, occasionally on coastal mudflats, around mainland and Tas.

SANDPIPERS

SHARP-TAILED

juv

non-br

br

PECTORAL

juv

non-br

br

COX'S

br

non-br

RUFF

♂

non-br

♀

br

juv

SANDPIPERS

Three of these medium-sized migratory sandpipers have noticeably downcurved bills distinguishing them from most other waders (but see Broadbilled p 110, Western p 108, and to lesser extent Cox's p 112). Only Curlew Sandpiper is common (one of commonest waders); the others are rarely seen. Stilt and Buff-breasted are so distinctive they are easy to recognise, but Dunlin in non-breeding plumage can easily be overlooked, can be confused with Cox's.

Curlew Sandpiper   *Calidris ferruginea* 21 cm

Medium-sized, rather slender sandpiper with curved bill, black legs, broad white wingbar and white rump. *Adult breeding:* head and underparts bright chestnut. *Non-breeding:* plain grey-brown above, white below with faintly streaked cloudy breast. *Juv:* neat-looking with pale edges to back and wing feathers, pale buff breast with faint streaking. Feeds belly-deep in water or on mud, probing to full bill length. *Voice:* liquid 'chirrip'. *Range:* common migrant from Siberia to mudflats and estuaries around mainland and Tas, common inland on migration Aug–Oct; many overwinter.

Dunlin   *Calidris alpina* 20 cm

Medium-sized, rather dumpy sandpiper with curved bill, black legs, white wingbar and dark centre to rump. *Adult breeding:* bright chestnut and black above, distinctive black belly. *Non-breeding:* mainly grey with white belly and undertail; upperparts like Curlew Sandpiper, darker streaks on sides of breast and flanks. *Juv:* brownish-grey with buff suffusion on breast and flanks; each back feather with fine black line inside white edge. Similar to non-breeding Curlew Sandpiper, but has dark centre to rump. *Voice:* high grating 'chrri' or 'cheerp'. *Range:* rare vagrant from Arctic; rarely migrates south of equator, most likely with Curlew Sandpiper on northern mudflats.

Stilt Sandpiper   *Micropalama himantopus* 21 cm

Medium-sized to large sandpiper with long slightly decurved bill, long yellow or greenish-yellow legs, white rump but no white wingbar. *Adult breeding:* distinctive with chestnut earpatch and barred underparts. *Non-breeding:* plain grey above, white below with faint streaking on sides of breast and faint barring on upper flanks; Wilson's Phalarope rather similar but much shorter legs. *Juv:* similar but with buff suffusion to breast and white or buff edges to feathers of back and wings. *Voice:* dry buzzing 'trrhuit' or 'zh-r-r-reet'; 'pit'; scolding 'dk-dk-dk'. *Range:* extremely rare vagrant from North America, most likely Aug–Apr on marshy or brackish ponds.

Buff-breasted Sandpiper   *Tryngites subruficollis* 21 cm

Medium-sized short-billed sandpiper with yellow legs, buff suffusion to plumage and strongly scalloped upperparts. *Adult:* breast and belly buff with faint spots on sides of breast, feathers of back and wings black edged buff. *Juv:* breast and belly pale yellowish buff with spotted sides to breast, feathers of back and wings black and buff with pale edges. *Flight:* no wingbar, sides of rump buff; underwing white with black bar on primary coverts and buff on lesser coverts. *Voice:* trilling 'prrrreet'; unmusical 'chwup . . .'; quiet 'tik' like tapping stones. *Range:* extremely rare summer vagrant from North America, most likely on damp grassy meadows.

SANDPIPERS

CURLEW

juv

non-br

br

DUNLIN

juv

non-br

br

STILT

juv

non-br

br

BUFF-BREASTED

ad

juv

TATTLERS, KNOT AND TURNSTONE

Grey-tailed Tattler
Tringa brevipes 27 cm

Plain-backed plump wader with yellow legs and distinctive call: 'pyuee-pyuee'. Very similar to Wandering, but paler with white eyebrow extending behind eye. *Adult breeding:* breast and flanks narrowly barred white and black, much of abdomen unbarred. *Non-breeding:* plain grey with white belly and dark underwing; forehead pale, looking white at distance, white eyebrow extending well behind eye; bill can be orange at base. *Juv:* grey centre to forehead, paler than Wandering; upperparts finely spotted white. *Range:* common summer visitor from Siberia (Aug–Apr) to northern coasts from about Shark Bay, WA, to Sydney, NSW, less common farther south to Tas.

Wandering Tattler
Tringa incana 27 cm

Plain-backed plump wader with yellow legs and distinctive call: 'ti-ti-ti-ti . . .'. Very similar to Grey-tailed but darker on upperparts, particularly mantle and crown, dark centre to forehead, and white eyebrow not extending behind eye. *Adult breeding:* breast flanks and abdomen heavily barred, broader and more extensive than Grey-tailed. *Non-breeding:* dark grey with dark centre to forehead; eyebrow white in front of eye only. *Juv:* upperparts finely spotted white. *Range:* uncommon to rare summer visitor from North America (Aug–Apr) to rocky shores and reefs in north-eastern Aust, particularly Barrier Reef islands, south to about Wollongong, NSW. Some overwinter.

Red Knot
Calidris canutus 25 cm

Stout wader with bill same length as head, crown unstreaked, narrow white bar in wing, pale rump with grey barring, shortish olive legs. *Non-breeding:* grey above with narrow pale edging to feathers, pale eyebrow, smudged sides to neck with faint spotting. Great Knot larger, has longer bill, striped brown, more heavily spotted breast. *Juv:* feathers of back edged white with dark subterminal bar, breast more heavily spotted pale buff and flanks barred, crown faintly streaked. *Breeding:* rufous underparts, feathers of back rufous patterned with black; similarly coloured breeding Curlew Sandpiper is more slender with downcurved bill, dark legs. *Voice:* 'knut-knut', 'nyut', high-pitched 'toowit-wit'. *Range:* uncommon summer visitor to tidal mudflats and beaches, also lakes, mainly northern Aust, less common south.

Great Knot
Calidris tenuirostris 29 cm

Stout wader with bill longer than head, crown streaked, distinct white bar in wing, white rump, shortish olive legs. *Non-breeding:* larger, more heavily spotted on breast and flanks than Red Knot and more streaked upperparts with white wing bar and rump more distinct, bill too short for Dowitchers. *Juv:* strikingly patterned on back, breast heavily spotted. *Breeding:* breast spotted and barred, sometimes so heavily it appears solid black, scapulars rufous with black markings. *Voice:* 'pyuee'. *Range:* common in north, uncommon farther south, mainly on tidal flats and beaches, occasionally inland, Oct/Nov–Mar, some overwinter.

Ruddy Turnstone
Arenaria interpres 23 cm

Dumpy short-billed sandpiper with short orange legs, dark bib, prominent white bars on wings and rump in flight, and habit of turning over stones and other debris. *Juv:* feathers of back with broader pale edges. *Breeding:* striking black, rufous and white pattern. *Voice:* 'titititititi', 'quitta-quitta-quit-it-it', 'kee-oo'. *Range:* common summer visitor to most rocky platforms, tidal flats, reefs and beaches with pebbles, shells and debris, around Aust, Aug–Apr, numbers overwinter.

TATTLERS

GREY-TAILED

non-br

br

br

WANDERING

non-br

br

KNOTS

non-br

juv

br

RED

GREAT

non-br

br

TURNSTONES

non-br

juv

RUDDY

br

SNIPE

Long-billed waders most often occurring in fresh grassy marshes, jumping high and flying rapidly when disturbed. Painted Snipe is not a true snipe, being less agile, flying more slowly with legs dangling rather like a rail – it is nomadic, polyandrous and breeds in Aust. True snipe are strictly migrants, generally arriving in late Aug. Most species are so similar they cannot be identified reliably in the field by most observers; pattern of feathers on sides of tail useful when visible, and calls when flushed may help; in the hand, number, shape and pattern of tail feathers is diagnostic. However, there is a high probability that birds seen in eastern Australia are Japanese, most in Northern Territory, Kimberley and north-west are Swinhoe's, while occasional birds in the 'Top End', Kimberley and Pilbara are Pintails. It is possible that other snipe visit northern and north-western Aust; most probable are Common Snipe (more vigorous in flight with more impetuous zig-zags than Japanese, shows wider white trailing edge to wing, has white edge to outer primary, flush call 'scaipe'), Jack Snipe (smaller with larger head, shorter bill, dark wedge-shaped tail, hesitant flight, usually silent when flushed, occasionally weak 'gagh'), and a remote possibility Great Snipe (prominent white edges to wing coverts, white outer tail feathers, heavy flight without zig-zags, flush call coughing 'heert').

Japanese (Latham's) Snipe *Gallinago hardwickii* 24–26 cm
The common snipe of eastern Aust, told in hand by 16–18 tail feathers, outer four narrower, barred black and white, not extremely narrow like Pintail and Swinhoe's. When flushed, wings longer, more erratic flight, more likely to tower up than Pintail and Swinhoe's, more white on belly than Swinhoe's. *Voice:* loud 'keow' like green tree frog; when flushed, loud rasping 'shack'. *Range:* common migrant from Japan and Kuril I to eastern and Tasmanian swamps and wet grasslands.

Swinhoe's Snipe *Gallinago megala* 24–26 cm
Northern and north-western snipe, told in hand by 20–24 tail feathers with three or four feathers barred, narrower than Japanese but broader, less stiff than Pintail. When flushed, white on underparts less extensive, heavier in flight than Japanese with shorter wings; less inclined to zig-zag away or tower up to escape. *Voice:* when flushed short 'shrek'. *Range:* uncommon regular migrant from eastern Asia Aug–Apr to northern Aust, from northern Qld to Pilbara, WA.

Pintail Snipe *Gallinago stenura* 23–25 cm
Rarely encountered snipe, told in hand by 24–28 tail feathers, outer four or five unbarred, stiff, very narrow (1 mm). Bill not as long as Japanese or Swinhoe's, less bulbous at tip; back colour duller, due to narrower buff edging to feathers. In flight, more ponderous than Japanese, not inclined to zig-zag on shorter, rounder wings, nor to tower into sky to escape. *Voice:* when flushed 'scaarp'. *Range:* rare migrant from northern USSR Aug–Apr to grassy swamps and wet grasslands in northern Aust, may also feed on drier pastures.

Painted Snipe *Rostratula benghalensis* 22–25 cm
Rail-like 'snipe' with prominent white ring around eye. *Female:* throat, breast and nape black with chestnut suffusion (less chestnut than overseas race), bright buff stripes down back, greenish finely-barred wings. *Male:* streaked greyish-brown throat, breast and nape, pale buff stripes on back, brownish wings with buff and black marbling. *Juv:* like male but lacking sharp demarcation between breast and abdomen. *Imm female:* wings and back like adult female, head and neck like adult male. First reaction to intruder is to freeze, then fly if pressed, fast and low with slow wingbeats and dangling legs. *Voice:* most calls by female – penetrating 'cook-cook-cook . . .' mainly dusk and dawn (Jan–Mar in Kimberley); soft mellow 'koo-oo koo-oo . . .' like blowing across mouth of bottle; when flushed 'kek', 'kak' or 'kit'. *Nesting:* lined scrape on ground under tussock or in samphire bush; 4 blotched and lined yellowish-olive eggs. *Range:* uncommon to rare nomad in marshes, mainly in north and east, and Tas.

SNIPE

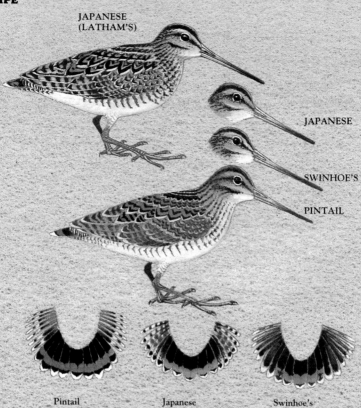

JAPANESE
(LATHAM'S)

JAPANESE

SWINHOE'S

PINTAIL

Pintail   Japanese   Swinhoe's

PAINTED

♂

imm ♀

♀

Painted 36 x 27

DOWITCHERS AND GODWITS

Medium-sized to large waders with long, straight or slightly upturned bills, usually longer in females, used for probing in mud for marine worms, crabs, etc. **Dowitchers:** related to snipe but look more like godwits. Only Asian so far recorded, often first noticed because of its 'sewing machine' method of feeding; two similar but smaller American species could turn up – only observers very familiar with them can separate them. **Godwits:** large waders with bicoloured bills. Bar-tailed is among commonest waders, ideal starting point for beginners to learn larger waders. Longer legged, more elegant Black-tailed is less common everywhere except south-west. Hudsonian is a rare vagrant, best told by dark underwing pattern.

Asian Dowitcher   *Limnodromus semipalmatus*   25–35 cm
Rare dowitcher, rather like a small Bar-tailed Godwit but with straight black bill slightly swollen at tip. *Adult breeding:* similar to Bar-tailed Godwit but less black on back. *Non-breeding:* like Bar-tailed but greyer with larger white eyebrow, more contrasting dark centres to back feathers, comparatively longer legs (body size of Great Knot but leg length of Bar-tailed). *Juv:* back feathers black with pale buff edges. *Flight:* rump and tail white with narrow bars, narrow white wing stripe. Characteristic feeding method, probing rapidly just in front of feet 'like a sewing machine'. *Voice:* quiet nondescript 'chewsk'; quiet plaintive 'miau'. *Range:* rare visitor from Asia (Aug–Apr) to mudflats around Aust, most common in north-west and Kimberley.

Bar-tailed Godwit   *Limosa lapponica*   38–45 cm
Very common godwit with slightly upturned bicoloured bill (substantially longer in female), barred whitish rump and tail, no white stripe in wing. *Adult breeding:* male: head, neck and underparts chestnut; female duller. *Non-breeding:* grey-brown above with darker central stripe to back feathers. *Juv:* similar to adult but with barred darker centres to back and wing feathers. Watch for European race with largely unbarred rump. *Voice:* quiet 'krrick' in flight; barking 'kit-kit-kit . . .'. *Range:* common migrant from Asia Sep–Apr to mudflats, estuaries and islands around mainland and Tas, more common in north and east; many overwinter.

Black-tailed Godwit   *Limosa limosa*   36–43 cm
Common godwit with straight or slightly downcurved bicoloured bill (longer in females), white rump and black tail, broad white stripe in wing. Longer-legged but more slender-bodied than Bar-tailed. *Adult breeding:* male: head and neck chestnut breaking into bars on breast and flanks, abdomen and undertail white; female has head and neck pale chestnut suffused with grey and buff. *Non-breeding:* uniform dark grey above, breast grey, belly white. *Juv:* head and neck buff, back strongly patterned. *Flight:* long legs trail well past black tail, rump white, broad white stripe in wing. *Voice:* quiet 'keek . . .'; sharp 'tititi'. *Range:* common to uncommon migrant from eastern Asia Aug–Apr to tidal mudflats and estuaries, also lakes and dams, perhaps in transit; less common in south than north, vagrant Tas; some overwinter.

Hudsonian Godwit   *Limosa haemastica*   37–45 cm
Rare godwit like Black-tailed in colour but Bar-tailed in shape, with slightly upturned bicoloured bill, less white on rump and wing than Black-tailed, mainly blackish underwing. *Adult breeding:* chestnut on underparts extends to undertail, with barring on belly; head with white eyebrow and streaked cheeks. *Non-breeding:* plain grey above like Black-tailed. *Juv:* darker above with pale buff edges to feathers, neck and breast grey suffused with buff. *Voice:* low 'ta-it'; sharp 'kit-it'. *Range:* rare vagrant from North America most likely Sep–Mar with other godwits on east coast.

DOWITCHERS

ASIAN

br

juv

non-br

GODWITS

BAR-TAILED

juv

non-br

br

BLACK-TAILED

juv

non-br

br

HUDSONIAN

non-br

br

CURLEWS

Curlews are large waders with long downcurved bills. Most have loud calls that epitomise salt marshes and mudflats. Eastern Curlew and Whimbrel are common, easily identifiable by their calls; Eurasian is rare vagrant, told from Eastern by overall paler colour and white lower back. Little Curlew occurs in flocks on northern plains, sometimes in enormous numbers (more than 500 000 at Derby, Feb 1955; 250 000 at Foggs Dam, Oct 1970).

Little Curlew *Numenius minutus* 31–36 cm
Small curlew with shortish, slightly downcurved bill, dark rump, stripe over eye, usually on northern tussock grassland, often in large flocks. Squats when startled; holds wings up briefly on alighting. In flocks unlikely to be misidentified, but solitary birds can suggest rarely-sighted Upland Sandpiper (overleaf) which also holds wings up on alighting but is more lightly built, has longer tail with white outer edges, paler underwings and abdomen, shorter less curved bill, yellow legs. *Voice:* when feeding in flock 'tee-tee-tee'; alarm note 'tchew-tchew-tchew'; in flight fluting 'klee-klee'. *Range:* common migrant from Asia Oct–Apr to grassland in northern Aust, most on (a) north-western plains between Fitzroy R floodplain and Anna Plains Oct–Apr, concentration depending on local conditions; (b) Top End Oct–Nov and Mar–Apr; and (c) on plains of south-eastern Gulf Dec–Apr; farther south uncommon to rare, usually in small flocks or singly, Dec–Mar.

Whimbrel *Numenius phaeopus* 38–43 cm
Medium-sized curlew with downcurved bill, stripe over eye, and pale rump and lower back (except rarely seen American race, which has brown rump and lower back). Smaller than Eastern with shorter bill; larger than Little which has much shorter, only slightly decurved bill, more likely away from coast. Two forms occur: Asian form (race *variegatus*) with pale lightly barred rump and lower back; and American form (Hudsonian Whimbrel, race *hudsonicus*) darker with brown barred rump and lower back. *Voice:* tittering 'ti-ti-ti-ti-ti-ti-ti' (sometimes called 'seven whistler'). *Range:* common migrant from eastern Asia (or very rarely from America) to coastal mudflats, estuaries, mangroves, islands, occasionally inland, around mainland and Tas; many overwinter. (Watch for Bristle-thighed Curlew *N. tahitiensis*, like Whimbrel but darker warm brown above, rusty tail, pale band across rump, migrates from America to Polynesia, could overshoot in Aug–Sep with adverse weather.)

Eastern Curlew *Numenius madagascariensis* 53–60 cm
Large dark curlew with long downcurved bill (shorter in male), dark rump and lower back. Smaller, rarely sighted Eurasian Curlew is paler, less heavily marked, has white rump and lower back visible in flight. Whimbrel smaller still, shorter bill, stripes over eye, pale rump (but rare American form, Hudsonian Whimbrel, has dark rump), rapid tittering call. *Voice:* loud 'curlee . . .'; sad 'carr-er'. *Range:* common migrant from eastern Asia to coastal mudflats, mangroves, islands, estuaries and sandpits around mainland and Tas; many overwinter.

Eurasian Curlew *Numenius arquata* 54–58 cm
Large pale curlew with long downcurved bill, white rump and lower back. Smaller and noticeably paler than Eastern which has dark rump; larger and longer-billed than Whimbrel, and lacks eyebrow pattern. Most likely form occurring here is Asian (race *orientalis*) with white axillaries; European form (race *arquata*) has barred axillaries. *Voice:* loud 'courleee'. *Range:* rare vagrant from western Asia, most likely with Eastern Curlews on mudflats on north or west coast.

CURLEWS

european form

asian form

Hudsonian american form

LITTLE

WHIMBREL

BRISTLE-THIGHED

EURASIAN

EASTERN

GREENSHANKS AND UPLAND SANDPIPER

Greenshanks: sandpipers with long greenish legs and extensive white on tail, rump and lower back, forming a white wedge between wings visible in flight. Common Greenshank has loud call, often first wader to rise when disturbed, alerting other species. **Upland Sandpiper:** American wader found mostly on grassy plains, recorded only once long ago.

Upland Sandpiper   *Bartramia longicauda*   28–32 cm
Vagrant long-tailed sandpiper with long yellow legs, slightly downcurved short bill, small head with large eye, whitish finely-barred underwings contrasting with buff breast and flanks, most likely to occur on grassy plains. *Adult:* strongly streaked neck and breast, breaking to 'arrowheads' on lower breast and flanks; edges of back feathers buff. *Juv:* paler, less strongly streaked on neck, backfeathers edged white. Unlike any other wader, but solitary Little Curlews could be misleading (p 122). *Voice:* bubbling 'peetlie-peet-peet . . .'; loud mellow 'qua-a-ily' or 'plut-plut'. *Range:* extremely rare vagrant from North America, remotely possible on eastern marshy grassland Sep–Apr.

Marsh Sandpiper   *Tringa stagnatilis*   20–23 cm
Small slender greenshank with needle-like bill. Long, dusky-olive legs trailing well beyond tail in flight; white rump and lower back prominent in flight; no white wingbar. *Adult breeding:* head and neck densely spotted black, breaking into chevrons on flanks. *Non-breeding:* back uniform grey with narrow white margins to feathers, darker on shoulder, underparts mainly white. *Juv:* back feathers notched and margined with buff. Wades up to belly, picks food from surface. Paler and longer-legged than Wood (p 128) with white extending further up back – Wood has darker-looking face, stouter shorter bill, squatter stance. *Voice:* metallic 'tew' when flushed; soft 'tee'; 'tu-ee-you'. *Range:* regular migrant from Asia, mainly to lakes and swamps but also mangroves, salt marshes and estuaries, common in north, uncommon to rare in south and Tas.

Common Greenshank   *Tringa nebularia*   31–35 cm
Noisy, large, rather heavy greenshank with long, slightly upturned bill pale at base, longish green legs trailing beyond tail in flight, white rump and lower back prominent in flight. No white wingbar. Head looks paler and more streaked than Marsh, with less distinct eyebrow. *Adult breeding:* head and neck heavily streaked, black feathers among grey feathers on back. *Non-breeding:* less heavily streaked neck, back grey with each feather with fine black line inside white margin. *Juv:* more heavily streaked neck, feathers on back more broadly edged white except at tip which is black; sometimes legs yellow, could be mistaken for Greater Yellowlegs, which has more speckled plumage and white on back confined to rump. *Voice:* loud, slightly hectoring 'tew-tew-tew', one of most familiar wader calls. *Range:* common to uncommon migrant from Asia to coastal mudflats, estuaries, salt marshes, mangroves, lakes and swamps throughout Aust; some overwinter.

Spotted Greenshank   *Tringa guttifer*   30 cm
Rare large greenshank with stout straight bill yellowish at base, greenish-yellow legs shorter than Common Greenshank, projecting just beyond tail in flight, lower back and rump white, underwing coverts white. *Adult breeding:* neck and breast heavily marked with large black spots. *Non-breeding:* similar to non-breeding Common Greenshank apart from differences noted above (but barring on underwing of Common not always visible in flight) – best characters are shorter, yellower legs, shorter, stouter bill, squatter stance. *Voice:* sharp 'kyew'. *Range:* very rare vagrant from Asia, most likely on northern swamps or mudflats Sep–Mar.

SANDPIPERS

UPLAND

juv

ad

MARSH

GREENSHANKS

COMMON

non-br

juv

br

SPOTTED

non-br

br

SHANKS

These waders with brightly coloured long legs are only rarely sighted in Australia, although Common Redshank may be regular visitor to the north-west coast, particularly on mangrove tidal marshes. The two yellowlegs are probably more easily distinguished by their size relationship to Greenshank and Marsh Sandpiper, with which they loosely associate while here.

Lesser Yellowlegs
Tringa flavipes   27 cm

Slender wader with long yellow or orange-yellow legs, fine straight bill, white rump patch not extending on to back; smaller than Greenshank but larger than Marsh. *Adult breeding:* back mottled black and white, below white with dark spots and streaks on breast. *Non-breeding:* back grey-brown with white spotting. *Juv:* above brown with buff spots. Looks larger and darker than Marsh Sandpiper (p 124) with which it loosely associates, wades up to belly, jerkily active, often feeds with sideways flick of bill. Similar but larger Greater has stouter, longer, slightly upturned bill, looks as big as Greenshank. *Voice:* usually silent on ground; in flight, single or double notes 'tew', 'tew-tew'. *Range:* very rare summer vagrant from North America, most likely on saltmarshes on east coast.

Greater Yellowlegs
Tringa melanoleuca   29–31 cm

Slender wader with long yellow or orange-yellow legs, long stout bill slightly upturned, white rump patch not extending on to back; about size of Greenshank. *Adult breeding:* back mottled dark brown and white, below white with darker streaks on breast and bars on flanks. *Non-breeding:* above grey-brown with white edges to feathers; below white streaked darker on breast. *Juv:* brown extensively spotted buff. Rather similar to Greenshank but has brighter legs, less white on rump, and wings spotted white. Lesser, very similar in plumage but is smaller, slimmer and with more slender bill. Wades up to belly, picks prey from surface, often flicks bill sideways through water. *Flight:* white rump, dark back, no wingbars. *Voice:* loud three-noted falling 'tew-tew-tew', slightly higher and quicker than Greenshank. *Range:* very rare vagrant from North America, most likely as summer visitor to saltmarshes on east or north coast.

Spotted Redshank
Tringa erythropus   30 cm

Slender wader with very long red legs, long slender bill slightly down-drooping at very tip and with red base, rump and lower back white. *Adult breeding:* mainly black with white edges and notches to feathers of back and wings; when moulting, black becomes increasingly mottled with white and grey non-breeding plumage. *Non-breeding:* above grey with white edges and notches to back and wing feathers; below white with grey clouding on breast and flanks. *Juv:* above dark brown heavily spotted white, below white with heavy mottling on breast and extensive barring on flanks. Common Redshank has prominent white trailing edge to wing, is smaller with shorter legs and shorter stouter bill with more red at base. *Flight:* white rump and lower back, barred tail, indistinct pale hind-edge to wing. *Voice:* sharp quick 'chew-it'. *Range:* very rare vagrant from Eurasia, most likely in summer on ponds and marshes on north-west coast.

Redshank
Tringa totanus   28 cm

Slender wader with long red legs, stout bill with red base, white rump and lower back and prominent white trailing edge to wings in flight. *Breeding:* above brown, barred darker, spotted and notched white; below white with breast barred darker. *Non-breeding:* above grey-brown with white tips to feathers, below white with breast clouded grey and streaked darker. *Juv:* above brown with buff edges, spots and notches to feathers, below white with breast clouded brown streaked darker; bill dull reddish-brown. Only other similar wader with prominent white trailing edge to wing is Terek, which has shorter orange-yellow legs, long upturned bill and more uniform grey-brown upperparts with no white on rump. Spotted Redshank has much less obvious pale trailing edge to wing, has longer finer bill and longer legs. *Flight:* rump and lower back white, tail barred, prominent white hind edge to wings. *Voice:* noisy; sad whistle 'tyoooo'; rapid 'tew-huku' or 'tew-hu'. *Range:* uncommon summer Eurasian visitor, most likely on fresh or salt marshes on north-west coast, but could turn up on any suitable marsh or pond.

SHANKS

LESSER YELLOWLEGS

br

juv

non-br

juv

non-br

GREATER YELLOWLEGS

br

SPOTTED REDSHANK

br

juv

non-br

COMMON REDSHANK

br

juv

non-br

'TEETERING' SANDPIPERS

Most small shorebirds bob their heads to aid vision at ground level, but in these birds the motion is exaggerated into a nervous 'teetering' action. They are active feeders, particularly Terek, which dashes about mudflats. Wood and Green are more likely to occur on fresh wooded swamps, both zig-zag off to some height when disturbed. Common is most likely on shores of estuaries and rivers, flies with characteristic stiff downturned wingbeats.

Common Sandpiper
Tringa hypoleucos 20 cm

Nervous, constantly teetering sandpiper with prominent white eye-ring, brown smudge on side of breast, faintly barred wing coverts, tail longer than folded wings, and shortish green legs. *Breeding:* arrow-shaped markings on back. *Non-breeding:* plain bronze-brown back, faint bars on coverts. *Juv:* buff edges to feathers of back and wings. *Flight:* dark rump, barred white edges to dark tail, white bar on wing meeting white trailing edge near body; characteristic flight alternating stiff wingbeats with short glides on downturned wings. (Watch for Spotted Sandpiper *T. macularia* from North America, almost identical in non-breeding plumage but with rather more bars on wing coverts, shorter tail (not much longer than folded wing) and white bar on wing not meeting white trailing edge; in breeding plumage unmistakable with spotted underparts.) *Voice:* high-pitched 'seep, seep, seep'. *Range:* common summer visitor to rocky shores, mudflats, rivers, lakes and sewerage farms throughout Australia, less common in south-east and Tas.

Wood Sandpiper
Tringa glareola 20–23 cm

Active, bobbing, mainly freshwater sandpiper with profusely spotted upperparts (much less obvious in worn plumage), long greenish-yellow legs protruding well beyond tail in flight, white rump, fine bars on largely white tail, underwing pale with darker barring. *Adult breeding:* neck and breast heavily streaked and spotted with black. *Non-breeding:* neck and breast clouded grey. *Juv:* spots on upperparts tinged buff. White spots in plumage abrade quickly, giving more uniform appearance to upperparts. Green is similar but darker, less spotted above, shorter green legs, dark underwings and broader bars on tail in flight. Lesser Yellowlegs similar but larger with bright orange-yellow legs. *Flight:* white rump, 6–8 fine bars on tail, no bars in wing but white shaft to first primary, usually takes off when disturbed with sharp call, zig-zags away in high semi-circle. *Voice:* shrill 'chiff-iff-iff'. *Range:* uncommon summer visitor to wooded swamps and lakes throughout mainland; vagrant Tas.

Green Sandpiper
Tringa ochropus 23 cm

Vagrant sandpiper similar to Wood but with longer bill, darker above with somewhat greenish cast to back feathers, smaller spots on upperparts, white rump and lower back, four broad bars on white tail, underwing dark with faint white barring, shortish green legs not protruding much beyond tail in flight. *Adult:* small white spots on back. *Juv:* Small buff spots on back. Do not confuse with Wood Sandpiper in worn plumage – dark underwing must be seen for positive identification. (Watch for Solitary Sandpiper *T. solitaria* from North America, almost identical but with dark rump and lower back.) *Flight:* less likely to flush than Wood, may fly only short distance but will also zig-zag off to some height like Wood; watch for dark underwing linings and broader bars on tail. *Voice:* musical 'tweet-weet-weet' similar to Wood but more liquid, less shrill. *Range:* extremely rare vagrant from northern hemisphere, most likely on ponds or wooded lakes.

Terek Sandpiper
Tringa terek 22–23 cm

Active, dumpy sandpiper with distinctive upturned bill, short orange-yellow legs, dark patch on shoulder and prominent white trailing edge to wing in flight. *Adult breeding:* black stripes on shoulder. *Non-breeding:* plain brownish-grey above with pale edges to feathers. *Juv:* buff edges to feathers of back and wing. Dashes jerkily about mudflats, reminiscent of wind-up toy, usually most active bird on shore. *Flight:* tail, rump and back dark, broad white trailing edge to wings. *Voice:* rapid high 'tee-tee-tee' or 'weet-weet-weet'; rippling 'hu-hu-hu-hu-hu' in flight. *Range:* common summer visitor from Asia to coastal mudflats, estuaries and islands in northern Aust; less common to rare in southern Aust; vagrant Tas.

SANDPIPERS

COMMON

br
juv
non-br

WOOD

br
juv
non-br

GREEN

br
non-br

SOLITARY

TEREK

br
juv
non-br

PRATINCOLES AND PHALAROPES

Pratincoles: dry-country waders with short bicoloured bills and very long wings. **Phalaropes:** unusual waders adapted for swimming with lobed toes. Females are larger than males, more brightly plumaged in breeding dress. Often spins while swimming.

Australian Pratincole   *Stiltia isabella*   22–24 cm

Long-legged, cinnamon-brown (isabelline) pratincole with deep rufous belly patch, short square tail. *Juv:* more greyish-brown with speckled neck, feathers of back edged pale buff. *Flight:* swift, erratic, swallowlike, often flushed from sides of outback roads; underwings black, rump and sides of tail white, centre of tail black. Runs gracefully, bobs hind end while head remains still. *Voice:* plaintive 'quirra-peet'. *Nesting:* scrape on ground near rock or stick; 2–3 blotched pale umber eggs. *Range:* common nomad on open semi-arid plains mainly in northern and central Aust, moves coastward in winter, migrates to Borneo, Java and New Guinea.

Oriental Pratincole   *Glareola maldivarum*   23–24 cm

Olive-brown pratincole with shortish legs, long forked tail, creamy buff throat edged with black. Most are moulting on arrival (mid Oct), showing mottled black throat band. *Juv:* feathers of neck, breast and back with fine black line inside narrow buff edge. *Flight:* rump and base of tail white, end of tail black; underwing chestnut. Usually in large flocks, often stands on ground, flying up to take grasshoppers. *Voice:* contact call in flocks 'churr' or 'churr-ah'; loud 'chick'. *Range:* common migrant from Asia, Oct–Apr, to dry open plains, mainly in north; less common in smaller numbers farther south.

Red-necked Phalarope   *Phalaropus lobatus*   17–20 cm

Slender phalarope with needle-like bill about as long as head, white stripe in upper wing, dark centre to rump. *Female breeding:* head and hind neck dark grey, sides of neck and upper breast chestnut, throat white, dark upperparts striped buff. *Male breeding:* duller on head and neck. *Non-breeding:* black stripe through eye, crown and hindneck grey, back dark grey with broad white edges to feathers, giving less uniform look than other non-breeding phalaropes. *Juv:* feathers of upperparts blackish-brown with buff edges forming obvious stripes, looks darker than adult. Often seen swimming high in water, spins while feeding; also feeds over mudflats. *Voice:* low-pitched 'twick'; querulous 'kerrek-kerrek'. *Range:* rare vagrant from flocks spending southern summers in seas north of New Guinea; sightings possible on any coastal wetlands.

Red Phalarope   *Phalaropus fulicarius*   19–22 cm

Tubby phalarope with stout bill shorter than head, narrow white stripe in upper wing, dark centre to rump. *Female breeding:* entire undersurface brick red, whitish cheeks, bill yellow with black tip. *Male breeding:* duller than female with white mottling on belly. *Non-breeding:* plain grey above, white below with black mark through eye bolder than Red-necked. *Juv:* above blackish-brown with buff edges to feathers, looking less distinctly striped than juvenile Red-necked; underparts white with pinkish-buff suffusion on face and breast; black stripe through eye. Swims high in water, spins while feeding; more marine than others. *Voice:* whistling 'wit'; chirruping 'zhit', both higher-pitched than Red-necked. *Range:* spends southern summer at sea off west coasts of Africa and South America, likely in Aust only as rare vagrant; one record near Swan Hill, Vic, 1976.

Wilson's Phalarope   *Phalaropus tricolor*   22–25 cm

Lanky phalarope with long needle-like bill longer than head, no white stripe in upperwing, white rump, legs trail beyond tail in flight. *Female breeding:* black stripe through eye extends down side of neck with chestnut suffusion, hindneck, throat, foreneck and underparts white. *Male breeding:* much duller with mottling on flank. *Non-breeding:* plain grey above, white below, stripe through eye least prominent of non-br phalaropes; legs become yellow. *Juv:* palest of juv phalaropes; pale grey above with buff edges to feathers, buff suffusion on breast. Swims lower than other phalaropes with longer profile, but more likely to feed on shore where it can be confused with sandpipers, closest to Marsh (p 124) which has white on rump extending up back, longer legs. *Voice:* usually silent but occasional soft 'aangh'. *Range:* less marine than other phalaropes, spends southern summer on wetlands in Argentina, thus sightings in Aust accidental; several recorded.

PRATINCOLES

AUSTRALIAN

juv

ad

ORIENTAL

ad

juv

31 x 24

Australian

moulting

PHALAROPES

RED-NECKED

♀

juv

♂

non-br

♀

RED

juv

♂

non-br

♀

♂

WILSON'S

juv

non-br

non-br

OYSTERCATCHERS, STILTS AND AVOCETS

Stilts: elegant long-legged waders with long straight needle-like bills and calls like puppy yapping. Length of leg enables them to wade in deeper water than other waders; also swim, Banded more often. Food is surface-swimming swamp life; Banded reliant on brine shrimps. Black-winged is almost world-wide in range, Banded confined to Aust, adapted to exploiting ephemeral salt lakes. Nest on ground, Black-winged singly or in small loose colonies, Banded often in very large colonies. **Avocets:** basically stilts with upturned bills, enabling more efficient gleaning of surface-swimming creatures. Nests on ground, usually in small loose colonies on islands or swamps. **Oystercatchers:** large dumpy waders of seashore with long blade-like bills adapted for opening bivalved molluscs. Pied mainly on sandy beaches and estuaries, Sooty on rocky beaches and islands. Solitary nesters, laying in scrape on ground.

Banded Stilt   *Cladorhynchus leucocephalus*   36–45 cm

Pied stilt with broad white bar in hindwing, legs shorter than Black-winged; head and neck wholly white, broad chestnut breastband shading to black centre of belly. *Juv:* back and underparts white with feathers edged paler, head and underparts white. *Imm:* similar to adult but lacking breastband. Often in large dense flocks, either wading or swimming in 'rafts' where brine shrimp are plentiful. Also fly in dense flocks; small groups, often immature, also seen. *Voice:* puppy-like bark. *Nesting:* scrape in sand on island in inland salt lakes, often in enormous colonies; 3–4 blotched whitish to pale umber eggs. *Range:* locally common nomad on salt lakes, salt marshes and estuaries, occasionally fresh or brackish lakes in southern mainland; vagrant Tas; regular visitor to Rottnest I, WA.

Black-winged Stilt   *Himantopus himantopus*   35–38 cm

Pied stilt with wholly black wings. *Adult:* head and neck white with black nape and hindneck. *Juv:* back and wings brown with pale edges to feathers, crown and nape dusky. *Imm:* back and wings black, head and neck white with dusky nape and hindneck. Usually in pairs or small groups, sometimes large loose groups on suitable wetlands. *Voice:* sharp yapping like small puppy; plaintive 'care-air-er' if disturbed at nest. *Nesting:* variable from barely-lined scrape in ground to substantial platform of waterweeds, may be built up if water level rises; 4 spotted and blotched umber eggs. *Range:* common nomad to wetlands throughout mainland; vagrant Tas.

Red-necked Avocet   *Recurvirostra novaehollandiae*   40–46 cm

Pied avocet with rufous head and upper neck. *Adult:* black and white marking in wings. *Juv:* brown and white markings in wings. Usually in pairs or small flocks; feeds by swinging bill from side to side while wading, or tip-tilting while swimming. *Voice:* liquid 'toot' or 'klute' in flight. *Nesting:* lined scrape or platform of waterweeds and mud on marshly ground; 4 blotched pale umber to olive eggs. *Range:* common nomad on suitable wetlands throughout mainland Aust; vagrant Tas.

Pied Oystercatcher   *Haematopus longirostris*   50 cm

Oystercatcher with white belly, white bar in wings and white rump. *Juv:* back and wing feathers brown with paler edges; eye brown, legs greyish-pink, bill reddish with dusky tip. *Imm:* back and wings sooty black, eye orange-red, legs pink. Usually in pairs or small groups. *Voice:* loud 'pit-a-peep, pit-a-peep'; piping 'kleep, kleep . . .' in flight. *Nesting:* scrape in sand; 2–3 spotted pale umber eggs. *Range:* common on sandy beaches, mudflats and estuaries around Aust.

Sooty Oystercatcher   *Haematopus fuliginosus*   50 cm

Sooty-black oystercatcher. *Juv:* back and wing feathers dark brown, edged pale, bill and eye-ring duller, legs greyish-pink. More solitary than Pied, usually in pairs or family parties on rocky shores. Two forms occur: southern form (race *fuliginosus*) with narrow eye-ring; northern form (race *ophthalmicus*) with broad eye-ring and longer bill. *Voice:* similar to Pied. *Nesting:* scrape in soil among rocks, pigface or shells; 2–3 spotted and blotched pale umber eggs. *Range:* uncommon resident on rocky shores around Aust.

STILTS

BLACK-
WINGED

ad

juv

BANDED

ad

juv

AVOCETS

ad

juv

RED-
NECKED

OYSTERCATCHERS

PIED

juv

ad

SOOTY

southern
form

juv

northern
form

45 x 30   55 x 40   50 x 34   63 x 42   63 x 42

Black- winged   Banded   Red- necked   Pied   Sooty

SKUAS AND JAEGERS

Aggressive seabirds pirating much of their food from other seabirds, also eat offal and food scraps. They have characteristic elongations to central tail feathers and white patches in outer wing which vary in size between species. Large species are known as skuas, smaller species, with longer tail extensions, as jaegers. Juveniles and immatures of jaegers are mottled brown, have shorter tail extensions. Skuas breed in Antarctica and are winter visitors; jaegers nest in northern hemisphere and visit mainly in summer.

Arctic Jaeger *Stercorarius parasiticus* 45–52 cm

Slender jaeger with pointed central tail feathers shorter than Long-tailed. *Adult light phase, breeding:* breast and abdomen whitish-buff, dark cap, diffuse collar; back and wings brown without distinct trailing edge; white window in wing more extensive than Long-tailed but less than Pomarine. *Non-breeding:* mottled on back and breast; white bars on rump and undertail coverts. *Dark phase:* dark brown all over except diffuse pale collar and white 'windows'. *Intermediate phase:* broad brown breast band. *Juv:* short pointed tail projections (blunt in Long-tailed and Pomarine); barred above and below with belly pale (light phase) or barred (intermediate) or mottled (dark); less barring on uppertail coverts than juv Pomarine or Long-tailed. More vigorous in flight than Long-tailed, less attenuated; Pomarine is more solid with broader-based wings, looks more powerful. *Range:* rare summer visitor Aug–June, most likely Oct–Apr, less common than Pomarine but more likely inshore.

Pomarine Jaeger *Stercorarius pomarinus* 47–55 cm

Bulky jaeger with twisted spoon-shaped central tail feathers. *Adult light phase, breeding:* throat and breast yellowish-buff with dark cap, diffuse yellowish collar, mottled dark breast band; back and upperwings brown with white windows in wings. *Non-breeding:* cap more diffuse, may join breast band; mottled brown and buff above and below, but not on underwings. *Dark phase:* sooty brown with paler face and collar, white windows; undertail and rump slightly more heavily barred than Arctic. *Imm:* heavily mottled brown and buff above and below, including underwings; short, blunt tail projections. Chases birds up to gannet size (Arctic up to Crested Tern size) *Range:* uncommon summer visitor Aug–June, mostly Oct–Apr, usually offshore, in seas off east and south and south-west coast.

Long-tailed Jaeger *Stercorarius longicauda* 51–56 cm

Small, slender jaeger with long pointed central tail feathers. *Adult light phase, breeding:* throat and breast nape pale buff with black cap and complete buff collar; back and upperwings grey with darker flight feathers and hind edge, white 'window' in wing restricted to shafts of outer primaries; *Non-breeding:* mottling on back and breast, white bars on upper tail and undertail coverts. *Juv:* tail projection short, blunt; head lacks cap; breast dusky; underwing, upperwing and back barred dark grey and buff; bases of primaries of underwing white, but only shafts white on upperwing. *Dark phase:* back and underparts dark brown, faint collar. If long tail feathers missing, best told by restricted white 'window' in wing, rather tern-like flight, narrower, longer wings. *Range:* rare summer vagrant (visitor) recorded Nov–Apr on eastern coast from Gulf of Carpentaria to about Robe, SA.

Southern (Great) Skua *Stercorarius skua* 61–65 cm

Uniformly-brown, winter-visiting skua with paler buff streaks on body and large white windows in wings; tail projections small, often not visible. *Juv:* similar. Dark phase Pomarine less powerful-looking, has distinct black cap, more obvious tail projections. Rare South Polar looks smaller-headed, has light and intermediate phases which are paler, particularly around head; dark phase is more similar but has pale feathers at base of bill and sometimes a pale collar giving black-capped appearance. *Range:* uncommon but regular winter visitor from Antarctic islands to southern waters north to about tropic, non-breeding birds may remain longer.

South Polar Skua *Stercorarius maccormicki* 53 cm

Rare winter-visiting skua with several phases, all having at least 'face' pale, all with white window in wings. *Light phase:* wings and back as for Southern (Great); head and underparts pale straw-coloured with buff collar. *Intermediate phase:* head and underparts dark straw-coloured with prominent pale brown collar. *Dark phase:* similar to Southern (Great) but feathers at base of beak ('face') white; some have pale collar; buff streaks on body not so obvious, looks darker and colder in tone. *Juv:* more heavily streaked on body. Large gulls are dark brown with pale face, but are different in shape, lack large white windows in wings. *Range:* very rare vagrant from Antarctica; winters in northern hemisphere, birds seen here probably on passage; records Nov and May.

SKUAS

ARCTIC

dark

light

inter

juv

juv

dark

light

POMARINE

SOUTHERN

juv

non-br

LONG-TAILED

br dark

br light

light

inter

dark

juv

SOUTH POLAR

LARGE GULLS

Gulls are gregarious web-footed seabirds mostly occurring on coasts or islands, usually not far from shore; some species live on or visit inland wetlands, and some are migratory. Most feed by scavenging, untidy nest on ground in colonies. All of the large species so far recorded in Australia have dark backs in adult plumage. Colour and shape of bill, pattern of white in wing tip and pattern of tail are important adult characters. Juveniles are brown; look for disposition of pale plumage on head and neck, size of bill, pattern of upperparts and tail.

Pacific Gull
Larus pacificus 63 cm

Adult: large black-backed gull with black band in tail (except in some very old birds 20+ years) and massive yellow bill with red tip; no mirrors in wings. *Juv:* mottled dark brown with pale face and black tip to stout pink bill. *Imm:* wings dark brown, body mottled whitish, bill yellow with black tip. Two forms occur: eastern form (race *pacificus*) east of SA/Vic border, with white eye and complete red tip to bill; and western form (race *georgii*, perhaps a separate species) west of SA/Vic border, with red eye and incomplete red tip to bill. *Voice:* 'ow-ow'; 'chuk-chuk . . .'. *Nesting:* flat cup of grass, twigs, on ground on island; 2–3 blotched olive eggs. *Range:* common resident (adult) or nomad (juv) around southern coast, very rarely on rivers, from Shark Bay, WA, to Hunter R, NSW; rare vagrant farther north.

Black-tailed Gull
Larus crassirostris 47 cm

Adult: large black-backed gull with black band in tail, no mirrors in wing, bill yellow, separated from red tip by black line. Back has greyish cast, not as black as Pacific, which has stouter bill without black subterminal strip. *Juv:* very similar to juvenile Pacific, paler below with more slender bill, pink with black tip. *Range:* very rare vagrant from Japan to north coast; recorded near Darwin, NT, and Melbourne, Vic.

Kelp Gull
Larus dominicanus 57 cm

Adult: large black-backed gull with white tail, small white mirror in outer primary, yellow bill with red spot on lower mandible. *Juv:* brown with paler mottling on hind neck and breast with black bill; juvenile Lesser Black-backed has paler rump and a little less black on tail. *Imm:* wings and back mottled brown, body whitish, bill yellow without black tip. *Voice:* 'cyop-cyop-cyop . . .'. *Nesting:* flat cup of grass, twigs on ground on island; 2–3 blotched olive eggs. *Range:* rare to uncommon, possibly increasing, around southern coast north to Perth, WA, and Newcastle, NSW; rare vagrant farther north; more common Tas, mainly south-east. (Almost identical Lesser Black-backed Gull *Larus fuscus* 60 cm, has been recorded doubtfully on north coast. Difficult to separate in field — Lesser Black-backed has paler back, slightly more white in wing tip.)

Silver Gull
Larus novaehollandiae 40–42·5 cm

Very common gull with white head lacking any dark markings (some birds may have irregular smudges of oil staining from polluted waters). *Adult:* red bill and legs. *Imm:* black bill and legs. *Juv:* brown edges to back and wing feathers, often dark subterminal band in tail. *Nesting:* small to large colonies on cays, islands, lakes, saltworks, etc; nest on ground lined with seaweed or on dead trees standing in lakes; 3–4 eggs. *Range:* very common around coastal mainland and Tas, as well as on inland waterways, increasing in numbers to nuisance level near rubbish dumps, airports and tern colonies.

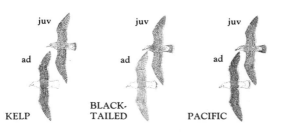

KELP

BLACK-TAILED

PACIFIC

GULLS

BLACK-
TAILED

juv

imm

ad

KELP

juv

imm

sub-ad

ad

PACIFIC

juv

imm

ad

eastern
form

western
form

SILVER

juv

ad

Silver 53 x 38

Kelp 71 x 49   Pacific 74 x 51

SMALL GULLS

All of these migratory gulls are smaller than Silver Gull and are very rare summer visitors, possibly blown off course in normal migration to west coast of South America, in case of Sabine's and Franklin's, and overshooting in Saunders'. These rarely-sighted gulls are all black-headed in breeding plumage, reduced to dusky smudges for most of the time they are in Aust, when they look much more like Silver and could be overlooked. Best distinguished by slightly smaller size and wing pattern which shows in flight. Another gull similar to Saunders', the Black-headed, is more likely to turn up, so is mentioned here for comparison. Usually nest colonially on ground or on vegetation in swamps, building untidy cup-shaped nest from grasses, sticks and weed; 4 eggs normal clutch. Gulls often congregate near colonies of other seabirds, raiding nests if disturbance causes adults to leave — no responsible birdwatcher ever enters seabird colonies if gulls are present.

Sabine's Gull
Larus sabini 34 cm

Small gull with yellow-tipped black bill, black wingtips, conspicuous white wedge in hindwing and forked tail. *Adult breeding:* head dark grey, narrow eye-ring red. *Adult non-breeding:* head white with smudgy black marking behind eye. *Imm:* differs from adult non-breeding by more extensive dusky markings on hind neck, variable amounts of juvenile feathers on wing, dark tail tip, moulting to all white tail Feb–Apr. *Juv:* upper parts scalloped brown with white wedge in hindwing, tail forked with dark band on tip, dusky bar on underwing. Very graceful almost ternlike gull breeding in Arctic, thought to migrate down west coast of America, to winter off north-western South America; takes food daintily from surface of water, runs after crabs, etc, on mudflats. *Range:* extremely rare accidental, recorded in Aust only once, near Darwin, NT.

Franklin's Gull
Larus pipixcan 35 cm

Small gull with red bill, hind-edge of wing narrowly edged white with two narrow white bars at wingtip separated by broad black bar, and square to rounded tail. *Adult breeding:* bright red bill, head and upper neck black with interrupted white eye-ring. *Adult non-breeding:* dull red bill, smudgy black marking surrounding eye; white eye-ring usually still visible. *Imm:* differs from non-breeding adult in variable amounts of juvenile feathers on wing, dark subterminal band in tail not extending to outer feathers, moulting progressively to all-white tail and all-grey wing Feb–Apr. *Juv:* browner upperparts with similar wing pattern to adult; tail square with dark subterminal band; immature Sabine's has forked tail and white wedge in hindwing, Saunders' has grey upperparts with black mottling on wing. *Range:* breeds on swamps in North America, migrates to South America; sightings on beaches in Aust in widely separated localities (Busselton, WA, Redcliffe, Qld) suggest it may be more common than suspected.

Saunders' Gull
Larus saundersi 32 cm

Small gull with stout bill, wing with four outer primaries white narrowly tipped black, remaining primaries grey in upperwing and black in underwing, tail square to rounded. *Adult breeding:* head black with narrow white eye-ring, short stout bill black. *Adult non-breeding:* head white with dusky smudges behind eye. *Imm:* differs from non-breeding adult in variable amount of juvenile feathers on wing, dark subterminal band in tail (outer tail feathers sometimes white) moulting Feb–Apr. *Juv:* upperparts grey with black freckling on shoulder, dusky smudges behind eye, black tip to tail, and black tipping on hind-edge of upperwing more extensive than adult (very similar to Black-headed Gull *Larus ridibundus*, which has longer thinner red bill, black on head ending further up the nape in breeding plumage, black tipping to flight feathers extending to secondaries; underwing shows all primaries blackish, ranges south to New Guinea). *Range:* extremely rare in Aust, normally occurs in north-east Asia, migrating as far south as Hong Kong.

Laughing Gull
Larus atricilla 38–43 cm (ws c. 100 cm)

Medium-sized gull with reddish or black bill, underwing with dark tip, dark grey upperwing with white confined to hind wing, primaries broadly tipped black, square to rounded tail. *Adult breeding:* black head with interrupted eye-ring, stout reddish slightly drooping bill, dark reddish to blackish brown legs. *Adult non-breeding:* black on head confined to dark smudge behind eye, often extending to back of crown. *Imm:* head and breast smudgy grey with darker patch behind eye extending around hind crown, mantle dark grey, wings with variable amount of juvenile feathering, dark subterminal band in tail, moulting Feb–Apr. *Juv:* head and breast dark brownish-grey, back and wings dark brown-grey with paler edges, moulting Sept–Oct.

1 SILVER	4 SABINE'S
2 LAUGHING	5 SAUNDERS'
3 FRANKLIN'S	6 SAUNDERS'

GULLS

SABINE'S

non-br

imm

juv

br

SAUNDERS'

non-br

juv

br

BLACK-HEADED

FRANKLIN'S

non-br

imm

juv

br

LAUGHING

non-br

imm

juv

br

TERNS

Terns are long-winged short-legged birds like slender gulls, which feed in salt or fresh water, either by plunging headlong after fish or by scooping up small creatures swimming just under the surface; some also take insects, particularly grasshoppers and dragonflies found in marshy areas. Often occur in flocks and usually breed in large colonies, mostly on the ground but occasionally in trees. Most terns are grey above and white below, with a black cap which is usually lost after breeding, when brightly coloured legs and bill become black, making recognition more difficult. A few species are dark in plumage, particularly the noddies which are reverse-coloured, with white caps and dark bodies. Those shown here include the largest species – colour of bill is easiest way to separate them. White Tern is a rare vagrant from Norfolk I, whose nesting habits are unique: lays egg on horizontal tree branch or in fork without any nest at all.

White Tern
Gygis alba 28–33 cm

Small white tern with large dark 'eye'. Told from pale terns by lack of black on crown or nape; in hand, blue base to bill, deeply indented webs on long toes, dark eyes distinguish it from albino dark noddies and terns (which have pinkish legs, bill and eyes). *Range:* rare vagrant to east coast from breeding colonies on Norfolk I, recently colonised Lord Howe I.

Gull-billed Tern
Sterna nilotica 35–43 cm

Adult breeding: large pale tern with neat black cap. *Adult non-breeding:* among palest of terns, with dusky smudge behind eye. *Juv:* mottled brownish scallops on back and leading edge of wing. Resident form (race *macrotarsa*) has long legs and most in summer have full black cap; migrant form (race *affinis*) with shorter legs, greyer above and on tail, smaller, most often seen in summer lacking black cap. *Nesting:* colonial; lined scrape on islands and spits in lakes; 2–3 blotched buff-white eggs. *Range:* uncommon nomad in wetlands, estuaries, mangroves, mudflats, islands throughout mainland, migrant race Sep–Apr mainly coastal, vagrant Tas.

Lesser Crested Tern
Sterna bengalensis 36–42 cm

Adult breeding: an orange-billed large tern (slightly smaller than Silver Gull) with shaggy black crest on crown and nape; sometimes forehead is black too. *Adult non-breeding:* black on head confined to nape; crown often speckled black and white. *Juv:* smudgy black nape; feathers of back and wings scalloped darker. Crested is larger, less brightly coloured lemon bill and has more black on head in non-breeding plumage, darker grey back. *Voice:* 'k-krrek'. *Nesting:* small, dense colonies in a clearing among low vegetation on comparatively few islands and cays; one egg laid in scrape in sand. *Range:* least common of large terns around northern Aust, from Point Cloates, WA, to southern Barrier Reef, Qld.

Crested Tern
Sterna bergii 41–52 cm

Adult breeding: large tern (slightly larger than Silver Gull) with lemon-yellow bill, shaggy black crest on crown and nape, often flecked with white. *Adult non-breeding:* black on head confined to nape and hind-crown, which is often flecked with black towards forehead. *Juv:* smudgy black nape; feathers of back and wings edged darker. *Voice:* grating 'krrreck', lower and less vigorous than Lesser Crested. *Nesting:* usually small dense colonies in clearing among low vegetation on island or cay; one, sometimes two eggs in scrape in sand. *Range:* very common around mainland and Tas coasts, may ascend rivers for considerable distance.

Caspian Tern
Hydropogne caspia 53–58 cm

Adult breeding: largest tern, with massive red bill and black cap. *Adult non-breeding:* cap streaked black and white. *Juv:* back and wing feathers edged darker, bill more orange. *Voice:* deep 'krraa'. *Nesting:* small compact colonies on sandy beaches, also nests singly; scrape in sand holds 1–3 eggs. *Range:* common around mainland and Tas coasts, but infrequent from Cape Howe, Vic, to southern Barrier Reef, Qld.

TERNS

GULL-BILLED

juv

non-br

br

WHITE

LESSER
CRESTED

non-br

br

CRESTED

non-br

br

CASPIAN

br

non-br

50 x 37

Gull-billed

51 x 36   60 x 42   64 x 45

Lesser Crested   Crested   Caspian

SMALL TERNS

Little Tern   *Sterna albifrons*   20–23 cm
Adult breeding: tiny tern with yellow black-tipped bill, black line
through eye reaching bill; at rest dark longer primaries contrast with
pale shorter primaries (paler Fairy Tern has all primaries same tone).
Non-breeding: bill black, forehead and crown white, nape black;
shoulder patch dark. *Juv:* crown streaked, back feathers edged brown,
dark shoulder patch. *Voice:* sharp 'creek'; 'peet–peet–peet'. *Nesting:*
scrape in sand or shell bank; 1–2 blotched pale umber eggs. *Range:* two
populations: (a) uncommon, resident population in Tas and on eastern
and northern coasts from Yorke Pen, SA, to Broome, WA; (b) common
summer migrant from Asia (Aug–Apr), from North West Cape, WA, to
Sydney, NSW, rarer farther south. (Race *saundersi* from Red Sea has
greyer rump and tail, more black in wing, recorded once Wollongong,
NSW.)

Fairy Tern   *Sterna nereis*   22–24 cm
Adult breeding: tiny tern with orange bill usually lacking black tip and
with black line through eye not reaching bill; at rest, primaries all same
tone (Little darker above, primaries two-toned). *Non-breeding:* similar
but blackish tip to bill, white forehead more extensive; non-breeding
Little has wholly black bill, black on head confined to nape, and dark
shoulders. *Juv:* bill black, crown streaked, back feathers edged brown,
no dark shoulder patch. *Voice:* 'chee-wick'; 'ket-ket-ket-ket . . .'.
Nesting: scrape in sand, often in large colonies; 1–3 eggs blotched pale
umber. *Range:* common resident in coastal waters, coastal salt lakes and
islands in Tas, and southern and western coasts from Port Albert, Vic,
to Derby, WA.

Black-naped Tern   *Sterna sumatrana*   31 cm
Very pale small tern with black nape extending just in front of eye;
underparts delicate rose, outer wing feather with dark edge. *Juv:* back
feathers pale brown edged darker, back of head streaked grey merging
into black nape. Little non-breeding is smaller, less definite black on
head, tail streamers shorter, darker grey on back, dark shoulders, more
likely to occur on beaches and estuaries. *Voice:* noisy 'shee-shee-ship'.
Nesting: small scrape in rock or coral rubble; often in compact colonies
but also solitary; 1–3 blotched umber eggs. *Range:* common on coral
cays and islands from Torres Strait to Lady Elliott I on Barrier Reef,
more rarely on mainland beaches.

Black Tern   *Sterna nigra*   25 cm
Adult breeding: black marsh tern with long black bill, grey wings, and pale grey, slightly forked tail;
underwing coverts white. *Adult non-breeding:* black patch at base of wings' leading edge. *Juv:* definite
collar on hindneck, dark patch on upper back shading into scalloped grey lower back, grey forked tail
with dusky subterminal band; black patch at base of wings' leading edge. *Range:* very rare vagrant,
likely to occur anywhere on estuaries, marshes or lakes.

White-winged Tern   *Sterna leucoptera*   22–24 cm
Adult breeding: black marsh tern with short reddish bill, white shoulder
on pale grey wing, white squarish tail, underwing coverts black. *Adult
non-breeding:* variable black horseshoe over top of crown from behind
eye. (Whiskered has black horseshoe from eye to eye around back of
neck.) Lacks black patch at base of wing found in Black. *Juv:* upper and
lower back dark grey. *Range:* irregular visitor to coastal and subcoastal
western, northern and eastern mainland; accidental Tas; usually seen in
small to large flocks hawking for dragonflies or grasshoppers on grassy
swamps or resting on estuary sandbanks, rarely at sea.

Whiskered Tern   *Sterna hybrida*   26 cm
Adult breeding: red-billed marsh tern with prominent white cheeks
('whiskers'), belly dark grey, underwing coverts white, slightly forked
tail. *Non-breeding:* bill blackish-red, grey above, white below with
either full black cap or cap reduced to nape with streaked crown. *Juv:*
bill black, forehead white, crown and nape black, back barred brown,
faint dark shoulder patch (juvenile White-winged darker above, larger
shoulder patch). *Voice:* staccato 'kit', 'kit-it', 'krrit'. *Nesting:* floating or
anchored heap of vegetation in shallow water or on marshy islets, loose
colonies; 3 blotched glossy brownish eggs. *Range:* common nomad in
wetlands throughout Aust; migrant in south, vagrant Tas.

TERNS

LITTLE

br

non-br

juv

FAIRY

non-br

juv

BLACK-NAPED

br

non-br

juv

38 x 29

Black-naped

WHITE-
WINGED

non-br

br

juv

BLACK

non-
br

br

WHISKERED

non-
br

juv

br

36 x 28

Whiskered

28 x 22

Little

30 x 23

Fairy

TERNS

WHITE-FRONTED

ROSEATE

COMMON

ARCTIC

ANTARCTIC

TERNS

These five similar graceful terns are found only in marine and estuarine environments. Identification presents such difficulties that some observers call them 'commic' terns (from Common and Arctic) and leave it at that. However, there are some pointers that help eliminate possibilities:

Range

Roseate: resident northern species seldom south of Tropic in east but reaches Cape Leeuwin in WA.
White-fronted: mainly visitor (May–Nov) to south-east and Tas, but a few breed in Tas.
Common: summer visitor (Oct–May) common on east coast, often overwintering, overlaps ranges of Roseate and White-fronted.
Arctic: rare visitor (Apr–Dec) mainly seen offshore in winter along south coast, but likely elsewhere.
Antarctic: vagrant, most likely on passage in spring and autumn.

In flight

Roseate: underwing with little translucence, no prominent dark trailing edge to primaries; rump and tail same colour as back, outer tail feathers edged white.
White-fronted: underwing with more translucence than Common but less than Arctic, no prominent dark trailing edge to primaries; rump and tail wholly white, but little contrast with back; in first-year birds, upperwing shows two dark triangles separated by a pale triangle, tail with prominent dusky tips.
Common: dark trailing edge to primaries, broader, more diffuse and shorter than Arctic; in non-breeding plumages rump and tail grey, in breeding plumage white; outer tail feather with dark edge; in first-year birds, upperwing shows dark secondaries.
Arctic: underwing with all primaries translucent, prominent dark trailing edge to primaries, narrower but longer than Common; rump and tail white, contrasting with back, outer tail feathers edged dark grey; in first-year birds, upperwing shows white secondaries.
Antarctic: underwing probably like Common but with narrower and longer dark trailing edge.

At rest

Roseate: Smallest and palest of these terns (smaller than Black-naped), tail much longer than wings, in non-breeding, carpal bar mid grey, darker than surrounding feathers but less dark than Common and Arctic.
White-fronted: larger than Roseate, Common and Arctic, paler than Common and Arctic, tail longer than wings; often with red legs; adult non-breeding very pale generally, with no dark carpal bar; first year, dark carpal bar, short dark-tipped tail, prominent white edges to primaries.
Common: larger than Black-naped, smaller than White-fronted; tail shorter than wings; often with red legs, longer than Arctic; moults while in Aust, old wing feathers look very dark in contrast to new feathers; dark carpal bar.
Arctic: same size as Common, darker than White-fronted and Roseate; tail slightly longer than wings; short legs; in non-breeding plumage, carpal bar paler than Common but darker than Roseate; adult crown completely white (eliminates all but Antarctic – but watch out for Little); first summer crown like Common (white in front of eye, black behind).
Antarctic: tail same length as wings, no dark carpal bar; bill and legs often red or reddish, adult non-breeding crown white like Arctic.

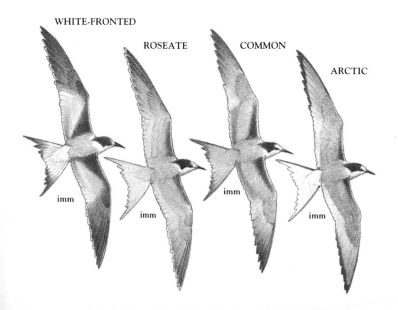

WHITE-FRONTED

ROSEATE

COMMON

ARCTIC

imm

imm

imm

imm

TERNS

Roseate Tern　　　　　　　　　　　*Sterna dougalli*　34–40 cm

Small pale slender resident tern based on islands of northern, north-
eastern and western coasts, with tail longer than folded wings, lacking
prominent dusky tips to primaries and lacking dark outer edge to tail.
Adult breeding: neat black cap, underparts faintly tinged pink (almost
Galah-coloured in some), bill black with red base (in some wholly red,
recorded at Ashmore Reef, Lacepede I and northern Barrier Reef), tail
much longer than folded wings. *Non-breeding:* forehead and fore-crown
white, hind-crown black, bill black, underparts lacking pink flush. *Juv:*
back barred black, grey and white; diffuse carpal bar. *Call:* soft 'chew-
ick'; grating 'sack'. *Nesting:* lined scrape in sand, coral rubble, in
colonies on islands; 1–3 blotched umber eggs. *Range:* common resident
or migrant around northern coast, from Cape Leeuwin, WA, to
southern Barrier Reef (Lady Musgrave I), vagrant farther south.

White-fronted Tern　　　　　　　　　*Sterna striata*　35–43 cm

Small, pale, winter-visiting tern with long tail, lacking dark outer edge,
and fine black bill; most are young birds from NZ, arrive May–June,
leave Oct–Nov, but some breed in summer in Bass Strait and around
Tas, probably locally resident. *Adult breeding:* narrow white forehead,
crown and nape solid black, bill black and legs red to black; lacks
prominent dusky tips to primaries; pink flush to underparts. *Non-
breeding:* forehead and fore-crown white, hind-crown black; similar to
non-breeding Roseate but larger, lacks dark carpal bar, ranges virtually
exclusive. *Imm:* bold dark carpal bar, outer primaries tipped dusky grey.
Juv: similar with barred and mottled brown back, tail tipped black.
Voice: high-pitched 'zrit'; harsher 'kee-at'. *Nesting:* lined scrape in shell
bank, sand or rock; 1–2 blotched umber eggs. *Range:* (a) common
winter migrant, mainly juveniles and immatures, from NZ May–Nov,
to south-east; uncommon north of Sydney and west of Melbourne; (b)
uncommon breeding resident in restricted areas in Bass Strait and
around Tas, may be increasing.

Common Tern　　　　　　　　　　　*Sterna hirundo*　32–38 cm

Small summer-visiting tern mainly to eastern estuaries and adjacent coasts, most birds in non-breeding
plumage with black carpal bar, grey upper tail, black bill, darker outer edge to tail. *Adult breeding:* tail
shorter than folded wings, bill black or red with black tip; in flight only inner primaries translucent,
rump white. *Non-breeding:* similar but with rump and upper tail grey, forehead white, hind-crown
black, dark carpal bar. *Imm:* dark carpal bar, dark secondaries in upperwing. Two forms occur:
commonly-sighted eastern form (race *longipennis*) with black bill in breeding plumage; and rarely-
sighted European form (race *hirundo*) with black-tipped red bill in breeding plumage. *Voice:* grating
'krrrick-krrrick'. *Range:* regular summer visitor from Asia or Europe to coast, islands and estuaries
around Aust, mainly in east; many first-year birds overwinter.

Arctic Tern　　　　　　　　　　　*Sterna paradisaea*　33–38 cm

Small mainly pelagic tern with short legs, tail just longer than folded wings; in flight long narrow dark
trailing edge to primaries, dark outer edge to tail, white upper tail, translucent area in wing extending
to outer primaries. *Adult breeding:* wholly red bill, complete black cap, grey underparts, white cheeks.
Non-breeding: bill black, crown white, nape black, dark carpal bar less obvious than Common, white
upper tail contrasting with back. *Imm:* white secondaries in upperwing, carpal bar less dark than
Common, white upper tail. *Voice:* nasal 'krrrick-krrrick'. *Range:* rare visitor, mainly Apr–Dec, to
southern coasts, most seen well offshore.

Antarctic Tern　　　　　　　　　　*Sterna vittata*　41 cm

Rarely-sighted small tern with heavy reddish bill and legs, translucent area more like Common, but
dark trailing edge to primaries more like Arctic, grey outer edge to tail paler than Common or Arctic.
Adult breeding: similar to Arctic but slightly larger, more solid and with stouter bill. *Non-breeding:* bill
and feet usually noticeably reddish, crown completely white like Arctic. *Juv:* prominent carpal bar,
back mottled brown, white and grey; underparts with brownish cast. *Range:* very rare vagrant from
Antarctic to south coast, most likely autumn or spring.

LARIDAE

TERNS

ROSEATE

imm

non-br

br

WHITE-FRONTED

juv/imm

non-br

br

COMMON

imm

non-br

br

ARCTIC

imm

non-br

br

ANTARCTIC

non-br

br

NODDIES AND DARK-BACKED TERNS

Noddies: dark-plumaged terns found widely in tropical seas; White-capped is one of world's most numerous species. Most are dark grey-brown to black with contrasting white caps. Feed on small fish and plankton, skimmed from the surface. All nest on islands. **Dark-backed terns:** very successful marine species with long tail streamers, found in many parts of world.

Sooty Tern   *Sterna fuscata*   42–46 cm

Adult: black-backed tern with white forehead not extending behind eye. Bridled is browner and has more extensive eyebrow; calls are different, Bridled yaps like puppy, Sooty gives 'wideawake'. *Juv:* unusual among terns, dark all over with white scallops on back and wings. *Voice:* sharp 'wideawake'. *Nesting:* scrape on ground in sandy cays, often in enormous colonies with Common Noddy, and surrounding smaller colonies of other terns. *Range:* common on cays in the north, from Houtman Abrolhos, WA, to about Cairns, Qld.

Bridled Tern   *Sterna anaestheta*   36–41 cm

Adult: brown-backed tern with white forehead extending as a line behind eye; much browner-looking than Sooty with underparts greyish-white rather than pure white. *Juv:* much more like adult than young Sooty: crown mottled brown, back and wing feathers edged paler. *Voice:* 'yap' like puppy. *Nesting:* colonies are much more diffuse than Sooty, with nest scrape usually placed under vegetation or overhanging rock. *Range:* common on islands and cays from Cape Leeuwin, WA, around north coast to southern Barrier Reef.

Common Noddy   *Anous stolidus*   37–41 cm

Large brownish ground-nesting noddy with sharply defined black lores. *Imm:* cap reduced. At sea difficult to separate from other dark noddies at distance; at close range, pale underwing coverts, brown tone to plumage, longer bifurcated tail may be obvious. *Voice:* 'karrk', 'kwok kwok'. *Nesting:* untidy shallow saucer of seaweed, twigs and grass, on ground or trampled low bush, occasionally no nest is constructed; 1 egg. *Range:* common in tropical seas around northern coast from Abrolhos, WA, to Lady Elliott I, southern Barrier Reef, but dispersing after breeding; rare vagrant farther south.

White-capped Noddy   *Anous minutus*   34–38 cm

Small black tree-nesting noddy with sharply defined black lores. Most common on Barrier Reef, particularly Bunker and Capricorn Groups. *Imm:* forehead only white. Difficult to separate at sea from Common, which is more abundant at northern end of Barrier Reef, whereas White-capped is more numerous at southern end; dark underwing coverts. *Voice:* querulous 'krrrk', rapid laughing 'k-k-k-k-k . . .'. *Nesting:* substantial arboreal platform of pisonia, pandanus or casuarina leaves or grass cemented with droppings, large colonies, nests occupying every tree fork; 1 egg. *Range:* common in north-eastern seas about Barrier Reef and Coral Sea from Bunker Group to north-eastern Torres Strait; occasional vagrant farther south.

Lesser Noddy   *Anous tenuirostris*   30–35 cm

Small greyish-black tree-nesting noddy with diffusely defined lores, more or less localised about Houtman Abrolhos islands near Geraldton, WA. Common Noddy, equally numerous in same area, is larger, browner, has longer wedged tail often divided in centre, has paler underwing coverts. *Voice:* soft 'churr'. *Nesting:* substantial arboreal platform of leaves and seaweed cemented with droppings, in large colonies, usually in mangroves; 1 egg. *Range:* common about Houtman Abrolhos, WA, occasionally blown farther afield by storms.

Grey Noddy (Ternlet)   *Anous cerulea*   25–30 cm

Small pale grey noddy with wedge-shaped tail and large dark 'eye'. *Imm and some adults:* darker grey. Underwing grey with white underwing coverts. Told from gulls by overall grey appearance, wedge-shaped tail and large dark 'eye'; from terns by tail shape, lack of black on crown or nape. *Range:* rare vagrant to east coast from Norfolk I and Lord Howe I.

TERNS

SOOTY

juv

ad

BRIDLED

juv

ad

LESSER

WHITE-
CAPPED

NODDIES

COMMON

GREY

50 x 35

46 x 33

Sooty

Bridled

53 x 37

45 x 31

45 x 31

Common

White-
capped

Lesser

BRONZEWINGS

Bronzewings are ground-feeding pigeons with iridescent patches in wing feathers, extensive in some species, restricted to only a few feathers in others. During bowing displays by male the iridescence is particularly striking. They fly rapidly, among fastest birds in bush, but Spinifex prefers to walk unless disturbed. Flock Bronzewing occurs in sometimes very large flocks, eerily silent on the wing; numbers vary according to climatic conditions. Nest is scanty platform of twigs in tree or on ground. Ground-nesting Squatter and Partridge are disappearing, but tree-nesting Crested is increasing.

Squatter Pigeon   *Petrophassa scripta* 26–31 cm
Brown bronzewing with white flank, pied pattern on head, bluish or tan bare skin around eye. Southern form (race *scripta*) has bare skin around eye blue-grey or blue-white; northern form (race *peninsulae*) on Cape York has tan skin around eye. Usually in small groups on ground in open woodland; flies into tree when disturbed. *Voice:* 'coo-coo'. *Nesting:* grass-lined scrape on ground; 2 white eggs. *Range:* locally common resident in open woodland near water, in interior of eastern mainland west of Divide from north-east NSW to Cape York; once common and widespread, now rare and patchily distributed in southern range (near Inverell, Louth, Cobar and west of White Cliffs).

Partridge Pigeon   *Petrophassa smithii* 25–28 cm
Brown bronzewing with white flank, and red or yellow skin around eye. Yellow-faced form (race *blaauwi*) occurs in northern Kimberley, WA; red-faced form (race *smithii*) occupies northern NT. Usually in small flocks on ground, 'explode' into flight when disturbed. *Voice:* 'croo-croo'. *Nesting:* grass-lined scrape in ground, usually near rock or tree; 2 white eggs. *Range:* rare to uncommon vanishing resident in grass-covered woodland in northern Aust, from west Kimberley to extreme north-west Qld.

Flock Bronzewing   *Phaps histrionica* 26–28 cm
Solid-looking bronzewing with reddish-brown upperparts, grey underparts and pattern on face, with white forehead. *Male:* face pattern black and white. *Female:* face pattern duller. Usually seen in flocks, sometimes of hundreds, in open country, flying silently and swiftly, or feeding on ground; visit water sunrise and sundown. *Nesting:* scrape on ground under bush or in grass, often close to others; 2 white eggs. *Range:* highly nomadic in open woodland and plains across northern Aust; once more abundant and widespread.

Crested Pigeon   *Ocyphaps lophotes* 31–35 cm
Brownish-grey bronzewing with fine, upright crest. *Juv:* duller in colour with less iridescence in wing. Flies with distinct whistle of wings, interspersing rapid flaps with long glides, cocks its tail on alighting. *Voice:* surprised 'wook'. *Nesting:* flimsy platform of sticks in tree; 2 white eggs. *Range:* common resident in all but forested country; spreading; requirements are open ground for feeding, trees for shelter and access to water.

Spinifex Pigeon   *Petrophassa plumifera* 20–22 cm
Small reddish bronzewing with upright crest, associated with spinifex on rocky hills. Birds in north-west (race *ferruginea*) and west Kimberley (race *mungi*) have reddish bellies; east of Hall's Creek, WA, bellies are white (race *plumifera*). *Voice:* 'cooroo-coo'; 'coo', hoarse 'coorr'. *Nesting:* scrape under spinifex bush; 2 white eggs. *Range:* locally common on spinifex-covered rocks and sandridges, usually near water, from north-west through central Aust to north-west Qld.

BRONZEWINGS

SQUATTER

red-faced form

PARTRIDGE

yellow-faced form

♀

♂ FLOCK

CRESTED

SPINIFEX

33 x 24	30 x 23	30 x 23	30 x 22	27 x 21
Flock	Squatter	Partridge	Crested	Spinifex

BRONZEWINGS AND OTHER GROUND-FEEDING PIGEONS

Two bronzewings shown here have more iridescence than other species; the sombrely-plumaged rock-pigeons are in essence bronzewings without any bronze, having instead white or chestnut patches in the wings, obvious only in flight, live in rocky gorges and on sandstone cliffs, are able to fly straight up cliff face on loudly clattering wings. Wonga is large pigeon, walks in stately fashion on forest floor, flies when startled with very noisy wingbeats, manoeuvres well between trees for such a big bird, alights on horizontal branch with back to intruder. Feral Pigeon is increasing, often found in bush away from cities and towns, nests in hollow branches and on ledges in open cuts.

Chestnut-quilled Rock-Pigeon　　*Petrophassa rufipennis*　28–31 cm
Dark brown rock-pigeon with chestnut patch in wing. Usually seen among rocks in gorges and on hillslopes, flying with loud wingbeats when startled, can fly staight up out of gorge. *Voice:* 'coo-caroo'. *Nesting:* bulky platform on rock ledge; 2 white eggs. *Range:* common resident on rocky hills and sandstone gorges in NT from Oenpelli to Katherine R.

White-quilled Rock-Pigeon　　*Petrophassa albipennis*　27–31 cm
Dark-brown rock-pigeon with white patch in wing. Birds in WA (race *albipennis*) have large white patch in wing; NT birds (race *boothi*) have small white patch. *Voice:* 'coo-caroo'. *Nesting:* substantial flimsy platform of twigs on rock ledge; 2 white eggs. *Range:* common resident on rocky hills and gorges in Kimberley, WA, and NT east to lower Victoria R.

Common Bronzewing　　*Phaps chalcoptera*　33–36 cm
Brown-backed bronzewing with pale forehead and chin. *Male:* forehead pale buff, purplish crown, white line under eye. *Female:* duller with forehead pale grey. Flies rapidly showing buff underwings. *Voice:* penetrating 'oom oom . . .'. *Nesting:* flat platform of twigs in tree among upright twigs; 2 white eggs. *Range:* common in forests and woodlands throughout Aust, where water available.

Brush Bronzewing　　*Phaps elegans*　28–33 cm
Rufous-backed bronzewing with rufous patch on chin. *Male:* forehead buff, crown and underparts grey; nape and chin rich rufous. *Female:* duller with greyish forehead, reduced patch on chin. Flies swiftly, often sits on horizontal branch. *Voice:* low far-carrying 'oom oom . . .'. *Nesting:* small platform of twigs on ground under bush, or low in bush or bushy tree, 2 white eggs. *Range:* uncommon resident in coastal heaths, woodlands with understorey, in southern Aust from Fraser I, Qld, to Tas, and south-west to Houtman Abrolhos islands.

Wonga Pigeon　　*Leucosarcia melanoleuca*　36–38 cm
Large slaty-grey pigeon with white forehead, white 'V' on breast and black spots on flank. Walks slowly and noisily on ground in forest and woodland, rises with loud clatter when disturbed, perches on horizontal branch. *Voice:* penetrating repetitive 'woo-woo-woo . . .'. *Nesting:* platform of twigs on horizontal fork; 2 white eggs. *Range:* locally common to uncommon in forest, woodland and brigalow scrub in eastern mainland from about Rockhampton, Qld, to Gippsland, Vic; once very numerous in northern range.

Feral Pigeon　　*Columba livia*　33 cm
Variable pigeon descended from domestic birds. Now widely spread in cities, also spreading into more natural habitats.

ROCK-PIGEONS

CHESTNUT-
QUILLED

WHITE-
QUILLED

western
form

NT
form

♀ COMMON

♂

BRONZEWINGS

♂ BRUSH

♀

PIGEONS

WONGA

38 x 29   33 x 24   33 x 24   38 x 29

Chestnut-quilled   White-quilled   Common   Brush   Wonga

DOVES

In general, small pigeons are called doves, although there is no arbitrary line that can be drawn to distinguish one from the other (in fact, Emerald Dove is often called Green-winged Pigeon). All feed on ground, eating mainly dry seed, congregate quickly on productive area, when disturbed fly up with whistling wingbeats, most show white outer feathers which are also used in displays. In brief showers, raise one wing to expose normally hidden feathers – also sunbathe in similar fashion. Nests are scanty platforms of twigs. Two introduced species are doing well; aviary escapees Namaqua Dove, Telpacoti Dove and Collared Dove have failed to establish themselves.

Diamond Dove   *Geopelia cuneata*   20 cm
Small grey dove with white spots on wings and red ring around eye. *Juv:* wings barred but not as regularly as Peaceful. Usually in small groups, occasionally larger flocks near water in hot weather. Rises with whistling whirr and waggling tail when disturbed, showing buff patch in wing, seldom flies far. *Voice:* 'crroo-coo crroo-coo'; 'goola-goo-goola-ga'. *Nesting:* flat flimsy platform in tree or bush; 2 white eggs. *Range:* common nomad in drier areas of Aust.

Peaceful Dove   *Geopelia striata*   19–21 cm
Small grey-brown dove with barred wings and neck, and with bluish ring around eye. Usually singly in small groups, rises with whistling whirr, showing dark brown wing lining. *Voice:* 'doodle-doo'; throaty 'crrr'. *Nesting:* flimsy platform of twigs in low tree or bush; 2 white eggs. *Range:* common resident in open woodland and tree-scattered plains throughout Aust where water available, but absent from south-west corner.

Bar-shouldered Dove   *Geopelia humeralis*   27–29 cm
Large brown dove with barred copper-coloured nape and barred wings. Flight direct with head up, rises with loud clatter, showing chestnut wing lining. *Voice:* alert 'wook-oo-coo'. *Nesting:* flat platform in tree, creeper or bush; 2 white eggs. *Range:* common resident in thick vegetation near water, including mangroves, forests and river margins in north-western, northern and eastern Aust (rarely recorded in Vic).

Spotted Turtle-Dove   *Streptopelia chinensis*   27–28 cm
Large introduced dove with spotted neck. Two forms were introduced: large Chinese form (race *chinensis*) and small South-East Asian form (race *tigrina*) with pale shoulder patch. In most areas both have hybridised, but pure *tigrina* occurs at Innisfail, Qld. *Voice:* 'coo-wook'. *Range:* common resident around cities and large towns in south-western, south-eastern, eastern mainland and Tas, spreading into natural bushland.

Laughing Turtle-Dove   *Streptopelia senegalensis*   24–27 cm
Small introduced reddish-brown dove with speckled breast and bluish shoulders, found in south-west. *Voice:* bubbling 'cooo-wook-coo'. *Range:* common resident around cities and towns in south-western Aust, spreading into natural bushland.

Emerald Dove   *Chalcophaps indica*   23–28 cm
Plump ground-feeding dove with iridescent green wings, red bill, reddish-brown head and body with delicate purplish-bloom, white patch on shoulder and two white bars on lower back. *Juv:* green on wing much duller, head and neck with faint buff bars. *Imm:* head and body more ashy than adult, wings less brilliant, white shoulder patch small or missing, bill grey or orange. Rises with loud whirr, flies relatively silently and swiftly through trees, showing buff underwing linings. *Voice:* monotonous, repetitive 'hoo-hoo-hoo . . .'. *Nesting:* flat platform of twigs, usually well hidden, on leafy branch; 2 white eggs. *Range:* common resident in forest, rainforest and riverside thickets in north and east from north Kimberley, WA, to southern NSW; rare vagrant north-east Vic.

DOVES

PEACEFUL

DIAMOND

ad   juv

LAUGHING

SPOTTED

BAR-SHOULDERED

northern form   ♂

EMERALD

eastern form

♂

♀

juv

21 x 17	20 x 16	26 x 21	29 x 23	28 x 21	29 x 22

Peaceful   Diamond   Laughing   Spotted   Bar-shouldered   Emerald

FRUIT-DOVES AND CUCKOO-DOVE

These forest pigeons are arboreal and feed almost entirely on fruit, having large gapes enabling them to swallow unexpectedly bulky items. **Fruit-doves:** brightly coloured pigeons usually observed with difficulty, in upper storey of rainforest, wet eucalypt forest or mangroves. Two small species are usually seen fleetingly as silhouettes flying from tree to tree on loudly whistling wings (caused by narrowed tip of outer primary); differences in pattern of spread tail aid identification with experience. Larger Wompoo is disappearing as rainforest disturbed, now common only in north. **Cuckoo-doves:** rusty-brown with long tails and barred plumage in at least some stage of development, thus looking rather like cuckoos. One Australian species, conspicuous through loud call.

Purple-crowned Pigeon (Superb Fruit-Dove)   *Ptilinopus superbus*
22–24 cm

Small colourful or plain green pigeon with white abdomen and grey or white tip to outer tail feathers. *Male:* purple crown, rufous nape and black breastband. *Female:* green with purplish patch on back of head and whitish abdomen. *Juv:* green with mottled green and yellow belly, similar to juvenile Red-crowned. Very difficult to spot in rainforest canopy where it feeds; falling fruit help locate feeding birds; flies swiftly with loud whistle of wings. *Voice:* 'coo-coo'; slow steady 'hoo-hoo . . .' not accelerating (Rose-crowned accelerates). *Nesting:* small platform of twigs in small bushy tree; 1 white egg. *Range:* common in north-eastern rainforest, forest and mangroves north of Cardwell, Qld; uncommon nomad or non-breeding migrant farther south to about Hunter R; rare vagrant farther south, with odd birds reaching Tas.

Rose-crowned Pigeon (Fruit-Dove)   *Ptilinopus regina*   18–24 cm

Small colourful or plain green pigeon with yellow abdomen and yellow tip to tail in adults or entirely dark tip in juveniles. *Male and female:* crown rose-pink, nape grey, no breastband. *Juv:* green with mottled yellow abdomen. Rose-crowned form (race *ewingi*) in Kimberley, WA, and NT; 'red'-crowned form (race *regina*) in Qld and NSW. Difficult to sight in canopy, falling fruit aids detection; swift flight with loud whistle of wings. *Voice:* accelerating 'hoo-hoo . . .' for 8–15 notes. *Nesting:* flimsy twig platform in bush or low tree; 1 white egg. *Range:* common to rare resident in rainforest, forest, mangroves and melaleuca forest from Cape Leveque, WA, to about Newcastle, NSW; rare vagrant farther south, odd records Tas.

Wompoo Pigeon (Fruit-Dove)   *Ptilinopus magnificus*   35–40 cm

Large colourful pigeon with purple breast. Difficult to spot in canopy, watch for falling fruit, listen for muttering 'pack-pack . . .'; rapid flight showing much yellow on underwing. Size decreases towards Cape York. *Voice:* loud 'wollack-a-woo', 'woo' or 'wompoo'; 'pack-pack-pack' while feeding. *Nesting:* flimsy platform of vine tendrils on slender horizontal branch; 1 white egg. *Range:* common to rare resident in rainforest and contiguous forest from Cape York to Hunter R, NSW; disappearing in southern range; once extended to Nowra, NSW.

Brown Pigeon (Cuckoo-Dove)   *Macropygia amboinensis*   38–43 cm

Large rusty-brown pigeon with long tail. *Male:* iridescence on neck. *Juv:* paler edges to feathers, faintly barred head, neck and breast. Noisy tame pigeon usually seen perched on high horizontal branch or feeding on low trees such as wild tobacco. Flies with lazy wingbeats, showing russet wing linings. *Voice:* 'did you walk'; hoarse 'croarrr'. *Nesting:* flimsy platform in low bush, vine or tree; 1 white egg. *Range:* common resident in rainforest and forest mainly near edges in eastern Aust, from Cape York to about Bermagui, NSW.

FRUIT-DOVES

PURPLE-CROWNED
(SUPERB)

♂

♀

ROSE-
CROWNED

rose-
crowned
form

'red'-crowned
form

juv

CUCKOO-DOVE

juv

BROWN

♀

♂

WOMPOO
(MAGNIFICENT)

30 x 21

33 x 22

34 x 23

42 x 28

Purple-crowned   Rose-crowned   Brown   Wompoo

FRUIT-DOVES AND IMPERIAL PIGEONS

These fruit-eating pigeons have a lot of white or grey in their plumage and are generally more obvious when feeding than the colourful species on the previous page. Banded is confined to western edge of Arnhem Land; Torresian occurs around northern coast, roosting in mangroves on islands, flying to the mainland each day to feed. Many were shot on islands in the past, resulting in reduced numbers. Topknot is also reduced in numbers, usually observed in flocks, often called Flock Pigeon. White-headed is not strictly a fruit-pigeon, but behaves like one – it is related to the common domestic pigeon. It too declined with shooting and clearing but has recovered by including fruit of the introduced camphor laurel in its diet. Black-collared has been recorded only in Torres Strait, unlikely on mainland.

White-headed Pigeon   *Columba leucomela*   38–42 cm
Dumpy long-tailed pigeon with white head and breast, and black upperparts with iridescent edges to feathers. *Juv:* head and breast smoky-grey, black back extending to hindneck. Flies swiftly, usually seen singly or in small groups. *Voice:* deep 'wook-wook'; low 'oom-oom . . .'. *Nesting:* twig platform in tall tree; 1 white egg. *Range:* common resident in rainforest and forest, visits isolated fruiting trees, eg introduced laurel; eastern Aust from Cooktown, Qld, to Bermagui, NSW; may be increasing in NSW.

Banded Pigeon (Fruit-Dove)   *Ptilinopus cinctus*   48–51 cm
White-headed pigeon with black band on breast, occupying relict rainforest in sandstone gorges in Arnhem Land. *Juv:* greyish wash to head and neck. Flies swiftly with whistling wings; usually seen in small groups in fruiting trees. *Voice:* low 'oom-oom-oom'. *Nesting:* flimsy twig platform in tree; 1 white egg. *Range:* uncommon resident in rainforest remnants and nearby eucalypts on western escarpment of Arnhem Land, NT, from Oenpelli to South Alligator R.

Torresian Imperial-Pigeon   *Ducula spilorrhoa*   39–44 cm
Mainly white pigeon with black outer wing and tail tip. Head and throat may be stained with fruit juices. Usually in flocks, Aug–Apr, feeding in rainforest on mainland, returning to breeding islands to roost. *Voice:* loud 'coo-hoo'. *Nesting:* substantial twig platform in colonies in mangroves on island or along river on Cape York; 1 white egg. *Range:* common but much reduced in northern Aust, migrant (Aug–Apr) over most of range but resident in west Kimberley, WA, from Napier Broome Bay to Kunmunya.

Black-collared Imperial-Pigeon   *Ducula mulleri*   38–40 cm
Large colourful pigeon with diagnostic black collar. Rare vagrant from New Guinea to Torres Strait islands.

Topknot Pigeon   *Lopholaimus antarcticus*   40–46 cm
Large grey pigeon with prominent topknot. Flies high over forest with swift powerful flight, often in flock, showing pale band in black tail and black flight feathers. Often seen hunched on perch, looking almost hawk-like. *Voice:* unusual unpigeon-like 'wirriga'. *Nesting:* bulky platform in leafy tree; 1 white egg. *Range:* common to uncommon nomad in rainforest and nearby eucalypts in eastern Aust from Cape York to about Bermagui, NSW.

PIGEONS

WHITE-HEADED

ad

juv

BANDED

ad

juv

TORRESIAN

BLACK-
COLLARED

ad

TOPKNOT

juv

| 40 x 30 | 32 x 27 | 43 x 30 | 40 x 30 |

White-headed   Banded   Topknot   Torresian

COCKATOOS

Cockatoos are large parrots with powerful bills and crests which vary from simple in Gang-gang to ornate. Like all parrots, powder-down feathers provide bloom on plumage. All have loud screeching calls and often occur in large noisy flocks. Much of their food is taken on the ground, but Gang-gang feeds in trees and shrubs. Nest in hollows, mainly in eucalypts, often chewed about entrance; 2–4 rounded white eggs.

Gang-gang Cockatoo *Callocephalon fimbriatum* 35 cm
Small grey cockatoo with paler edges to feathers and wispy crest. *Male:* head and crest red; *Female:* head and crest grey, bars on breast and abdomen reddish. Flies with loose deep wingbeats; usually seen feeding quietly in outer leaves of eucalypts or in hawthorn hedges. *Voice:* distinctive rasping screech with upward inflection. *Nesting:* deep hollow; 2–3 white eggs. *Range:* uncommon in forest and woodland in south-east, limited range in Tas.

Pink Cockatoo (Major Mitchell) *Cacatua leadbeateri* 36 cm
White, pink-tinted cockatoo with prominent orange and red crest. *Male:* brown eyes; *Female:* red eyes. Flies singly or in small flocks with hesitant wingbeats, revealing pink underwing linings. Feeds on ground and in trees on green seed pods. *Voice:* quavering 'quee-err'; harsh alarm screech. *Nesting:* deep hollow in eucalypt; 2–3 white eggs. *Range:* uncommon nomad in mulga, mallee and cypress woodlands and grasslands near tree-lined watercourses, mainly in 250–400 mm rainfall area of southern Aust.

Galah *Cacatua roseicapilla* 36 cm
Small pink and grey cockatoo. *Male:* brown eyes; *Female:* red eyes. *Juv:* pink areas suffused with grey. Northern and eastern birds (race *roseicapilla*) have eye-ring reddish; western birds (race *assimilis*) are paler with eye-ring grey or whitish. Usually in loud, noisy flocks. *Voice:* pleasant 'czzk-czzk'; loud screech. *Nesting:* much-chewed hollow in eucalypt, lined with eucalypt sprigs, ground below usually scattered with sprigs; 3–4 white eggs. *Range:* common, widespread and increasing in woodlands, tree-scattered grasslands, agricultural clearings throughout Aust except forested areas; increasing in south-east Tas, possibly from escapees.

Sulphur-crested Cockatoo *Cacatua galerita* 49 cm
Large white cockatoo with long yellow crest. *Male:* eye black. *Female:* eye ruby-red. Usually in noisy flocks, showing yellow underwing linings in flight. *Voice:* loud raucous screech. *Nesting:* hollow in eucalypt; 2–3 white eggs. *Range:* common in wide variety of wooded habitats, northern and eastern mainland and Tas.

Little Corella *Cacatua pastinator* 38 cm
Large white cockatoo with short white crest, long or short bill and no obvious pink on throat. Several forms occur: widespread form (race *gymnopis*); with short bill and pink face; northern and north-western form (race *sanguinea*) with short bill and white face; and south-western form (race *pastinator*); with long bill and pink face. Usually in large noisy flocks. *Voice:* very vocal, most often-heard calls are 'ee-ee-ar' and loud screech, deafening where large flocks occur. *Nesting:* hollow in eucalypt or baobab; 2–3 white eggs. *Range:* common in eastern, western and northern Aust, rare and vanishing but with local concentrations in south-west.

Long-billed Corella *Cacatua tenuirostris* 38 cm
Large white cockatoo with short white crest, long bill and obvious pink feathers on throat. *Voice:* noisy screeching; quavering 'currup'. *Nesting:* hollow in eucalypt; 2–3 white eggs. *Range:* after decline, now increasing in woodland and agricultural clearings in south-eastern SA, western and central Vic and south-western NSW.

COCKATOOS

GANG-GANG

♂

♀

PINK

SULPHUR-CRESTED

western
form

GALAH

♀

♂

eastern
form

CORELLAS

LITTLE

northern form

widespread form

south-western
form

35 x 26

Little

36 x 28   39 x 30   47 x 34   35 x 26

Gang-gang   Pink   Sulphur-crested   Galah

LONG-BILLED

36 x 27

Long-billed

BLACK COCKATOOS

Large cockatoos, most having coloured panels in tail; fly with slow, deep wingbeats, often in flocks, keeping contact with loud sad wailing. The two white-tailed species in south-west are very similar, differing mainly in shape of beak. Two forms of Yellow-tailed Black-Cockatoo occur: a long-tailed form (race *funereus*) in NSW and eastern Vic; and smaller short-tailed form (race *xanthonotus*) in western Vic, SA and Tas — regarded by some as separate species.

Yellow-tailed Black-Cockatoo   *Calyptorhynchus funereus*   60–69 cm

Large black cockatoo with yellow tail panels, yellow earpatch and pale edges to feathers. *Male:* eye-ring pink, bill dark grey. *Female:* eye-ring grey, bill whitish, yellowish edges to breast feathers, freckling in tail panels. *Juv:* broader pale edges to feathers. Flies in small flocks with slow measured wingbeats, calling frequently. *Voice:* plaintive 'plee-erk'. *Nesting:* hollow in tree; 1 white egg. *Range:* common resident and nomad in woodland and forest in south-eastern Aust, from Rockhampton, Qld, to Eyre Peninsula, SA, and Tas.

Long-billed (Baudin's) Black-Cockatoo   *Calyptorhynchus baudinii*
53–60 cm

Large black cockatoo with white tail panels and long bill, confined to south-western forested area from Perth to Albany, WA. Feeds mainly on marri nuts and wood-boring grubs, also apples and pears. Similar Carnaby's has shorter bill, occupies drier range but moves to coast Dec–June, visiting pine plantations in large numbers, feeds on hard seeds such as hakea and pine. *Voice:* plaintive 'plee-erk'. *Nesting:* upright hollow in eucalypt; 1–2 white eggs. *Range:* locally common.

Carnaby's Black-Cockatoo   *Calyptorhynchus latirostris*   53–60 cm

Large black cockatoo with white tail panels and short bill. *Male:* bill dark grey, eye-ring pink. *Female:* bill whitish, eye-ring dark grey; earpatch larger and whiter than male. *Juv:* broader pale margins to feathers; dependent on adults for many months, keeps up incessant grating call. Feeds on hard seeds such as hakea and introduced pines. *Voice:* plaintive 'plee-erk'. *Nesting:* deep hole in eucalypt; 1–2 white eggs. *Range:* common migrant or nomad in woodlands in south-west from about Murchison R to Esperance, WA.

Glossy Black-Cockatoo   *Calyptorhynchus lathami*   46–50 cm

Large black cockatoo with small crest on brown or brown and yellow head, and red or barred red and yellow tail panels, bill dark brownish grey in both sexes. *Male:* dark brown head; red tail panels. *Female:* brown head splotched yellow; tail panels red barred with black and edged yellow; *Juv:* yellow or orange spotting on ear-coverts and underwing coverts, yellow/orange bars on belly. *Voice:* soft repeated 'tarr-ed'. *Nesting:* hollow in dead tree; 1 white egg. *Range:* uncommon to rare in open casuarina woodland mainly on highland slopes, in east from about Mackay, Qld, to eastern Vic; isolated population on Kangaroo I.

Red-tailed Black-Cockatoo   *Calyptorhynchus magnificus*   50–61 cm

Large glossy black cockatoo with large crest on black head and red or barred orange tail panels. *Male:* black all over with red panels in tail; bill dark grey. *Female:* black with yellow spots and yellow bars on underparts; orange-yellow panels in tail with narrow black bars; bill whitish. *Juv:* similar. Flies with slow funereal wingbeats in small to large flocks. *Nesting:* hollow in dead tree; 2 white eggs. *Voice:* harsh metallic 'krurr'. *Range:* common in woodlands, and along wooded watercourses in arid country.

Palm Cockatoo   *Probosciger aterrimus*   60 cm

Large black cockatoo with enormous bill, red face, many-feathered upright crest and wholly black tail. *Female:* bill and red face smaller. *Juv:* yellowish bars on underparts. Singly or in small parties, flies with slow deep wingbeats, gliding with downcurved wings. Feeds in palms and pandanus. *Voice:* loud whistling 'kweet-kweet', second note high; harsh 'shaark'. *Nesting:* hollow with bed of wood chips; 1 white egg. *Range:* common in rainforest and nearby eucalypt and paperbark woodland on northern Cape York, Qld.

COCKATOOS

LONG-BILLED

CARNABY'S

YELLOW-TAILED

PALM

RED-TAILED

GLOSSY

50 x 35	50 x 36	51 x 37	50 x 37	45 x 34
Yellow-tailed	Long-billed	Red-tailed	Palm	Glossy

LORIKEETS AND SWIFT PARROT

Lorikeets are small to medium-sized parrots with long necks, pointed tails and wings, and brush-tipped tongues for gathering nectar and pollen from flowers. They fly swiftly and noisily in flocks searching for flowering trees, mainly eucalypts, also ripening sorghum, maize and soft fruit in some areas. Several species may feed together in blossom-rich trees. Nests are typical of parrots, hollows in trees, usually with small opening such as a knot hole. Some species also roost in hollows, but most roost in large noisy congregations in tall trees. Larger species will come into gardens to take nectar from birdtables. Many species have brilliant flashes of colour under the wings, possibly as warning of approaching predator. The **Swift Parrot** is not a lorikeet but behaves as if it were, breeds in Tas and nearby islands and moves to mainland in Feb–Mar to wander over south-east looking for blossoms.

Rainbow Lorikeet *Trichoglossus haematodus* 25–30 cm

Large blue-headed lorikeet with orange or red breast. Two forms occur: eastern form, also occurs as escapee in south-west (race *novaehollandiae*) with red breast and yellow-green collar; and red-collared form (race *rubritorquis*) from northern Aust, with orange breast and collar. *Voice:* noisy screech while in flight, more pleasant single note while feeding. *Nesting:* unlined hollow in tree, usually at some height, 2–3 eggs. *Range:* common in all forest, heath and woodland where blossoms occur in northern and eastern Aust from Broome, WA, to Kangaroo I, SA, and introduced in south-west in vicinity of Perth, WA.

Scaly-breasted Lorikeet *Trichoglossus chlorolepidotus* 23 cm

Only lorikeet with plain green head and red bill. Yellow markings on breast and hindneck vary in size according to way bird holds its feathers, may disappear altogether with feathers sleeked down. Underwing coverts red. Often in flocks with Rainbow. *Voice:* similar to Rainbow but higher-pitched. *Nesting:* unlined hollow in eucalypt, usually with small entrance, 2–3 eggs. *Range:* common in forests, woodlands and heaths where blossoms occur in eastern Aust, from about Cooktown, Qld, to about Thirroul, NSW; as aviary escapees elsewhere.

Swift Parrot *Lathamus discolor* 24 cm

Lorikeet-like parrot with long thin tail, red face and red patches on shoulder, outer edges of secondaries and underwing coverts. *Voice:* distinctive piping 'pee-pit, pee-pit . . .', tinkling chatter. *Nesting:* unlined hollow in eucalypt, sometimes in small colonies, in Tas and Furneaux group; 3–5 eggs. *Range:* breeding in Tas, Sep–Feb; winter nomadic visitor (Mar–Nov, most common May–Sep) to sclerophyll forests and woodlands where trees are blossoming in south-east from south-eastern Qld to Mt Lofty Ranges, SA.

Varied Lorikeet *Trichoglossus versicolor* 18 cm

Small tropical lorikeet with red cap and white eye-ring. *Female:* slightly duller than male. *Imm:* red cap reduced to forehead. *Voice:* thin screech. *Nesting:* unlined hollow, often in horizontal limb, in eucalypt, also used as roost; 2–3 eggs. *Range:* fairly common nomad in woodlands and riverine vegetations in northern Aust from about Broome, WA, to Gulf drainage, Qld, north to Jardine R, south to Mt Isa, and occasionally into Channel Country in wet years.

red-collared form

LORIKEETS

RAINBOW

eastern form

SCALY-BREASTED

PARROTS

SWIFT

VARIED

ad

juv

27 x 23

25 x 20

25 x 20

24 x 19

Rainbow

Scaly-breasted

Swift

Varied

SMALL LORIKEETS AND FIG-PARROTS

Lorikeets: these small lorikeets have high-pitched screeching call, usually fly in small flocks; nest in hollows in trees usually considerable distance from ground, lay white eggs. Young birds similar to adults but with coloured pattern on heads duller. **Fig-parrots:** small dumpy large-billed parrots with short tails, usually feeding on fruit in the canopies of trees such as figs and elaeocarpus in rainforest; generally difficult to observe in upper storey of rainforest trees; quiet high-pitched calls and falling fruit may alert observer, but more likely to be seen in flight when dumpy short-tailed silhouette is diagnostic. Massive clearing of rainforest in southern range has caused Coxen's form to become extremely rare; northern forms are more common, may visit fruiting figs in towns. Nest in hollows in rainforest trees, 2 white eggs, young are similar to female but duller.

Musk Lorikeet   *Glossopsitta concinna* 22 cm
Small lorikeet with red forehead, cheek and bill-tip and green underwing coverts. Often flies with Little and Purple-crowned, larger and stockier. *Voice:* a rolling screech midway in pitch between Rainbow and Little. *Nesting:* unlined hollow in eucalypt; 2 eggs. *Range:* uncommon nomad in woodlands and drier forests in south-east mainland, mainly west of Divide, and Tas.

Little Lorikeet   *Glossopsitta pusilla* 15 cm
Very small lorikeet with red face, black bill and green underwing coverts. *Voice:* high-pitched screeching 'zzet'. *Nesting:* unlined hollow in eucalypt usually with small opening; 2 eggs. *Range:* common nomad in sclerophyll forests in eastern and south-eastern Aust from about Cairns, Qld, to Mt Lofty Ranges and Yorke Peninsula, SA, and Tas.

Purple-crowned Lorikeet   *Glossopsitta porphyrocephala* 16 cm
Small lorikeet with orange forehead and earpatch, purple crown, black bill and red underwing coverts. *Voice:* pleasant harmonious screech, lower-pitched than Little but higher than Musk. *Nesting:* unlined hollow in eucalypt; 3–4 eggs. *Range:* scarce to common nomad in woodland, dry sclerophyll (wandoo) and scrubland (mallee) in south-eastern, southern and south-western Aust, from southern NSW (on rare occasions to south-east Qld) to Shark Bay, WA.

Double-eyed Fig-Parrot   *Psittaculirostris diophthalma* 13–15 cm
Small dumpy large-billed parrot with red or blue on cheeks, usually in fig trees. Sometimes flies with lorikeets but usually in pairs or small flocks with high-pitched call. Three distinct forms: extremely rare Coxen's (race *coxeni*) in south-eastern Qld and north-eastern NSW, with small blue patch on forehead (smaller in female) and patchy red cheek; Macleay's (race *macleayana*) in north-eastern Qld from about Cooktown to Townsville, with small red patch on forehead, and red cheek (male) or bluish-grey cheek (female); and Marshall's (race *marshalli*) on tip of Cape York about Lockhart R and Claudie R, with large red patch on forehead and cheeks (male), or blue forehead and blue and brownish cheeks (female). *Voice:* high-pitched 'zeet-zeet'. *Nesting:* in decayed tree hollows; 2 eggs. *Range:* in fig and elaeocarpus trees in three separate areas on east coast from north-eastern NSW to Cape York.

LORIKEETS

MUSK

LITTLE

PURPLE-CROWNED

FIG-PARROTS

Marshall's form ♂

♀

Coxen's form ♂

Macleay's form ♀

♂

♀

DOUBLE-EYED

25 x 20

22 x 19

Musk

Little 20 x 17

Purple-crowned 20 x 17

Double-eyed

TYPICAL PARROTS AND LONG-TAILED PARROTS

Typical parrots: only two Australian species belong to widespread family of 'typical parrots'. Both occur also in New Guinea and are confined to forest and woodland on northern Cape York Peninsula.
Long-tailed parrots: King and Red-winged are similar in many ways, having similar calls, flying in small flocks with leisurely wingbeats, feeding mainly in trees on green seeds. Cockatiel is quite different, having different tail shape with unique pattern, occurring in sometimes large, fast-flying flocks, feeding mainly on dry seed on ground.

King Parrot   *Alisterus scapularis*   43 cm
Large, long-tailed parrot with red belly. *Male:* head, breast and underparts red, back bright green with paler patch, dark blue tail. *Female:* head and breast brownish-green, eye yellow. *Imm:* similar to female, eye brown. Usually in small flocks mostly of green-headed birds, often perching on high dead branches above forest canopy. *Voice:* metallic 'chack'; 'sweeeee'. *Nesting:* unlined hollow in tree, usually deep, stout upright trunk; 3–5 eggs. *Range:* fairly common in rainforest, wet sclerophyll and eucalypt woodland, orchards, parks in eastern Aust, from Cooktown, Qld, to Otway Ra, Vic.

Red-winged Parrot   *Aprosmictus erythropterus*   32 cm
Red-billed parrot with red patch in wing. *Male:* red wing patch, back black, concealed blue patch on rump. *Female and imm:* red wing patch small, back green. Other parrots with red in wing (Paradise, female Mulga, Regent, Blue Bonnet) have long slender tails and dark bills. *Voice:* metallic 'chiac'. *Nesting:* unlined hollow in tree, usually deep stout upright branch or hollow trunk; 3–6 white eggs. *Range:* common in open woodland, brigalow and tree-lined watercourses through grasslands in northern and eastern Aust, from Broome, WA, to central NSW and north-eastern SA.

Red-cheeked Parrot   *Geoffroyus geoffroyi*   22 cm
Dumpy bright green parrot with short square yellowish tail, confined to rainforests in north-eastern Cape York, Qld. *Male:* face red, crown and nape blue, bill red. *Female:* head brown, bill grey. Flight is direct and swift, showing blue underwing linings. Some lorikeets have red faces, but they have pointed tails and don't range so far north. *Voice:* metallic 'hank'; high screeching. *Nesting:* unlined hollow in tree or palm, sometimes excavated in rotting wood; 2–3 eggs. *Range:* uncommon in dense rainforest and nearby woodland on Cape York between Pascoe R and Rocky R, inland to McIlwraith Ra.

Cockatiel   *Nymphicus hollandicus*   32 cm
Only crested parrot. *Male:* face yellow with orange earpatch, plain tail. *Female:* face paler, tail and abdomen barred. White patch in wing obvious in flight, often in large flocks. *Voice:* plaintive 'queel', 'weero'. *Nesting:* unlined hollow in tree; 4–7 eggs. *Range:* increasingly common in woodland, plains, roadsides, clearings over most of drier areas of mainland Aust; some records in Tas.

Eclectus Parrot   *Eclectus roratus*   43 cm
Male: green short-tailed parrot with red bill and sides, confined to rainforests in north-eastern Cape York, Qld. *Female:* red, black-billed parrot with blue belly and shoulders. Rather cockatoo-like in behaviour: noisy screeching when disturbed, slow wingbeats, roosting in large flocks. *Voice:* rolling 'kraa kraa'; flutelike whistle. *Nesting:* unlined hollow in tree, usually at considerable height; 2 eggs. *Range:* uncommon in rainforest and nearby woodland on Cape York from Pascoe R to Massey Cr, inland to Iron Ra and McIlwraith Ra.

PARROTS

KING

RED-WINGED

♂

♀

♂

♀

COCKATIEL

♀

♂

RED-CHEEKED

♂

♀

ECLECTUS

♀

♂

33 x 27

King

31 x 26

Red-winged

26 x 19

Cockatiel

30 x 25

Red-cheeked

43 x 34

Eclectus

LONG-TAILED PARROTS AND BROAD-TAILED PARROTS

Long-tailed parrots: elegant swiftly-flying birds, usually encountered in small flocks, feeding in foliage or on ground (where they do not often appear to be encumbered by tails). Nest in hollow limbs of eucalypts, Princess sometimes in loose colonies. **Broad-tailed parrots:** the two remaining parrots on this page and all those on the following plates are classified as broad-tails. The tail feathers are not particularly broad (and some are quite narrow, eg Elegant p 176), but they are often spread in display, giving a characteristic tail-heavy appearance. They usually move in pairs or small noisy parties. Most nest in hollow limbs, but a few nest on the ground and some dig burrows into termite mounds. Widespread Ringneck has been regarded as two or three species in past; observers may prefer to use old names Twenty-eight, Port Lincoln, Mallee Ringneck and Cloncurry Ringneck for the different forms except when intergrades are encountered.

Ringnecked Parrot (Ringneck)   *Platycercus zonarius*   34–38 cm
Long-tailed green parrot with yellow ring on hindneck. Four forms occur, with areas of intergradation: 'Twenty-eight' form (race *semitorquatus*) with black head, red patch on forehead, green belly, calls 'twenty-eight', confined to forested south-west; Port Lincoln form (race *zonarius*) and more turquoise-green north-western *occidentalis*, with black head, little or no red patch on forehead, yellow belly, calls 'twenty', widely spread in west from south-east Kimberley, WA, to west Flinder's Ra, SA; Mallee form (race *barnardi*), with green head, indigo back, found in eucalypt, mulga and callitris woodland east and north-east of Flinder's Ra to about Windorah, Qld; and pale green Cloncurry form (race *macgillivrayi*) mainly in river red gums, from Boulia, Qld, north through Cloncurry to about Camooweal, NT. *Nesting:* hollow in eucalypt; 4–7 white eggs. *Range:* common in forest and woodland, mallee, mulga and tree-lined watercourses in arid interior throughout Aust except wetter areas in north and east.

Red-capped Parrot   *Purpureicephalus spurius*   36 cm
Colourful red-capped parrot with yellow earpatch and rump, purple breast. *Female:* duller. *Juv:* cap green with mottled red forehead, breast greyish-brown. Associated with marri trees; bill adapted for removing seeds. Flight undulating with rapid fluttering wingbeats followed by glide with closed wings, revealing yellowish rump. *Voice:* grating metallic 'getacheck'. *Nesting:* hole in eucalypt; 4–9 white eggs. *Range:* common in forest and woodlands in south-west.

Princess Parrot   *Polytelis alexandrae*   40–47 cm
Long-tailed pastel-olive parrot with lilac forehead and rump, pink throat and grass-green shoulder. *Female and juv:* duller with shorter tail. Extremely elegant, flies with lazy grace in small flocks, rises when disturbed with waggling tail. *Voice:* 'krrrt'; 'queet-queet'. *Nesting:* hollow in eucalypt, often in small colonies; 4–6 white eggs. *Range:* rare, highly nomadic in central Aust, mainly along timbered watercourses near spinifex or in mulga woodland; no precise limits can be placed on its range, nor can prediction be made where it may be seen.

Superb Parrot   *Polytelis swainsonii*   40 cm
Long-tailed green parrot with yellow face and red throat, or with blue cheeks. *Male:* face yellow, throat red, undertail black. *Female:* all green with bluish cheeks, undertail black with red edges to feathers. Very elegant in flight, with effortless wingbeats, usually in small groups. *Voice:* grating 'crrrack crrrack'. *Nesting:* hole in eucalypt; 4–6 white eggs. *Range:* locally common in open woodland and riverine forest in inland NSW.

Regent Parrot   *Polytelis anthopeplus*   40 cm
Long-tailed smoky-yellow parrot with black tail and flight feathers and red patch on wing. *Male:* bright smoky-yellow with broad red wing patch. *Female:* dull with small red wing patch. Western birds (race *westralis*) are greener, particularly female. Fly in often large noisy flocks (>50) with swift graceful flight. *Voice:* grating 'krrrack'. *Nesting:* hole in eucalypt; 4–6 white eggs. *Range:* common resident or nomad in woodland near agricultural clearings in south-west and locally common partial nomad in riverine forests in south-east.

PARROTS

'28'

Port Lincoln form

RINGNECK

mallee form

Cloncurry form

ad

RED-CAPPED

juv

PRINCESS ♂ ♀

SUPERB ♀ ♂

REGENT ♀

31 x 25

Ringneck

26 x 23

Red-capped

28 x 22

Princess

28 x 24

Superb

31 x 25

Regent

ROSELLAS

Colourful broad-tailed parrots with distinctive cheek patches white, yellow or blue. Because forms with the same cheek colour intergrade where ranges overlap, there is a move to think of rosellas as three species, White-cheeked, Blue-cheeked and Yellow-cheeked (Western).

Eastern Rosella
Platycercus eximius 30 cm

White-cheeked rosella with red head and breast. *Juv:* smaller, duller and with patchy green on head. Two forms: green-mantled form (race *eximius* in south-east and race *diemenensis* in Tas) with green edges to back feathers; and golden-mantled form (race *splendidus*) on tablelands in north-east NSW and south-east Qld, intergrading with Pale-headed in south-east Qld with yellow edges to back feathers. *Voice:* 'tink-tink'; 'pseet-it'; 'tock-swit-it'. *Nesting:* hole in tree stump; 4–7 white eggs. *Range:* common in open woodland in south-east and Tas.

Pale-headed Rosella
Platycercus adscitus 30 cm

White-cheeked rosella with yellow head. *Juv:* duller, often with red feathers on crown and cheeks, moulted within three months. Two forms: white-cheeked form (race *palliceps*) in southern range, north to about Cairns, Qld; and blue-cheeked form (race *adscitus*) with blue and white cheek, rather paler yellow head, blue more extensive north of Cooktown, Qld. *Range:* common resident in open woodland near clearings in eastern Qld.

Northern Rosella
Platycercus venustus 29 cm

White-cheeked rosella with black cap. *Juv:* red patches in black cap, moulted within four months. Towards Kimberley, WA, cheeks become more blue on lower edge. Usually seen in small groups, more retiring and quieter than Pale-headed, flies in typical rosella fashion. *Voice:* 'pseet-ee pseet'. *Nesting:* hollow in eucalypt or paperbark; 3–4 eggs. *Range:* uncommon resident in woodland, often near watercourses, in northern Aust, from Nicholson R, Qld, to about Derby, WA.

Western Rosella
Platycercus icterotis 25–30 cm

Yellow-cheeked rosella confined to south-west. *Male:* head and breast red. *Female:* head and breast green with variable mottled red feathers. Two forms: green-backed form (race *icterotis*), with feathers of back edged green, found in wetter areas; and red-backed form (race *xanthogenys*) with red edges to back feathers, found in drier areas. *Voice:* quiet 'tink-tink'. *Nesting:* hollow in eucalypt; 3–6 white eggs. *Range:* common in forest, woodlands and orchards in south-west.

Green Rosella
Platycercus caledonicus 32–36 cm

Blue-cheeked, yellowish-green rosella confined to Tas and Bass Strait islands. *Juv:* duller, more olive. *Voice:* two-noted metallic call; piping alarm; flute-like whistle. *Nesting:* hollow in tree or stump, 4–8 white eggs. *Range:* common resident in wide range of wooded habitat in Tas.

Blue-cheeked Rosella
Platycercus elegans 32–36 cm

Blue-cheeked rosella with red, yellow or orange body and blue wings and tail. *Juv:* in south, mottled red and green on head, throat and undertail coverts; in north, much more red or entirely red on body. Two basic forms: Crimson Rosella (races *elegans* and *nigrescens*) crimson body, darker in north, occurring from Kangaroo I to Cairns, Qld; and Yellow Rosella (race *flaveolus*), yellow body, found in riverine forests of Murray, Lachlan, Murrumbidgee and lower Darling Rs. Intergrades between Crimson and Yellow are known as Adelaide Rosellas and have decreasing intensity of red from south to north in SA between Fleurieu Peninsula, Flinders Ra and Murray R, but 'regressives' occur so yellowish birds may appear in reddish populations and vice versa. *Voice:* 'psit-a-see'. *Range:* common in rainforest and forest in east and south-east.

ROSELLAS

WHITE-CHEEKED

eastern

pale-headed

blue-cheeked

northern

ad

juv

BLUE-CHEEKED

green

ad

juv

♂

♀

WESTERN

crimson

ad

juv

yellow

Adelaide

ad

juv

27 x 23

White-cheeked

26 x 22

Western

29 x 24

Blue-cheeked

PSEPHOTUS PARROTS

Psephotus means 'inlaid with small pebbles', referring to the mosiac-like cheek feathers of these small broad-tailed parrots. Plumage has almost iridescent appearance, resulting in great demand as cagebirds. Some are now very rare, others common and increasing.

Red-rumped Parrot *Psephotus haematonotus* 27 cm
Male: brilliant-green parrot with red rump, small yellow patch on shoulder and yellow lower breast and abdomen. *Female:* plain brownish-green. Usually in pairs or small groups, flies swiftly, often along roadsides. *Voice:* trilling song; shrill two-noted whistle. *Nesting:* hollow in eucalypt; 4–6 white eggs. *Range:* common resident in open woodland in inland south-east, spreading towards coast and into southern Qld.

Mulga Parrot *Psephotus varius* 27–31 cm
Male: brilliant-green parrot with yellow lower back, small red patch on rump and on nape, large orange-yellow patch on shoulder, lower abdomen yellow and red. *Female:* dull brownish-green with reddish-brown patch on nape and shoulder. Usually in pairs or small groups, flight swift and direct. *Voice:* 'swit swit'. *Nesting:* hollow in eucalypt; 3–6 white eggs. *Range:* uncommon resident in mallee and mulga woodland, along wooded watercourses in arid south.

Paradise Parrot *Psephotus pulcherrimus* 30 cm
Male: brown-backed parrot with red shoulder, forehead and undertail coverts, black crown, brilliant blue and turquoise cheeks and breast. *Female:* duller, with whitish face, small reddish patch on shoulder, bluish-grey underparts with reddish mottling under tail. (Red-shouldered form of Blue Bonnet, sometimes mistaken for Paradise, has no blue or green on underparts, lacks red forehead and black crown.) *Voice:* metallic 'queek'. *Nesting:* tunnel in terrestrial termite mound or earth bank; 3–5 white eggs. *Range:* rarest Aust bird, likely only in isolated areas of southern Qld west of Divide where habitat is unchanged by stock.

Golden-shouldered Parrot *Psephotus chrysopterygius* 26 cm
Golden-shouldered parrot found on western Cape York, Qld. *Male:* back brown, forehead and shoulder yellow, crown black, underparts turquoise and undertail coverts red. *Female:* dull yellowish-green with bronze crown and bluish undertail coverts with some salmon feathers. Usually in pairs or small parties, flight swift and undulating. *Voice:* 'cluk-cluk'. *Nesting:* tunnel in termite mound; 4–6 white eggs. *Range:* rare but locally common in paperbarks and eucalypt woodlands where termite mounds occur on western side of Cape York north to Watson R and Coen, Qld.

Hooded Parrot *Psephotus dissimilis* 26 cm
Golden-shouldered parrot found in north-eastern NT. *Male:* forehead and crown black, back dark brown with broad yellow patch on wing, underparts turquoise, small red patch under tail. *Female:* similar to female Golden-shouldered but more salmon under tail. Usually in pairs or small parties. *Voice:* 'chillick'. *Nesting:* tunnel in termite mound; 4–6 white eggs. *Range:* uncommon resident in open woodland with spinifex and termite mounds, from western Arnhem Land to Mataranka, NT, and formerly on Macarthur R, near Qld border; some early reports of annual movements to Melville I.

PARROTS

RED-RUMPED
♂
♀

MULGA
♂
♀

GOLDEN-
SHOULDERED
♀
♂

PARADISE
♂
♀

HOODED
♀
♂

21 x 17
Paradise

23 x 19
Red-rumped

24 x 19
Mulga

21 x 19
Golden-shouldered

21 x 19
Hooded

GRASS-PARROTS

Small 'grass' parrots found in a variety of habitats providing seed-bearing ground-cover throughout southern Aust. Feed on ground, in small flocks, fly high when disturbed. Cheek feathers often fluffed out to cover sides of beak; juveniles have pale orange bills that darken once they become independent of parents. Some are now very rare, others increasing. Rarest is Orange-bellied, some others have belly orange (Rock, Elegant, Blue-winged) but differ in pale blue, not green, central tail feathers.

Rock Parrot   *Neophema petrophila* 22 cm
Dull brownish-olive grass-parrot with blue forehead and face, and shoulder narrowly edged light and dark blue; slender tail pale blue above, yellowish below. Usually seen in coastal vegetation among rocks along south and south-west coast, flies high, zig-zagging, when disturbed. *Voice:* rather metallic 'zit-zit . . .'. *Nesting:* on ground in hollow among rocks; 4–5 white eggs. *Range:* common resident or partial nomad on rocky coastline from Robe, SA, to Shark Bay, WA.

Elegant Parrot   *Neophema elegans* 22–24 cm
Bright golden-olive grass-parrot with yellow and blue facial band and narrow blue edge to wing; slender tail blue above, yellow below. Usually seen in grasses in wide variety of habitats, zig-zagging like snipe when flushed, flying fast and high in great arc. *Voice:* grating 'zit'. *Nesting:* hollow in dead eucalypt or mulga spout; 4–5 white eggs. *Range:* common nomad in open woodland and agricultural clearings in south-west; uncommon nomad in south-east about junction of NSW, Vic and SA.

Blue-winged Parrot   *Neophema chrysostoma* 21 cm
Dull olive-green grass-parrot with yellow and blue facial band and broad blue edge to wing; slender tail blue above, yellow below. Blue frontal band less extensive than Elegant, and breast more grey-green. Usually seen in grasses in wide variety of habitats; when flushed either flutters ahead to alight or zig-zags up and away like snipe. *Voice:* high-pitched 'zzt-zzt'; soft warbling tinkle. *Nesting:* hollow in eucalypt or stump; 4–6 white eggs. *Range:* common migrant or nomad in Tas and south-east coastal areas, moving approximately Mar and Sep; uncommon to rare inland as far as southern Qld.

Orange-bellied Parrot   *Neophema chrysogaster* 20 cm
Bright grass-parrot with blue frontal band and narrow blue edge to wing; tail green above, yellow below. (Orange patch on belly in common with other grass-parrots). Similar Elegant and Blue-winged have yellow lores, Rock duller, more olive, has thinner tail bluish above, more westerly range. *Voice:* buzzing rapid chatter. *Nesting:* hollow in tall eucalypt in south-west Tas; 4–6 white eggs. *Range:* rare, vanishing migrant, winters Mar–July in coastal samphire flats and subcoastal wooded paddocks with capeweed from Westernport, Vic, to Robe, SA; breeds in south-west Tas around Port Davey.

Turquoise Parrot   *Neophema pulchella* 20 cm
Blue-faced grass-parrot with red shoulder patch (male) or with white lores (female). *Juv:* similar to female but duller. Usually in flocks or small parties; flight rather erratic but fast. *Nesting:* hollow in eucalypt or stump; 4–5 white eggs. *Range:* uncommon to rare in open woodland or grassland with scattered trees in south-eastern mainland, increasing after long decline.

Scarlet-chested Parrot   *Neophema splendida* 20 cm
Blue-face grass-parrot with scarlet breast (male) or with yellow breast and bluish lores (female). Female differs from female Turquoise only in more bluish lores; habits very similar. *Voice:* soft twittering. *Nesting:* hollow in eucalypt or mulga, several pairs often in neighbouring trees; 3–5 white eggs. *Range:* rare nomad in open woodland, mallee and mulga in southern inland Aust.

PARROTS

ELEGANT

BLUE-WINGED

ROCK

ad

juv

ad

juv

ORANGE-BELLIED

juv

ad

SCARLET-CHESTED

TURQUOISE

♀

♂

♂

♀

Scarlet-chested 23 x 19

24 x 20

21 x 18

22 x 19

23 x 20

22 x 18

Rock

Elegant

Blue-winged

Orange-bellied

Turquoise

PARROTS

Blue Bonnet is a psephotus parrot of arid country with a number of readily recognisable forms.
Budgerigar, among our commonest birds, is related to two of our rarest, Night and Ground Parrots.
Although brightly plumaged, they are cryptically coloured and are difficult to see in long grasses,
where they feed. Bourke's Parrot, an aberrant grass-parrot, is coloured to suit mulga woodland.

Blue Bonnet   *Psephotus haematogaster*   28–34 cm

Variable olive-brown parrot with blue face. Several forms occur, with
intergradations between some: yellow-vented form (race *haematogaster*)
has red belly, yellow vent, no red on shoulder, ranges from Eyre
Peninsula, SA, to north-west Vic and south-west Qld; red-vented form
(race *haematorrhous*) has red belly, vent and shoulder, ranges in central
NSW and southern central Qld; pale form (race *pallescens*) in Lake Eyre
Basin, SA; and Naretha Blue Bonnet (race *narethae*) with yellow belly,
red vent and small red patch on shoulder, found in myall woodland
fringing Nullarbor Plain. *Voice:* harsh 'cluck-cluck' or 'cloot-cloot';
high-pitched whistle. *Nesting:* hollow in eucalypt or she-oak; 4–7 white
eggs. *Range:* uncommon to rare resident in woodland and wooded
watercourses in drier areas of the south-east and the Nullarbor Plain.

Budgerigar   *Melopsittacus undulatus*   18 cm

Small bright-green parrot with yellow face and yellow edging to back
feathers. Occurs in small to large flocks mainly in interior of Aust.
Voice: continuous pleasant warbling and scolding. *Nesting:* hollow with
small entrance in eucalypt, she-oak or mulga; 4–8 white eggs. *Range:*
common nomad in interior, concentrating near water and in areas
blooming after rain; perhaps regular north–south movement in SA.

Bourke's Parrot   *Neophema bourkii*   19–23 cm

Brown 'grass-parrot' with delicate pink and blue underparts. *Male:*
forehead blue, brownish underparts. Usually in small groups in mulga
woodland; generally tame and inconspicuous, when flushed from
ground flies short distance into nearby mulga. *Voice:* mellow 'chu-ee';
shrill alarm 'kleet-kleet'. *Nesting:* hollow in mulga or she-oak; 3–6 white
eggs. *Range:* uncommon to rare nomad in mulga woodland in mid-
western and central Aust, east to about Cunnamulla, Qld.

Ground Parrot   *Pezoporus wallicus*   30–33 cm

Long-tailed green parrot with green and yellow pattern on feathers,
found in coastal heaths. *Adult:* red forehead, face and upper breast
unmarked green. *Juv:* green forehead, face and upper breast spotted and
blotched with black. Rarely seen unless flushed from heaths and buttongrass plains; flies like Common Sandpiper, on stiff downtilted wings,
dives back into heath. *Voice:* mainly dawn and evening, three or four
thin sweet notes 'tsee-tsee-tsee-tsit'. *Nesting:* hollow under grass tussock
or low bush lined with chewed grass stems; 3–4 white eggs. *Range:* rare
in limited areas of coastal heaths on south coast WA, south-east from
Fraser I, Qld, to south-east SA; more common Tas, mainly west and
south-west.

Night Parrot   *Pezoporus occidentalis*   23–26 cm

Short-tailed green parrot with black and yellow pattern on feathers,
found in arid grasslands in central Aust. Similar to Ground but shorter
tail, dumpier shape, lacks red forehead, has more yellowish cheeks.
Rarely sighted; when flushed flutters short distance 20–30 m. *Voice:*
short sharp repeated mournful whistle; frog-like croak. *Nesting:* scrape
in spinifex or samphire bush; 4 white eggs. *Range:* rare nomad in
spinifex, samphire and bluebush plains or rocky hillsides in arid interior;
no precise limits known; recent sightings in eastern Lake Eyre Basin.

PARROTS

BLUE BONNET

Nullarbor form

pallid form

yellow-vented

red-vented form

juv

BUDGERIGAR

BOURKE'S

juv

ad

NIGHT

GROUND

juv

ad

23 x 19

20 x 17

19 x 14

26 x 21

Blue Bonnet

Bourke's

Budgerigar

Ground

CUCKOOS

Cuckoos are loosely-feathered birds with two toes on each foot directed forwards and two backwards; unlike parrots, which have similar toes, cuckoos do not use their feet for holding food. Beaks are slightly downcurved and have unusual, slightly-raised circular nostrils. Wings fit loosely, appear to droop away from body. Many eat hairy caterpillars, others feed on fruit. Most cuckoos lay eggs in other birds' nests; eggs are often similar to host's.

Chestnut-breasted Cuckoo   *Cuculus castaneiventris*   24 cm
Blue-grey cuckoo with rich chestnut breast and yellow eye-ring; outer webs of tapered tail not notched. Similar to Fan-tailed but brighter in colour with upper tail lacking visible notches. *Juv:* rich rufous above, pale cinnamon below. Probably not uncommon summer migrant in rainforest on Cape York, but little known. No definite breeding records in Aust, but may parasitise Tropical Scrubwren. More information on range, habits, plumage and calls required.

Fan-tailed Cuckoo   *Cuculus pyrrhophanus*   24–28 cm
Blue-grey cuckoo with dull chestnut breast and yellow eye-ring. *Juv:* brown above with rufous edges to feathers, but no notches, underparts obscurely barred darker; graduated tail with buff notches. Usually sits on prominent perch, often quite low. *Voice:* far-carrying trill with downward inflection; mournful whistle 'wh-phweee', first note often not heard unless close. *Nesting:* parasitises mainly dome-shaped nests of scrubwrens, warblers and wrens, less commonly cup-shaped nests. *Range:* common resident or migrant in forest and woodland in south-west, south-east and east; common migrant in Tas.

Brush Cuckoo   *Cuculus variolosus*   24 cm
Grey-headed cuckoo with pale buff breast and grey eye-ring. *Juv:* heavily mottled brown and buff on upperparts, barred on underparts; square tail broadly notched buff, begins moult by Apr. *Imm:* similar to juv, but rufous notches and edges on back and wing feathers richer and smaller. Usually sits unobtrusively, attracts attention by loud call. *Voice:* commonest call ascending 'ph-ph-phew, ph-ph-phew . . . ' becoming louder, more persistent. *Nesting:* in south, parasitises mainly cup-shaped nests of flycatchers, occasionally domed nests of wrens and warblers; in north, mainly Bar-breasted Honeyeater. *Range:* common migrant Sep–Apr or resident in northern and eastern Aust, from Derby, WA, to about Melbourne, Vic.

Oriental Cuckoo   *Cuculus saturatus*   28–34 cm
Large greyish cuckoo with barred abdomen, yellow eye-ring and orange feet. Rare rufous phase (female only) barred reddish-brown and dark brown above, bars on underparts extending to throat. *Juv:* dark grey above with white edges to feathers, bars on underparts extending to throat, moults mid Nov–Mar. *Voice:* usually silent; low trill; harsh 'gaak-gaah-gah-ak-ak-ak'. *Range:* uncommon non-breeding migrant Sep–Apr in forests, woodlands and thick riverine vegetation in northern and eastern Aust from Derby, WA, to about Nowra, NSW, some overwinter; periodic irruptions of juveniles.

Pallid Cuckoo   *Cuculus pallidus*   30–33 cm
Large greyish-brown cuckoo with white spot on nape and yellow eye-ring. *Juv:* whitish-buff with black centres to feathers of crown, back and wings and black stripe through eye to side of breast. *Imm:* mottled brown and buff, barred below (often confused with Oriental). *Subadult:* similar above, grey below, with rufous patch on neck. *Voice:* male, loud whistle ascending scale; female, hoarse ascending 'wh-wh weeya wh-wh weeya'. *Nesting:* parasitises mainly cup-shaped nests of honeyeaters, flycatchers, orioles, magpie-larks etc. *Range:* common in variety of habitats (except forests) throughout Aust; regular migrant Tas.

CUCKOOS

FAN-TAILED

juv

ad

juv

ad

CHESTNUT-BREASTED

juv

BRUSH

ORIENTAL

juv

hepatic ♀

ad

ad

Brush 19 x 14

PALLID

ad

♀ sub-ad

imm

juv

21 x 15

Fan-tailed

25 x 16

Pallid

BRONZE-CUCKOOS

Small cuckoos with iridescent plumage and barred underparts; juvenals generally lack bars. (Black-eared has juvenal-like plumage without iridescence or bars). Distinctive pattern of bars and colours on tail varies with species. Most have similar but distinctive plaintive calls. All are parasitic in nesting behaviour, generally choosing birds with domed nests, but Shining (Golden) and Horsfield's also lay in cup-shaped nests.

Shining (Golden) Bronze-Cuckoo *Chrysococcyx lucidus* 17–18 cm
Iridescent green bronze-cuckoo with complete bars on underparts extending to chin. Two forms occur: Golden form (race *plagosus*) with head and bars on throat coppery-bronze, and with little white freckling on forehead, breeding in Aust; and Shining form (race *lucidus*) with head and bars on throat iridescent green, with extensive white freckling on forehead, breeding in New Zealand, recorded on migration during Feb–Apr and Sep–Nov. *Voice:* series of whistles with upward inflection, followed by a few downward trills. *Nesting:* parasitises mainly dome-shaped nests of warblers, thornbills and scrubwrens, less often nests of fantails, robins, etc. *Range:* common migrant mainly Aug–Apr in forests and woodland in eastern and south-western mainland and Tas.

Black-eared Cuckoo *Chrysococcyx osculans* 19–20 cm,
Large brown bronze-cuckoo with black ear patch, pale rump and grey-brown tail. Only slightly glossed with iridescence on back; somewhat similar to juvenile Horsfield's, which is smaller with dark rump, rufous sides to tail and with pale edges to wing feathers. *Voice:* series of whistles with downward inflection, slower and an octave lower than Horsfield's. *Nesting:* parasites mainly Fieldwren and Redthroat, also Speckled Warbler in drier areas of its range. *Range:* uncommon to rare nomad or migrant in open eucalypt woodland, samphire flats and wooded watercourses in drier areas of Aust; breeding range coincides with Fieldwren and Redthroat, wanders into northern Aust in winter.

Horsfield's Bronze-Cuckoo *Chrysococcyx basalis* 17 cm
Dull bronze-cuckoo with dark earpatch, narrow incomplete bars on each side of breast and flanks and rufous base to sides of tail. *Juv:* similar to Black-eared, but with rufous in tail and pale edges to back-feathers. *Voice:* series of whistles with downward inflection, faster and higher than Blacked-eared. *Nesting:* parasitises dome-shaped and cup-shaped nests, particularly wrens, thornbills and robins. *Range:* common nomad or resident in woodland throughout mainland and Tas, mainly inland; absent from heavily forested areas.

Little Bronze-Cuckoo *Chrysococcyx minutillus* 15–16 cm
Small bronze-cuckoo with red eye-ring, dull bronze-green above, broad incomplete bars on underparts, some rufous in outer tail. Two forms occur: resident rufous-breasted form with rufous outer tail feather (race *russatus*), in coastal north-east Qld from Bowen to Cape York, parasitising warblers of mangroves and rainforest; and migratory white-breasted form with white outer tail feather (race *minutillus*), in woodlands of northern and eastern Aust, Sep–Mar, parasitising mainly White-throated Warbler. *Voice:* distinctive extended downward trill; four-noted descending quail-like call 'tew-tew-tew-teeew'. *Nesting:* parasitises warblers. *Range:* uncommon migrant or resident in rainforest, forest, woodland and mangrove in northern and eastern Aust.

CUCKOOS

SHINING

Aust form

in sun

juv

in shade

NZ form

HORSFIELD'S

ad

BLACK-
EARED

ad

juv

juv

rufous-breasted
form

LITTLE

white-breasted
form

juv

♂

♂

♀

juv

ad

juv

white-
breasted
form

rufous-
breasted
form

18 x 12

20 x 14

Horsfield's 18 x 12

Little 19 x 13

Shining

Black-eared

LARGE CUCKOOS, COUCAL AND KOOKABURRAS

Large cuckoos eat mainly fruit, particularly native figs. Two are summer-breeding migrants, attracting attention by loud calls; Long-tailed breeds in New Zealand, is seen here very rarely on migration. **Coucals** belong to the cuckoo family, but care for their own young; they inhabit long grasses in open woodland. **Kookaburras:** large kingfishers with loud laughing calls; feed mainly on reptiles and large invertebrates, generally occur and nest in small parties. Nest in tree hollow or tunnel dug into termite mound.

Pheasant Coucal   *Centropus phasianinus*   60–80 cm
Large long-tailed cuckoo with body black (summer) or brown (winter and juvenile) and rufous barred wings and tail. Usually seen running across road or perched (particularly on wet days) on fencepost or dead tree near long grass; when flushed flies heavily with laboured wingbeats. *Voice:* loud booming 'coot coot coot coot . . .'; hissing 'skeowk'. *Nesting:* concealed grass cup in long grass; 3–5 white eggs. *Range:* common resident in long grass in woodland in north-western, northern and eastern Aust.

Channel-billed Cuckoo   *Scythrops novaehollandiae*   60–67 cm
Large grey cuckoo with long wings and tail and heavy bill. *Juv:* head and back suffused with buff, moults within two months. *Flight:* looks like flying walking-stick, usually in small very noisy groups, often chased by crows. *Voice:* loud raucous 'kork ork ork ork . . .'. *Nesting:* parasitises crows, magpies and currawongs; young reared with host's chicks. *Range:* common migrant Sep–Apr in forest and woodland with native figs in northern and eastern Aust; vagrant Vic, ACT and Tas.

Common Koel   *Eudynamis scolopacea*   39–46 cm
Long-tailed cuckoo either black (male) or barred brown (female). *Female and imm male:* barred and spotted brown with head, nape and stripe on cheek black. *Juv:* head and nape buff, black line through eye; moults at 2–3 months into female-like plumage. *Voice:* various raucous notes, most noticeable loud 'koo-ee'. *Nesting:* parasitises cup-shaped nests of large honeyeaters, orioles, magpie-larks. *Range:* common migrant Sep–Apr in rainforest, forest and woodland in northern and eastern Aust; rare vagrant elsewhere.

Long-tailed Koel   *Eudynamis taitensis*   38 cm
Similar to female Common Koel, but has white stripes above and below eye, and streaked, not barred underparts; tail brown with dark bars (female Koel has white bars on black tail). *Voice:* loud 'zweeesht'. Very rare vagrant to east coast during migration from NZ to islands to north of Aust.

Blue-winged Kookaburra   *Dacelo leachii*   40–46 cm
Large kingfisher with streaked head, white eye, large blue patch on shoulder. *Male:* tail and rump blue. *Female:* rump blue, tail reddish-brown, barred darker. *Voice:* extended maniac laughter. *Nesting:* hollow in tree or hole in termite mound; 2–4 white eggs. *Range:* common to uncommon in open woodland and wooden watercourses in north-western, northern and north-eastern Aust.

Laughing Kookaburra   *Dacelo novaeguineae*   46 cm
Large kingfisher with dark patch behind dark eye and limited silver-blue on shoulder. In fresh plumage (Feb–Mar) underparts have faint dark barring which wears off towards breeding season. *Juv:* bill dark; underparts and head faintly barred brown. *Voice:* loud chuckling laugh often in chorus. *Nesting:* hollow in tree or tunnel in arboreal termite mound. *Range:* common resident in woodland, forest and wooded watercourses in eastern Aust; introduced into south-west, Tas, Flinders I, Kangaroo I.

CUCKOOS

CHANNEL-BILLED

ad

juv

PHEASANT COUCAL

br

non-br

LONG-TAILED KOEL

COMMON KOEL

♂

♀

juv

LAUGHING

KOOKABURRAS

eastern form

BLUE-WINGED

northern form

♂

♀

western form

38 x 29	48 x 32	34 x 24	45 x 35	46 x 36
Pheasant	Channel-billed	Common	Blue-winged	Laughing

FOREST KINGFISHERS

Kingfishers have large heads with long heavy bills and weak feet, having two outer front toes joined for most of their length; some small species only have three toes, two directed forwards and one back. Those shown here are forest kingfishers, with broad black band through eye, feeding on dry land on grasshoppers, lizards, etc; they have longer tails and heavier bills than the two fishing kingfishers shown overleaf. Nests are holes in earthen banks, trees or termite mounds.

Collared (Mangrove) Kingfisher *Halcyon chloris* 25–28 cm
Green-rumped 'forest' kingfisher with white underparts, green and blue back (more green than Sacred), small white spot before eye, usually found in mangroves. Larger-billed and less colourful-looking than Sacred. *Voice:* loud 'ke-kek', second note higher and more emphatic. *Nesting:* hollow in mangrove or arboreal termitarium; 3–5 white eggs. *Range:* common resident in mangroves on north-west, north and east coast; rare south of Brisbane.

Sacred Kingfisher *Halcyon sancta* 19–23 cm
Blue-rumped 'forest' kingfisher with pale buff underparts, peacock blue and green upperparts and small buff spot before eye. *Female:* duller and more green above, less buff below, some almost white. *Juv:* buff edges to wing feathers, black scallops on breast. *Voice:* loud four or five notes slightly descending 'kik-kik-kik-kik . . .'; chuckling 'ch-rrr-k' when nesting; loud 'skreeek' in nest defence. *Nesting:* hole in tree in west, tunnel in arboreal termite mound in east. *Range:* common migrant in most wooded habitats except rainforest throughout Aust, resident in north; limited in Tas.

Forest Kingfisher *Halcyon macleayii* 19–22 cm
Blue-rumped forest kingfisher with two-toned blue upperparts, white 'windows' in wing, large white spot before eye, white underparts. *Male:* complete white collar. *Female:* incomplete white collar. *Juv:* incomplete collar, buff on flanks. Two forms occur: eastern form (race *incincta*) with turquoise back and small white wing-patch; and northern form (race *macleayi*) with blue back and large white wing-patch. *Voice:* loud scissor-like 'scissor-weeya scissoraweeya'. *Nesting:* tunnel in arboreal termitarium; 4–6 white eggs. *Range:* uncommon resident in north; irregular migrant in eastern Aust, leaving some years, remaining others.

Yellow-billed Kingfisher *Syma torotoro* 18–21 cm
Yellow-billed kingfisher with dark green back, buff head and underparts, black patch in front of eye and on hindneck. *Male:* crown buff. *Female:* crown black. *Juv:* bill black. Keeps under cover in rainforest, raises tail when calling. *Voice:* pervasive ascending trill. *Nesting:* tunnel in arboreal termitarium or in tree hollow; 3–4 white eggs. *Range:* common resident in rainforest, particularly edges, in Cape York south to Chester R.

Red-backed Kingfisher *Halcyon pyrrhopygia* 20–24 cm
Red-rumped 'forest' kingfisher with streaked crown. *Female:* duller green on back. Usually seen in open on dead branch, on even hottest day. *Voice:* loud mournful 'ter-ep'. *Nesting:* tunnel in termite mound, earth bank or soil in uprooted tree; 4–5 white eggs. *Range:* common resident or migrant in open woodland, plains with scattered trees, wooded watercourses in inland Aust, moving towards coast in winter.

KINGFISHERS

COLLARED

SACRED

♀

♂

juv

eastern
form

northern
form ♂

♀

FOREST

juv

RED-
BACKED

♂

YELLOW-BILLED

♀

♂

♀

32 x 26

25 x 22

25 x 22

25 x 22

25 x 22

Collared

Sacred

Forest

Yellow-billed

Red-backed

KINGFISHERS, ROLLER AND BEE-EATER

Kingfishers: two small species fish for food, have long bills and short tails: Azure is common along eastern and northern streams; Little is less common in tropical mangroves. Paradise-kingfishers have long central tail feathers, feed mainly on snails and insects: Buff-breasted is common migrant to lowland rainforest in north; Common is rare vagrant to Torres Strait islands. Nests are burrows in termitaria or in creek banks. **Rollers:** similar to kingfishers but have short broad bills, take large flying insects in midair. Nest in horizontal tree hollows, usually at considerable height. **Bee-eaters:** have finer bills than kingfishers, are more slender, have projecting feathers in tail, take food, mainly stinging insects and dragonflies, in midair. Dig tunnels in vertical banks or gentle sloping sandy soil.

Buff-breasted Paradise-Kingfisher   *Tanysiptera sylvia* 30–35 cm
Red-billed kingfisher with buff breast and long white tail. *Female:* tail shorter. *Juv:* duller, lacking long tail feathers, bill black. *Imm:* buff breast barred darker, pale edges to wing coverts, central tail feathers longer than others but much shorter than adults. Surprisingly difficult to spot in rainforest, catches eye with slowly raised and lowered white tail. *Voice:* repeated 'kiu-kiu-kiu-kiu . . .'; soft liquid downward trill; high-pitched 'see'; trill with upward inflection 'ch-kow ch-kow . . .'. *Nesting:* tunnel in small terrestrial termitarium; 3–4 white eggs. *Range:* locally common migrant Oct–Apr in mainly lowland rainforest in north Qld, south to Mt Spec; vagrant farther south.

Common Paradise-Kingfisher   *Tanysiptera galatea* 32–42 cm
Red-billed kingfisher with white breast and long blue spatulate tail. Very rare vagrant to islands in Torres Strait; being sedentary in New Guinea, little likelihood of records in north Qld.

Little Kingfisher   *Ceyx pusilla* 12–13 cm
Tiny short-tailed fishing kingfisher with white breast. Birds in NT and northern Cape York have head turquoise blue; south of Cooktown, Qld, head is royal blue. Usually seen in mangrove and pandanus along watercourses or flying just above water. *Voice:* high-pitched 'tsee'. *Nesting:* tunnel in creek bank or rotten tree trunk; 4–5 white eggs. *Range:* uncommon resident along heavily vegetated rivers, creeks and tidal channels in NT and north Qld, south to about Ayr.

Azure Kingfisher   *Ceyx azurea* 17–19 cm
Short-tailed fishing kingfisher with orange breast; northern birds have richer plumage than eastern birds. *Juv:* duller blue above, paler orange below. Usually seen on prominent perch over water or flying rapidly just above water along creek. *Voice:* high-pitched piping 'pseet-pseet'. *Nesting:* hole in earthen bank, usually on creek bank; 5–7 white eggs. *Range:* uncommon resident along creeks, rivers and mangroves in northern and eastern mainland and Tas.

Rainbow Bee-eater   *Merops ornatus* 23 cm
Colourful, blue, green and orange bee-eater with extended shafts to central tail feathers. *Juv:* duller, lacks black patch on throat. Usually in small groups, sometimes roosts in hundreds in small leafy tree. *Voice:* pleasant 'prrrp-prrrp'; single 'plk' when alarmed. *Nesting:* long tunnel in sand bank or sloping sandy soil; 4–7 white eggs. *Range:* common migrant Sep–Apr in woodland and timbered plains throughout Aust; resident population in north.

Dollarbird   *Eurystomus orientalis* 27–30 cm
Red-billed roller with bluish-white window ('dollar') in wing. *Juv:* bill blackish with yellow gape. Usually seen sitting on tall dead tree or telephone wires, frequently flying out on windowed wings after insects. *Voice:* harsh 'krak-kak-kak'; faster 'kek-ek-ek-ek . . .'. *Nesting:* horizontal hollow in tree; 3–5 white eggs. *Range:* common migrant Sep–Apr in forest and woodland in northern and eastern mainland; vagrant Tas.

KINGFISHERS

BUFF-BREASTED

imm

juv

ad

COMMON

AZURE

eastern

northern

north-eastern

LITTLE

northern

ROLLERS

DOLLARBIRD

juv

ad

BEE-EATERS

RAINBOW

ad

juv

Rainbow
23 x 19

37 x 29

7 x 14

25 x 22

22 x 19

Little

Buff-breasted

Azure

Dollarbird

BARN OWLS

Nocturnal birds with flat heart-shaped facial disks and dark eyes. Usually roost during day in hollow or hole in rocks, in dense vegetation or in long grass. Most feed on rodents, have supersensitive asymmetrical ears to help locate prey in grass. Nest in tree or rock hollow or in grass (Grass Owl). Some Masked Owls are very similar in coloration to Barn Owls, but can be distinguished by the more heavily feathered legs. Some regard the two forms of the Sooty Owl as separate species, others consider them races.

Barn Owl *Tyto alba* 32–35 cm

Small barn owl with buff and grey upperparts. Usually seen at night flying in car headlights or on roadside fence; often heard flying overhead uttering rasping hiss; if flushed from hollow during day, immediately mobbed by other birds. *Voice:* rasping hiss like tearing calico. *Nesting:* hole in tree or stump; 3–9 eggs. Common resident or nomad in woodlands throughout Aust, numbers subject to wild fluctuations depending on rodent numbers.

Masked Owl *Tyto novaehollandiae* 38–50 cm

Large barn owl with blackish-brown to speckled buff upperparts. Generally darker above than Barn, some quite similar but more heavily spotted below, darker margin to face. Females from Tas and south-east forests have chestnut face (race *castanops*); elsewhere variable, generally paler in north except rusty form on Melville Island. *Voice:* rasping hiss, louder than Barn. *Nesting:* hollow in tree or in cave; 2–3 white eggs. *Range:* uncommon to rare resident in forest, woodland, and in caves on Nullarbor Plain, in south-western, northern and eastern Aust; common resident in Tas.

Sooty Owl *Tyto tenebricosa* 35–45 cm

Very dark barn owl with face sooty grey, eyes very large. *Adult:* black around eye reduced, underparts pale grey with limited dark mottling. *Juv:* extensive black around eye, underparts extensively mottled darker. *Voice:* loud whistle like bomb falling; chirruping cricket-like duet. *Nesting:* hollow in tall eucalypt; 1–2 white eggs. *Range:* rare resident in rainforest and wet sclerophyll forest in south-eastern and north-eastern Aust.

Grass Owl *Tyto longimembris* 33–36 cm

Ground-loving barn owl with chocolate and buff upperparts, small eyes, long bare legs. Usually flushed from grass tussocks (Barn Owl occasionally roosts in grass – much paler above). *Voice:* rasping hiss; thin whistle. *Nesting:* trodden-down grass under bush or tussock; 4–6 white eggs. *Range* rare nomad, locally common during mouse plagues, in grass tussocks, lignum, canegrass and heaths, often in swampy ground, mostly north-eastern and northern Aust, but likely anywhere in correct habitat.

Lesser Sooty Owl *Tyto multipunctata* 32–35 cm

Very dark barn owl with bicoloured face, black around eye surrounded by white or pale grey; eye very large. *Adult:* outer face white. *Juv:* outer face grey. Smaller and more heavily spotted above than Sooty and with different facial pattern. *Voice:* similar to Sooty but higher-pitched. *Nesting:* deep hollow in eucalypt; 1–2 eggs. *Range:* uncommon resident in rainforest and wet sclerophyll forest in north-eastern Qld from Mt Spec to Cooktown.

OWLS

MASKED
Tasmanian form

light phase

♂

♀

BARN

42 x 32

Barn

GRASS

43 x 32

45 x 35

Masked

Grass

LESSER SOOTY

SOOTY

juv   ad

46 x 40   46 x 38

Sooty   Lesser Sooty

HAWK-OWLS

Nocturnal birds with rather hawk-like faces and large yellowish eyes. Usually roost by day in trees, often choosing leafy branch close to trunk. Size varies from sparrowhawk-size to eagle-size; prey ranges from beetles to possums. Nest in hollows in trees. Juveniles retain down on underparts for considerable time after leaving nest, feathers gradually appear through down from flanks towards centre of breast. Males are larger than females, unusual among owls.

Southern Boobook   *Ninox novaeseelandiae*   30–35 cm
Small brown hawk-owl with greenish-yellow eyes in large facial mask with pale margins. In Tas, eyes yellow; in arid areas, upper parts much paler; in north-eastern rainforests, very dark (race *lurida*). Although variable, never grey-headed like Barking. *Voice:* loud 'book-book', perhaps best-known nocturnal sound. *Nesting:* hollow in tree; 2–3 white eggs. *Range:* common resident throughout mainland and Tas.

Oriental Boobook (Brown Hawk-Owl)   *Ninox scutulata*   25–29 cm
Small brown hawk-owl with staring yellow eyes, heavily streaked on underparts, head brown, lacking definite 'mask' of Boobook, looks more like small brown Barking but has no spotting on upperparts. Very rare vagrant from Indonesia, recorded once on north-west coast.

Barking Owl   *Ninox connivens*   38–43 cm
Large greyish-brown hawk-owl with white spots on back, grey head with large staring yellow eyes, sparse brown to rusty streaks on underparts. Variable in plumage, but with pale face, not heavily masked like Boobook. *Voice:* deep 'wook-wook', like dog barking; quavering scream. *Nesting:* hollow in tree; 2–3 white eggs. *Range:* uncommon to rare resident in open forest, woodland and wooded watercourses in arid country throughout Aust, mainly west of Divide.

Rufous Owl   *Ninox rufa*   44–50 cm
Large reddish-brown hawk-owl with yellow eyes, narrow rufous bars on underparts and darker bars on upperparts. Rare dark form from near Mackay, Qld (race *queenslandica*), has darker brown bars on underparts. Usually seen perched in leafy tree in dense vegetation. *Voice:* loud slow 'hoo-hooo'. *Nesting:* hollow in tree; 2–3 white eggs. *Range:* rare resident in rainforest, dense river margins and paperbark forest in northern and north-eastern Aust.

Powerful Owl   *Ninox strenua*   60–65 cm
Very large dark hawk-owl with yellow eyes, coarse V-shaped bars on underparts and broad bars on upperparts. Usually seen perched in tall tree, often with half-eaten possum in claws. *Voice:* loud 'hoo-hooo', louder than Rufous. *Nesting:* hollow in tree; 1–2 white eggs. *Range:* uncommon resident in sclerophyll forest in south-eastern Aust, from Dawson R, Qld, to south-east SA.

dark form

light phase

SOUTHERN BOOBOOK

typical phase

juv

OWLS

BARKING

typical phase

light phase

juv

ORIENTAL

POWERFUL

RUFOUS

ad

juv

Southern 42 x 35

47x41

54 x 45

55 x 46

Barking

Rufous

Powerful

NIGHTJARS AND OWLET-NIGHTJARS

Nightjars: small nocturnal birds with long wings and weak feet, roosting by day on ground; plumage cryptically coloured like dead leaves. Feed on wing, catching insects with enormous gape. Two species lack long bristles around gape, live on hillsides; the other species. with long bristles, lives in rainforest. Often sit on roads at night. Lay eggs on ground. **Owlet-nightjars:** small active birds with short wings and long tail, roosting by day in tree hollow or rocks. Often call during day. Feed on wing at night, like a nocturnal Willy Wagtail. Nest in hollow limb of tree.

Large-tailed Nightjar   *Caprimulgus macrurus*   25–27 cm
Mottled brown rainforest nightjar with resonant 'chof-chof' call, white window in wings and white in tail. Usually roosts on ground among leaves, at night perches on ground or on branch. *Voice:* incessant resonant 'chof-chof-chof'. *Nesting:* 2 clouded pinkish eggs laid on ground. *Range:* common resident in rainforest, monsoon forest, bamboo thickets and paperbark swamp in northern and north-eastern Aust, vagrant to Qld–NSW border.

Spotted Nightjar   *Eurostopodus guttatus*   29–30 cm
Mottled reddish-brown nightjar with large white window in wing, no white in tail. *Juv:* freckled rusty-brown, Usually roosts on ground on stony hillsides, often flushed from roads by night, eye reflects bright red. *Voice:* eerie accelerating 'caw-caw-cook-ook-ook . . .', all on same note. *Nesting:* 1 spotted pale green egg laid on bare ground. *Range:* uncommon to common in woodland, particularly stony hillsides, throughout Aust.

White-throated Nightjar   *Eurostopodus mysticalis*   33–35 cm
Dark grey-brown nightjar with little white in wing and no white in tail. *Juv:* freckled rusty-brown, similar to juv. Spotted. Usually roosts in undergrowth on hillsides or on rocky slopes in open woodland. *Voice:* eerie accelerating 'caw caw cook-ook-ook . . .' similar to Spotted but accelerating up-scale, rather like Whistling Kite. *Nesting:* 1 blotched cream egg on ground. *Range:* common migrant (Sep–May, south-east) or resident (north-east) on hillsides or ridges in forest and woodland in eastern Aust.

Australian Owlet-nightjar   *Aegotheles cristatus*   22 cm
Small grey or rusty-grey owlet-nightjar with large black eyes and double collar on hindneck. Birds in southern Aust have grey bellies (race *cristatus*); in northern Aust belly is white (race *leucogaster*) and two phases occur, one reddish, often flushed from rocks, the other grey. *Voice:* pleasant chuckle uttered night or day. *Nesting:* hollow in tree or fencepost, lined with leaves; 3–4 white eggs. *Range:* common resident in forest, woodland and wooded watercourses throughout Aust and Tas.

NIGHTJARS

LARGE-TAILED

ad

imm

SPOTTED

ad

juv

WHITE-THROATED

ad

ad
cryptic
posture

OWLET-NIGHTJARS

AUSTRALIAN

reddish
phase

grey
phase

juv

ad

30 x 21	34 x 25	41 x 29	29 x 22
Large-tailed	Spotted	White-throated	Australian

FROGMOUTHS

Nocturnal birds with powerful frog-like beaks, fiery eyes usually closed when disturbed in daytime, long tails and cryptic coloration. Roost during day on branches close to treetrunk, usually in pairs or family groups; when disturbed, raise beak to heighten resemblance to broken branch. At night hunt like kookaburras, taking similar prey in similar way. Most have grey and brown phases, with females more often brown. Nests are flimsy stick platforms in fork of tree. High mortality on roads at night.

Tawny Frogmouth *Podargus strigoides* 32–46 cm

Widespread frogmouth with yellow or orange-yellow eyes. *Adult male*: mainly grey with black streaks; in north silvery-grey. *Adult female*: plumage suffused with brown, particularly on back and shoulders; in north, rufous; in central Aust, more like male. *Imm*: darker than adult with more marbled plumage on back (often mistaken for Marbled Frogmouth, which is richer rufous-brown) and has a prominent ear patch. *Juv*: paler with black and white mottling on back, wings and breast, soft streaked feathers on abdomen, moult shortly after fledging. Two basic forms: large eastern and southern birds with greyish males and brownish females (races *strigoides* in east, smaller *brachyurus* in south, paler *centralia* in centre); and small northern birds (race *phalaenoides*) with silvery-grey males and often rufous females. *Voice*: soft but penetrating 'oom-oom . . .'; loud 'grr-er' when disturbed. *Nesting*: flimsy stick platform in tree; 2 white eggs. *Range*: common resident in forest, woodland and wooded watercourses throughout mainland and Tas.

Marbled Frogmouth *Podargus ocellatus* 40–48 cm

Greyish- or reddish-brown frogmouth with orange eyes, prominent pale eyebrow, found in mid-eastern and north-eastern rainforest. *Male*: greyish with black and white marbling. *Female*: less heavily marbled rusty-brown. *Juv*: brown eyes. Two forms occur: northern form (race *ocellatus*) with lightly barred wings; and southern form ('Plumed Frogmouth', race *plumiferus*) with heavily barred wings. *Voice*: descending 'coop coop coop gobble gobble gobble . . .'; toad-like 'dugger-dugger-dugger . . .'; 'koo-er-loo, koo-er-loo'. Female has harsher croaking quality than male. *Nesting*: scant platform of twigs in tree, often high; 2 white eggs. *Range*: uncommon to rare resident in rainforest on Cape York Peninsula, Qld, south to McIlwraith Ra; and in south-eastern Qld and north-eastern NSW from Conondale Ra, Qld, to Grafton, NSW.

Papuan Frogmouth *Podargus papuensis* 45–54 cm

Very large brown (female) or grey (male) frogmouth with red eyes and white mottling on underparts, inhabiting rainforest and woodlands in north-eastern Qld. *Juv*: similar to juv Tawny but much larger. *Imm*: mottled blackish-brown, darkest of all frogmouths. *Voice*: low 'oom-oom . . .'; laughing 'hoo-hoo . . .'. *Nesting*: platform of twigs; 1–2 white eggs. *Range*: common migrant or resident, numbers decreasing southwards, in rainforest and nearby woodland, mangroves and riverine forest in north-east from Cape York to about Townsville.

FROGMOUTHS

TAWNY — northern form

red
phase

♀

♂

♀

MARBLED

**TAWNY — southern
and eastern
forms**

rufous
phase

♀

juv

imm

♂

PAPUAN

♀

imm

♂

juv

44 x 31

40 x 29

49 x 34

Tawny

Marbled

Papuan

SWIFTS

Aerial birds that spend most of their waking hours in flight on long scythe-like wings. Feed on small flying insects, have large gape surrounded by bristles that act as 'scoop'. Most are summer migrants; large swifts spend most of their time ahead of storm fronts; updraughts and thermals allow them to fly in perpetual shallow dive, enabling them to move very quickly; wing designed to change shape for soaring, so very little energy expended in flight. Small swallow-sized species known as swiftlets have more batlike flight; look for notched hindwing to differentiate them from swallows, but very difficult to tell from one another, some among hardest birds to identify correctly. White-rumped breeds in caves in north Qld; isolated population near Chillagoe may be a separate species.

Glossy Swiftlet   *Collocalia esculenta*   10 cm
Tiny glossy bluish-black swiftlet with dark rump, white belly and mottled greyish throat and breast. Smaller than other swiftlets, but valid only for direct comparison; diagnostic white belly not always easy to see. *Voice:* soft twittering. *Range:* rare vagrant or visitor around hilly rainforest and clearings in north-east from Iron Ra to Eungella Ra, Qld.

White-rumped Swiftlet   *Collocalia spodiopygia*   11 cm

Small dark grey swiftlet with greyish-white rump. Usually in flocks around ridges and cliffs, when pale rump difficult to see, but often low over canefields when flying ants swarm. *Voice:* high-pitched 'cheep'; metallic clicking in nest caves. *Nesting:* only swift nesting in Aust, tiny basket of grass and twigs cemented with saliva to cave wall in colonies; 1 white egg. *Range:* common resident about caves in rainforest areas in north-east Qld from Claudie R to Eungella Ra; vagrants farther south.

Uniform Swiftlet   *Collocalia vanikorensis*   12–13 cm
Large glossy blue-black swiftlet with dark rump and belly. Very difficult to tell from almost identical Mountain Swiftlet *C. hirundinacea* and Whitehead's Swiftlet *C. whiteheadi*, which has grey throat and brown abdomen, both equally likely to wander over Torres Strait from New Guinea. *Range:* very rare vagrant from New Guinea, recorded at Peak Point, a few kilometres from Cape York, 1913; four birds at Atherton 1971 probably Uniform or Mountain.

Fork-tailed Swift   *Apus pacificus*   17·5 cm

Large slim black swift with deeply-forked tail, white rump and white throat; underparts sometimes scalloped paler but rarely visible in field. Tail forked, even when fanned, More slender than House and White-throated Needletail. *Voice;* short trill like hoarse Rainbow Bee-eater. *Range:* common migrant Oct–Apr throughout mainland Aust, mostly west of Divide; uncommon Tas.

White-throated Needletail   *Hirundapus caudacutus*   20 cm

Large stocky swift with short square tail with protruding spines (not visible in flight), white throat and undertail, smoky-grey back, dark rump. Much heavier-looking than Fork-tailed, almost falcon-shaped, usually arrives ahead of storms in large, wheeling flocks, mostly over or east of Divide; attracted to low-pressure weather along east coast. *Voice:* churring twitter. *Range:* common migrant Oct–Apr, mainly in east and Tas; scattered records elsewhere.

House Swift   *Apus affinis*   15 cm
Small black stocky swift with square or slightly forked tail, white throat and white rump. Tail looks square when fanned, slightly forked when closed. Probably reasonably regular visitor in small numbers, but overlooked because of its similarity to Fork-tailed, which is larger, has longer finer wings, more cigar-shaped body, deeply forked tail, less fluttering flight. *Range:* rare vagrant or visitor, probably likely anywhere in Aust.

SWIFTLETS

UNIFORM

WHITE-RUMPED

GLOSSY

SWIFTS

WHITE-THROATED

FORK-TAILED

HOUSE

18 x 13

White-rumped

SWALLOWS AND MARTINS

Small long-winged birds mostly seen chasing insects on the wing, sitting on telephone wires or on twigs on dead trees. Generally swallows have long forked tail, martins have short square tails. Nests are either bottle-shaped or cup-shaped mud structures, holes in trees or tunnels in sandy banks. So often seen that they are usually ignored; nevertheless, flocks in north should be examined for rarer migrant species. Swiftlets are not unlike martins and may be overlooked, just as higher-flying swifts may be dismissed as swallows.

Welcome Swallow   *Hirundo neoxena* 15 cm
Rusty-throated swallow with grey underparts and no breastband; white spots in tail. Usually seen on utility wires or flying about towns and cities. (Watch for Pacific Swallow *H. tahitica*, 13 cm, very similar but lacks long outer tail feathers – easily confused with juveniles and adults in moult; recorded doubtfully in Torres Strait.) *Voice:* pleasant twittering, piercing 'sierp' in alarm. *Nesting:* mud cup usually among rafters, under eaves, under bridges or under overhanging rocks; 2–5 spotted whitish eggs. *Range:* common around habitations in southern and eastern mainland and Tas, vagrant elsewhere.

Barn Swallow   *Hirundo rustica* 14–17 cm
Rusty-throated swallow with white underparts and black breastband; white spots in tail. Most birds moult while in Aust, losing long outer tail feathers. Usually seen on utility wires in northern towns. Uncommon visitor Oct–Apr to towns in north from Derby, WA, to Innisfail, Qld; vagrants elsewhere.

Red-rumped Swallow   *Hirundo daurica* 18·5 cm
Stripe-breasted swallow with rusty eyebrow and rusty rump; no white spots in tail. Rare vagrant or visitor, likely in small flocks anywhere in north, may perch with other swallows; although distinctively marked, easily overlooked. Some birds in South-East Asia lack striped breast, have uniform chestnut underparts, could possibly turn up here.

Fairy Martin   *Cecropis ariel* 11–12 cm
White-rumped martin with clean white underparts and rusty head. Looks cleaner than Tree, has churring call; usually seen near culverts Aug–Jan, in flocks often on dead trees at other times. *Voice:* soft 'churr'. *Nesting:* large bottle-shaped mud nest under culvert or eaves; 4–5 freckled whitish eggs. *Range:* common partial migrant, most common in south Aug–Feb, throughout Aust; uncommon south-west; vagrant Tas.

Tree Martin   *Cecropis nigricans* 12–13 cm
White-rumped martin with cloudy white underparts and rusty spot on forehead. Not as clean-looking in flight as Fairy, has sharper call. *Voice:* sharp 'chk'; pleasant warble. *Nesting:* small hollow in tree, sometimes plastered with mud; 3–5 freckled pinkish eggs. *Range:* common nomad or migrant in open woodland throughout Aust; large flocks winter in north and in New Guinea.

White-backed Swallow   *Cheramoeca leucosternum* 14–15 cm
Black swallow with white cap, throat and back; no white patches in deeply-forked tail. Looks darker in flight than square-tailed martins; usually flies higher than other swallows; once call learnt, found to be more common overhead than expected. *Voice:* quiet but far-carrying, 'check'; loud hawk-alarm 'pseeoo' pleasant pipit-like warble; juvenile call 'churr'. *Nesting:* deep tunnel in sandy bank, also used for roosting; 4–6 white eggs. *Range:* locally common near breeding sites in open woodland, tree-scattered plains and grassland in southern Aust.

SWALLOWS

WELCOME

PACIFIC

BARN

RED-RUMPED

PACIFIC

BARN

RED-RUMPED

WELCOME

WHITE-BACKED

juv

TREE

FAIRY

ad

19 x 13

17 x 12

17 x 11

18 x 13

Welcome

Fairy

White-backed

Tree

PITTAS AND GROUND-THRUSHES

Blue-winged Pitta
Pitta moluccensis 17–20cm

Buff-breasted pitta with white throat. Very similar to Noisy but rather paler throat and considerably more white in wing. *Range:* very rare vagrant to north-west coast, recorded at Derby, WA, 1931, and Mandora Station, near Eighty Mile Beach, 1927. (Watch for similar Fairy Pitta *P. brachyurus* smaller with whitish lower edge to buff eyebrow, smaller white patch in wing and with pale trailing edge to wing.)

Noisy Pitta
Pitta versicolor 17–20cm

Buff-breasted pitta with black throat. *Juv:* duller with throat greyish, no pale blue on shoulder, breast with dark scallops on feather edges. Rather difficult to sight in forest, attracts attention by loud call, busy tossing aside of litter; rapid flight on broad wings with white window in tip. *Voice:* piercing 'walk to work'; single mournful whistle. *Nesting:* large dome of sticks between tree buttresses; 3–5 blotched bluish eggs. *Range:* common resident or migrant in rainforest and nearby eucalypt forest in east, from Cape York to Hunter R; vagrant farther south; some altitudinal migration in winter.

Red-bellied Pitta
Pitta erythrogaster 16–19cm

Blue-breasted pitta with red belly. *Female:* duller than male with brownish face. *Juv:* bronze-brown with throat and breast streaked pale buff, moults into immature plumage May–June. *Imm:* similar to female but with buff tips to blue breast feathers. *Voice:* mournful whistle. *Nesting:* large dome of sticks; 3 blotched whitish eggs. *Range:* common migrant Nov–Apr in rainforest on northern Cape York south to Rocky R and Wenlock R.

Rainbow Pitta
Pitta iris 16–20cm

Black-breasted pitta. Usually seen on ground in thick vegetation, flies rarely showing white window in wing; ascends trees to call and roost. *Voice:* 'want a whip'. *Nesting:* bulky dome on ground or in low fork or clump of grass or bamboo; 4 blotched creamy eggs. *Range:* uncommon resident in monsoon forest, bamboo thickets, river margins and mangroves in northen Aust from Prince Regent R, WA, to about Victoria R, NT.

Russet Ground-Thrush
Zoothera heinei 25–29cm

Rusty-backed ground-thrush with reduced black scallops on back and rump. Similar to Australian; rustier on back than south-eastern form, more white in shorter tail, less complex song, generally lower on mountains where they meet in northern NSW; similar in colour to Atherton form but smaller, lower down mountain, and smaller scallops on back. *Voice:* high-pitched 'tseep'; strong sweet 'tooweeya tooa'. *Nesting:* similar to Australian. *Range:* common in rainforest and wet sclerophyll forest in north-eastern Aust from northern NSW to Cape York.

Australian Ground-Thrush
Zoothera lunulata 25–29cm

Olive-backed (south-east) or large rusty-backed (Atherton Tableland) ground-thrush with prominent black scallops on back and rump. Two forms occur: south-eastern form (race *lunulata*) and Atherton form (race *cuneata*). Russet Ground-Thrush very similar but with reduced black scallops on back and rump, different song. *Voice:* high-pitched 'seep'; extended warbling song like Blackbird's. *Nesting:* untidy decorated bowl of bark and rootlets in low fork of tree or on stump; 2–3 freckled greenish eggs. *Range:* common in damp gullies, forest, rainforest, moving to woodlands in winter in south-east from north-eastern NSW to Mt Lofty Ra, SA, Bass Strait islands, Tas, Kangaroo I.

Song Thrush
Turdus philomelos 23cm

Introduced ground-thrush with black 'arrowhead' markings on underparts. *Voice:* beautiful extended song 'did he do it did he do it Judy did'; harsh 'took took', 'come out come out'; quiet 'chick' in flight. *Range:* uncommon to rare in suburban gardens, woodlands and forest around Melbourne, from Warragul to Lorne.

Blackbird *Turdus merula* — see p 210.

PITTIDAE, MUSCICAPIDAE

PITTAS

BLUE-WINGED

NOISY
ad
juv

FAIRY

RED-BELLIED
juv

ad

RAINBOW

SONG

THRUSHES

AUSTRALIAN

RUSSET

Atherton form

south-eastern form

33 x 25	28 x 23	30 x 24	34 x 23	34 x 23	29 x 21
Noisy	Red-bellied	Rainbow	Russet	Australian	Song

SCRUB-BIRDS AND BRISTLEBIRDS

Scrub-birds: small primitive ground birds with loud voices, living in dense undergrowth; usually keep under cover, very difficult to see well (tape recording of call may bring one close). Fly only rarely, usually move along ground with tail erect. Nests are bulky domes on or near ground, lined with woodpulp. **Bristlebirds:** not related to scrub-birds but included here because of similarity in appearance. Probably called 'bristle' birds because of three fine whiskers on each side of gape, absent in scrub-birds (*Atrichornis* means 'bird without bristles') but present in many wrens and warblers. Nests are loose domes of grass placed in clumps of grass or heath. Bristlebirds and scrub-birds must be regarded as endangered; under no circumstances should they be disturbed while nesting.

Rufous Scrub-bird *Atrichornis rufescens* 16 cm

Finely-barred rufous-brown scrub-bird of highland rainforest in central eastern Aust. Among most difficult of birds to see, keeps to undergrowth in areas where thinner canopy allows ground cover to flourish; most likely to be seen fleetingly while crossing fallen logs. *Voice:* powerful accelerating 'chip-chip . . .'; mimicry; harsh churring; sharp 'tic'. *Nesting:* similar to Noisy, but more lining; 2 spotted pink eggs. *Range:* rare and localised resident in scattered highland rainforest above 400 m, from Barrington Tops, NSW, to MacPherson Ra, Qld.

Noisy Scrub-bird *Atrichornis clamosus* 21 cm

Finely-barred scrub-bird of coastal heaths near Two People Bay, WA. Feeds on ground mainly in more densely vegetated gullies. Western Bristlebird inhabits same area, lacks white throat, black bib and barred plumage, often moves into shorter heath between gullies. *Voice:* powerful, varied, including mimicry of local birds. *Nesting:* dome of rushes lined with pulp of decayed rushes, on or near ground in clump of rush, with runway; 1 freckled pale buff egg. *Range:* locally common resident in very restricted area of coastal heath, mainly in gullies with some eucalypts, near Two People Bay, WA.

Eastern Bristlebird *Dasyornis brachypterus* 21–22 cm

Grey-brown bristlebird found in widely scattered heaths in south-east. Usually keeps well hidden, moving on ground with tail cocked when disturbed, occasionally emerges to sing on top of bushes; flies seldom, just over heath for short distances. *Voice:* musical 'it-wooa-weet-sip'; 'zip'; soft 'tuck'. *Nesting:* dome of grass, bark and twigs in grass tussock or low bush; 2 pale buff eggs. *Range:* uncommon to rare resident in heath, sedges, blady grass and overgrown watercourses in few remaining localities from Cunningham's Gap, Qld, to Wingan Inlet, Vic.

Western Bristlebird *Dasyornis longirostris* 18–22 cm

Grey-brown bristlebird found in coastal heaths on south coast of WA. Similar to Brown, but with longer bill, more dappled in colour. *Voice:* male, variable 'chip-pee-tee-peetle-pet . . .'; female, whistling 'quick more beer'; both, loud 'gee-ick', soft 'zit'. (G. Smith). *Nesting:* similar to Eastern. *Range:* scarce resident in few suitable remaining coastal heaths between Two People Bay and Fitzgerald R, WA.

Rufous Bristlebird *Dasyornis broadbenti* 23–27 cm

Reddish-brown bristlebird with rufous crown, becoming richer in colour from east to west. Birds in south-west (race *litoralis*) are smaller with bright rufous crown, possibly extinct. Usually heard more often than seen; runs swiftly with cocked tail when disturbed. *Voice:* rising 'chip chip chip cheweee'; sharp 'tewick'. *Nesting:* coarse twig and grass dome hidden in long grass growing in low bush; 2 freckled whitish eggs. *Range:* locally common resident in long grass in tangled shrubbery from mouth of Murray R, SA, to Torquay, Vic; race possibly extinct in south-west, formerly between Cape Naturaliste and Cape Leeuwin.

SCRUB-BIRDS

RUFOUS

♂ ♀ imm ♂

NOISY

♂ ♀ juv

BRISTLEBIRDS

EASTERN

WESTERN

eastern form south-western form

RUFOUS

23 x 18	29 x 20	26 x 19	26 x 19	29 x 21
Rufous	Noisy	Eastern	Western	Rufous

WAGTAILS

Brightly coloured ground birds with long tails. Wag tail up and down constantly, often sit on fence posts and sometimes in trees (do not confuse with Willy Wagtail, which is a misnamed fantail, wags tail from side to side.) All are irregular migrants or vagrants from northern hemisphere, probably overflying normal wintering grounds in South-East Asia and islands north of Aust. Usually seen in non-breeding, juv and imm plumages in Aust, Aug–Apr, making identification more difficult. Local birds in unusual situations sometimes confused with wagtails, eg White-winged Triller feeding on ground identified as White Wagtail.

Yellow Wagtail
Motacilla flava 16–19 cm

Relatively short-tailed wagtail with black legs; rump same colour as back (olive-green or brownish-grey). *Breeding:* yellow breast and throat; olive-green back, female duller. *Non-breeding:* brownish-grey above with pale eyebrow, buff below with whitish throat. *Juv:* brownish-grey above with pale eyebrow, buff below with dark breastband. Several races probably winter in Aust, only breeding males can be reliably identified. Race *taivana*, breeding: yellow eyebrow, crown same colour as back (non-breeding: eyebrow yellowish). Race *simillima*, breeding: white eyebrow, bluish-grey crown (non-breeding: buff eyebrow). Race *tchutchensis*, breeding: white eyebrow, blue crown, white throat (non-breeding: identical to *simillima*). Seldom perches in trees. *Voice:* 'jijit, jijit'; 'zweep' when flushed. *Range:* uncommon, possibly regular migrant to marshes, damp paddocks, airfields in northern and eastern Aust, recorded from Derby, WA, to Hawkesbury R, NSW.

Grey Wagtail
Motacilla caspica (cinerea) 20 cm

Long-tailed wagtail with pale legs, yellow rump and at least some yellow on underparts; white stripe in wing; actually most are more yellow in non-breeding plumage than Yellow Wagtail. *Breeding:* upperparts blue-grey with yellow rump and white eyebrow, underparts yellow with black throat edged with white stripe (male) or white throat (female). *Male non-breeding:* similar to female. *Imm:* upperparts like adult, underparts whitish with at least some yellow on undertail coverts. Often perches in trees. *Voice:* metallic 'chirrick chirrick'. *Range:* very rare vagrant to northern Aust.

Yellow-headed Wagtail
Motacilla citreola 16·5 cm

Relatively short-tailed wagtail with black legs and yellow face. *Male breeding:* upperparts dark grey with black nape and white edges to wing feathers; head and underparts yellow. *Male non-breeding and female:* similar but with dark nape extending to crown; eyebrow and face yellow; grey clouding on sides of breast. *Juv:* similar to juv Yellow but more extensive white edges to wing feathers. *Voice:* wheezy 'zweep' similar to Yellow. *Range:* very rare vagrant, recorded from Woolooware Bay, near Sydney, 1962.

White Wagtail
Motacilla alba 21 cm

Long-tailed pied wagtail with large white patch in wing. *Male breeding:* throat and broad breastband black. *Male non-breeding:* throat white, broad breastband black. *Female:* narrow breastband black. *Imm:* less white in wing, confined to shoulders. *Juv:* similar to juv Yellow but more white in wing. Some races likely to occur in Aust are identifiable in adult non-breeding plumage (ie with wing mainly white in flight), as follows. Race *lugens:* grey back, dark crown, dark line through eye. Race *ocularis:* grey black, grey crown, dark line through eye. Race *leucopsis:* black back and crown, no dark line through eye. Race *baicalensis:* grey back, dark crown, no line through eye, grey secondaries in flight. *Voice:* 'churruck churruck'. *Range:* very rare vagrant, recorded at Katanning and Broome, WA; likely in marshy grasslands, seashore and estuaries anywhere in northern Aust Sep–Apr.

WAGTAILS

YELLOW

taivana

simillima

br

non-br

juv

br

tchutch-ensis

non-br

YELLOW
non-br

GREY
non-br

GREY

imm

br

br
♀

br
♂

YELLOW-
HEADED

imm

♀

♂

WHITE

ocul1aris

leucopsis

baicalensis

lugens

juv

br
♂

br
♀

non-br ♂

non-br
♀

LARKS AND PIPITS

Small ground birds living in open spaces, seen in aerial display flights with beautiful songs, also given from perches such as fenceposts. Nests are grass cups, usually in a small depression under grass tussock. **Larks:** differ from all other passerines in having tarsus round. Hind toe has elongated claw, head slightly crested, bill conical rather finch-like, eat seed as well as insects; vary in colours according to soil. **Songlarks:** aren't larks but Old World warblers with lark-like habits; lack elongated hind toe and crest and have no white in tail; perch in trees rather more often. **Pipits:** are not larks either, related to wagtails (see p 206), wag tail up and down between spurts of running on ground.

Skylark　　　　　　　　　　*Alauda arvensis*　19 cm
Large streaked brown lark with small crest, light trailing edge to wing. Usually seen in open grassland; male has undulating display flight while singing. Larger than Singing, which is more reddish with rufous shoulders, lacks crest. *Voice:* beautiful trilling either in air or on perch; liquid 'chirrup'. *Nesting:* hidden grass cup, sometimes sparsely domed, under tussock; 2–4 freckled or blotched pale umber to whitish eggs. *Range:* common introduced nomad in grassland, pastures, in south-east and Tas.

Singing Bushlark　　　　　　*Mirafra javanica*　13 cm
Small streaked brown lark with reddish shoulders, lacking crest. Eastern birds are migratory, basically similar greyish-brown; northern and north-western birds often resident, vary considerably according to soil colour, reddest birds in west Kimberley, WA. Usually flutter on triangular-looking wings with drooping white-edged tail when disturbed, using hesitant fluttering flight. *Voice:* melodious tinkling song in display flight or on perch. *Nesting:* hidden grass cup; 2–4 blotched or freckled pale umber eggs. *Range:* patchily distributed migrant, nomad or resident in grasslands in north-western, northern and eastern Aust.

Rufous Songlark　　　　*Cinclorhamphus mathewsi*　16–17 cm
Streaked songlark with rufous rump. *Male:* bill black. *Female:* bill pale brown. *Juv:* spotted breast. Males very obvious during breeding season, flying around territory with loud incessant song, some years more abundant than others. *Voice:* rich melodious 'twitchy-tweedle . . .'; sharp 'tlik'. *Nesting:* well-hidden cup in grass sometimes with sparse dome; 3–4 speckled white eggs. *Range:* nomad or migrant in south Aug–Apr, common in open grassy woodland throughout Aust, mainly west of Divide.

Richard's Pipit　　　　*Anthus novaeseelandiae*　15–17 cm
Slender streaked brown pipit with slender pale bill, varying from greyish-brown in humid areas to pale cinnamon in desert, wags tail up and down. Paler than larks, more slender with longer tail, finer bill. Among commonest Aust birds. *Voice:* trilling 'prrrt' in undulating display flight. *Nesting:* well-hidden grass cup under bush or tussock; 3–4 freckled whitish-buff eggs. *Range:* common resident in grassland, paddocks, pastures, coastal dunes, road verges throughout mainland and Tas.

Brown Songlark　　　　*Cinclorhamphus cruralis*　18–24 cm
Brown songlark with brownish rump. *Male:* dark brown with fawn crown, dark breast and belly, black bill. *Female:* much smaller, similar to Rufous but colder in colour, dark patch on belly, no rufous rump, pale bill. Male often sits on fencepost with tail cocked, flies up 45° with quivering wings, planes back to earth. *Voice:* creaky, like sound of stretched wire when getting through fence. *Nesting:* well-hidden grass cup in long grass; 3–4 speckled pinkish eggs. *Range:* common nomad or migrant (Aug–Apr in south) in grasslands with scattered trees throughout Aust, mainly west of Divide.

LARKS

SKYLARK

SINGING

PIPIT

RICHARD'S

SONGLARKS

RUFOUS

♀

♂

♂

BROWN

♀

juv ♂

24 x 17	15 x 13	23 x 17	23 x 17	23 x 17
Skylark	Singing	Rufous	Richard's	Brown

CUCKOO-SHRIKES

Cuckoo-shrikes are birds with soft-feathers, particularly on rump. All Aust species are basically grey in colour, with more or less black about eye. Sexes are similar. (Cicadabird, with very different sexes, is better thought of as a large triller.) Feed on insects and small soft fruit such as native figs, mulberries. Nest is small bark and grass cup bound by cobwebs, usually placed on horizontal fork in tree.

Barred (Yellow-eyed) Cuckoo-shrike   *Coracina lineata*   24–29 cm
Dark grey cuckoo-shrike with yellow eye and heavily barred belly. *Juv:* lacks barred belly; head and breast mottled and streaked grey; wing feathers edged paler. Mostly seen in native small-fruited fig trees, also eats insects of outer foliage, eg phasmids. *Voice:* plaintive 'wheee' or 'whee-uh'. *Nesting:* cup of twigs, bark, grass and cobweb on horizontal fork, more bulky than Little; 2 blotched white eggs. *Range:* uncommon migrant (south Sep–Mar) or nomad (north) in rainforest and sclerophyll forest in north-east, from Cape York to north-east NSW, vagrant south to Sydney.

Ground Cuckoo-shrike   *Coracina maxima*   36 cm
Pale grey cuckoo-shrike with pale yellow eyes, black wings and tail, barred abdomen and rump. *Juv:* heavily barred on head, back and underparts. Usually seen in small parties walking slowly on ground or flying, showing deeply forked tail, or sitting on dead branch. *Voice:* loud 'kree-el'; 'cheer-cheer'; harsh 'coo-look'; 'hic-o-weeyit'. *Nesting:* bulky cup of twigs, bark, grass and cobwebs on horizontal fork; 2–3 blotched olive-green eggs. *Range:* uncommon nomad in open woodland and tree-scattered paddocks through drier areas of Aust, spasmodically visits open country in wetter areas.

Black-faced Cuckoo-shrike   *Coracina novaehollandiae*   33 cm
Mid-grey cuckoo-shrike with face black extending behind eye (imm) and to upper breast (adult). *Adult:* face, forehead, throat and upper breast black. *Imm:* face black, extending behind eye. *Juv:* similar but with barring and mottling on head, breast and flank. One of commonest Aust birds. White-bellied similar to imm, but black 'face' does not extend behind eye. *Voice:* plaintive 'plee-urk'; pleasant churring; 'plurr peh'. *Nesting:* flat cup of twigs and grass on horizontal fork; 2–3 blotched green eggs. *Range:* common migrant (south), nomad or resident in most habitats with trees throughout mainland Aust and Tas.

White-bellied (Little) Cuckoo-shrike   *Coracina papuensis*   28 cm
Pale-grey cuckoo-shrike with black between eye and beak; eastern birds have breast grey (race *robusta*); northern birds have breast white (race *hypoleuca*). Some birds in east have head and breast black, breaking into bars on abdomen, never cleancut as Black-faced; often confused with juv and imm Black-faced, which are best told by black on face extending behind eye. *Voice:* 'kisseek'; 'quee-erk' like Black-faced. *Nesting:* flat platform of twigs and cobwebs; 2–3 blotched green eggs. *Range:* common resident (north) or nomad (south-east), possibly migrant (south) in forest, woodland, mangrove and riverine forest in northern and eastern Aust.

Blackbird   *Turdus merula*   25 cm
Male: sweet-singing long-tailed black thrush with bright yellow bill and eye-ring (dusky in immature). *Female:* dark brown, more rufous below with paler streaked throat and yellow-brown bill. *Juv:* rufous brown with fine pale streaks on upperparts and wings, paler mottling and barring below, buff stripe below eye. *Range:* Introduced. Common resident in thick understorey in variety of habitats including gardens and orchards in south-east and Tas.

THRUSHES   *juv*   ♀   BLACKBIRD   ♂

CUCKOO-SHRIKES

GROUND

BARRED
(YELLOW-EYED)

juv

ad

ad

juv

BLACK-FACED

juv

imm

humid
form

arid
form

WHITE-BELLIED
(LITTLE)

Papuan
form

eastern
form

dark
phase

imm

northern
(white-
breasted)
form

31 x 22	34 x 24	30 x 22	34 x 24
Barred	Ground	White-bellied	Black-faced

TRILLERS AND BULBULS

Trillers: small cuckoo-shrikes with different male and female plumages. Two Aust species have pied males and brownish females; White-winged is very obvious, particularly in breeding season when male sings vigorously; Varied and Cicadabird more secretive, each with distinctive, far-carrying call. Usually feed on insects, fruit and nectar in trees, but White-winged also feeds on ground. Some birds may show orange or yellow foreheads from pollen. Male White-winged has eclipse plumage after migrating to northern Aust. Nest is small cup of grass, bark and cobwebs on horizontal fork of tree.
Bulbuls: tropical birds with fine bristles around bill and on nape; rump feathers are loose and puff-like. Mostly gregarious, they feed on insects, fruit and berries. Nests are cup-shaped, well-hidden, made from bark, grass and fine rootlets. Introduced into Aust.

White-winged Triller *Lalage sueurii* 18 cm
Obtrusive small slender cuckoo-shrike without prominent white eyebrow. *Male breeding:* shining black above with grey rump, white shoulder and black brow; white below. *Male non-breeding:* black back and crown replaced with brown. *Female:* similar to non-breeding male, but tail paler, rump browner. *Juv:* similar with greyish-brown streaks in head and breast. *Voice:* male, rich vigorous extended 'ch-ch-joey-joey-joey . . .'; female, usually silent; juv, rusty 'chick-airp'. *Nesting:* flat cup in fork; 2–3 blotched green eggs. *Range:* common irregular migrant (south Aug–Mar) or resident (north) in open woodland, particularly mulga, throughout Aust, vagrant to Tas, occasionally breeding.

Varied Triller *Lalage leucomela* 18–21 cm
Unobtrusive small plump cuckoo-shrike with prominent white eyebrow. *Male:* black above with white eyebrow, white edges to wing feathers, underparts pale grey to buff, sometimes faintly barred, and with buff undertail coverts. *Female:* similar but greyer on back, underparts buff with faint bars. Usually attracts attention by distinctive churring call. Several forms occur, differing in males only: eastern form (race *leucomela*) has only vent buff; Cape York form (race *yorki*) has vent and belly buff; and northern form (race *rufiventris*) has vent, belly and breast buff. *Voice:* 'chee-urr'; 'drr-eea, drr-eea'. *Nesting:* small flat cup in horizontal fork; 1 blotched green egg. *Range:* uncommon nomad in rainforest, sclerophyll forest, mangroves and riverine forest in northern and eastern Aust, from north-west Kimberley to about Taree, NSW.

Cicadabird *Coracina tenuirostris* 24–26 cm
Dark blue-grey (male) or barred brownish-grey (female) cuckoo-shrike, with far-carrying cicadalike call. *Male:* dark blue-grey with black face. *Female:* brownish-grey above with darker streaked earpatch, and buff edges to wing feathers, underparts buff with faint bars – similar to female Varied Triller but much larger, paler on back, eyebrow less prominent. More often heard than seen. *Voice:* loud extended cicadalike 'kree-kree . . .', with slight downward trend; 'clewk clewk' sometimes continuing for six or seven flutelike notes; harsh 'check-up'. *Nesting:* small flat cup on horizontal fork; 1 blotched green egg. *Range:* common migrant Sep–Apr/May in forest and woodland in northern and eastern Aust.

Red-whiskered Bulbul *Pycnonotus jocosus* 23 cm
Introduced bulbul with conspicuous black crest, white earpatch, red bristles behind eye, plain brown back, red undertail coverts. *Juv:* no red on cheek, undertail pink. Usually singly or in pairs, but in winter may congregate. *Voice:* jaunty falling whistle, 'wee-whit-h-h-h-who'. *Nesting:* untidy cup of bark, leaves and rootlets in thick bush; 2–4 spotted or blotched whitish eggs. *Range:* common around Sydney, less common Coff's Harbour, NSW; Melbourne (South Yarra); some sightings Adelaide.

Red-vented Bulbul *Pycnonotus cafer* 20 cm
Introduced bulbul with vestigial crest on black head, scaly smoky-brown back, undertail coverts red. *Range:* small population in Melbourne may now be extinct.

TRILLERS

WHITE-WINGED

juv

♀

non-br ♂

br ♂

northern form

VARIED

juv

♀

eastern form

Cape York form

CICADABIRD

BULBULS

♀

♂

RED-VENTED

RED-WHISKERED

22 x 17

21 x 15

25 x 17

22 x 17

32 x 22

Red-whiskered

White-winged

Varied

Red-vented

Cicadabird

ROBINS

Thickheads are insectivorous birds with large heads and generally stout bills with bristles around gape, and include robins, scrub-robins, whistlers and shrike-thrushes. Australian robins are small thickheads, with flycatcher appearance but take most of their prey on ground, flying down from prominent perch; particularly adept at spotting caterpillars. These species have red-breasted males, superficially similar but easily identifiable. Nests are deep cups placed in tree forks, decorated with lichen; nests of Rose and Pink are particularly beautiful. Juveniles have head and breast streaked dark grey, and all look much like the juv Red-capped illustrated.

Pink Robin   *Petroica rodinogaster* 13 cm

Black-backed, pink-breasted or brownish robin with no white in tail, confined to south-east and Tas. *Male:* pink breast, black above. Similar Rose has white in tail; female Rose and Flame are paler and have wing patches less buff. *Voice:* weak warble; 'tick-tick' like snapping twig. *Nesting:* beautiful rounded cup covered with lichen; 3–4 speckled greenish-white eggs. *Range:* locally common migrant or nomad in forest, woodlands and tea-tree scrub in Tas and south-east.

Rose Robin   *Petroica rosea* 11 cm

Slender grey-backed pink-breasted or brownish-grey robin with white in tail. *Female:* brownish-grey with small buff spot on forehead, two pale buff bars in wing, with or without pink clouding on breast. *Voice:* tinkling whistle 'we-are, we-are, we-are the champions' or 'a-weet a-weet-a rrrreep rrreep'; faint nasal 'neep' – once learnt reveals many more in forest than suspected. *Nesting:* possibly most beautiful Aust nest; rounded cup covered with lichen on lichen-crusted branch; 2–3 freckled pale green or pale blue eggs. *Range:* uncommon altitudinal migrant in rainforest and wet sclerophyll forest, more open forest in winter but usually along densely vegetated creeks, in eastern Aust.

Red-capped Robin   *Petroica goodenovii* 11·5 cm

Male: black-backed, red-breasted with white in tail and wing, red cap. *Female:* brown with faint pink flush on forehead (and breast in some). *Imm:* similar to female but lacks pinkish forehead and breast. *Juv:* mottled and steaked on head, back and breast (all juvenile robins are similar), moults within weeks. Very widespread in inland, particularly mulga (2–3 pairs per hectare). *Voice:* reeling metallic 'trr-trr-derradee dee'; 'toc-toc' like tapping stones. *Nesting:* deep grass and bark cup in fork, sparsely decorated with lichen; 2–4 spotted green or blue-green eggs. *Range:* common resident in woodland in drier areas of Aust, generally west of Divide except in extreme south-east corner.

Flame Robin   *Petroica phoenica* 13·5 cm

Male: grey-backed robin with orange-red breast, small white patch on forehead, white in wing and tail. *Female:* brown upperparts with small white spot on forehead, white in wing, brown clouding on breast (faint orange in some). *Imm:* wing patch more buff. *Juv:* streaked and mottled. *Voice:* pleasant piping 'you may come if you wish to the sea'. *Nesting:* bulky cup, untidier than other robins, in wide variety of situations; 3–4 freckled greenish-white eggs. *Range:* common migrant, nomad or resident in forest and woodland in Tas and south-east.

Scarlet Robin   *Petroica multicolor* 13 cm

Male: black-backed red-breasted robin with white forehead, white in wings and tail. *Female:* upperparts brown (in east) or dark grey (in west) with white spot on forehead, white eye-ring, pink breast, white in wing and tail. *Imm:* lacks pink breast, patches in wing buff. *Juv:* streaked and mottled. *Voice:* warbling 'sh-sh-sha-weeya'. *Nesting:* untidy cup of bark and grass in fork; 2–4 blotched whitish or pale bluish-green eggs. *Range:* common altitudinal migrant or nomad in east, resident in west in open forest and woodland in south-west and south-east.

PACHYCEPHALIDAE

ROBINS

ROSE

PINK

♀

♂

♀

♂

RED-
CAPPED

imm

♀

juv

♂

SCARLET

FLAME

♂

♀

♂

♀

18 x 14	17 x 14	15 x 13	18 x 14	18 x 14
Pink	Rose	Red-capped	Flame	Scarlet

ROBINS

These 'robins' are large rather inactive thickheads usually taking insect prey on ground or from treetrunks. They sit quietly on branches close to ground or sideways on treetrunk. Some spend more time on ground, hopping rather dumpily. Nests are neat cups decorated with lichen, moss or spiders' eggsacs, usually in low fork of tree, on palm frond (Grey-headed) or well-concealed (Hooded, Dusky).

White-browed Robin   *Poecilodryas superciliosa* 14–18 cm
Brownish robin with prominent white eyebrow found in undergrowth along tropical creeks and rivers. White-sided form (race *superciliosa*) on Cape York south to Rockhampton, Qld; and buff-sided form (race *cerviniventris*) with darker face in NT and Kimberley, WA. Active robin, often raising tail, opening wings, hopping through undergrowth and on ground. *Voice:* loud piping whistle 'peet-peet-peet-peet'. *Nesting:* unsubstantial cup lined with grass decorated with scraps of whitish bark, lichen; 2 spotted green to blue eggs. *Range:* uncommon in dense tropical vegetation mostly along rivers and creeks in northern Aust from Rockhampton, Qld, to Derby, WA.

Mangrove Robin   *Eopsaltria pulverulenta* 14–17 cm
Greyish robin with black lores, white bases to outer tail feathers and no obvious white in wing, found in mangroves. *Imm:* mottled brown and rufous with paler streaks similar to immature Yellow Robin (p 218). Often feed on ground among mangrove roots. *Voice:* mournful distant 'pee-peee'; harsh 'chit'; vigorous song. *Nesting:* small grass-lined bark cup with hanging strips in mangrove fork; 2 spotted pale green eggs. *Range:* locally common resident in mangrove around northern Aust from Exmouth, WA, to Ayr, Qld.

Hooded Robin   *Melanodryas cucullata* 14–17 cm
Male: pied robin. *Female:* greyish robin with white bases to outer tail feathers and white patches in wing. *Imm male:* similar to female but breast darker and more defined, whitish patch on shoulder. Rather stolid robin, spends long periods sitting motionless before flying down to ground for food; larger territory than other robins, +2 hectares. *Voice:* metallic 'peet peet peet . . .' before dawn. *Nesting:* neat cup of grass bark and rootlets well hidden in deep fork, often among dead twigs, ground level to 2 m; 2–3 olive or greenish eggs. *Range:* locally common resident in dry woodland, mallee and mulga throughout Aust.

Dusky Robin   *Melanodryas vittata* 16·5 cm
Dusky brown robin found in Tas and Bass Strait islands. Females of other robins found in Tas are much smaller, have pale lores and whiter bellies; Scarlet and Flame have more white in wing and tail; Pink has two bars in wing and no white in tail. Usually quiet and inactive like Hooded; in small family parties. Often sits on stumps and fence posts. In flight, white bar in outer wing. *Voice:* 'tu-wer tu-wer'. *Nesting:* well-concealed cup of bark and root fibres, 1–6 m from ground; 3–4 spotted or smudged pale green eggs. *Range:* common resident or nomad in Tas and Bass Strait islands, mainly in wooded habitats, clearings and coastal heath.

Grey-headed Robin   *Poecilodryas albispecularis* 16–18 cm
Large tortoiseshell robin with grey head, found in north-eastern upland rainforests south to Townsville, Qld. Usually unobtrusive but often flushed from ground along roadsides and tracks; hops along ground with upright stance. *Voice:* pervasive loud 'peee-pi-pi-pi'. *Nesting:* untidy decorated cup usually in lawyer vine, 1–3 m; 1–2 freckled and blotched pale green or pale blue eggs. *Range:* common resident in upland rainforests above about 250 m, usually seen alongside roads and clearings, in north-east Qld from Bloomfield R to Mt Spec.

ROBINS

WHITE-BROWED

white-sided form

buff-sided form

MANGROVE

imm ♂

♀

HOODED

♂

DUSKY

GREY-HEADED

19 x 15	20 x 15	21 x 16	21 x 16	26 x 19
Mangrove	White-browed	Hooded	Dusky	Grey-headed

YELLOW ROBINS

Dumpy, yellow-breasted, relatively inactive 'robins' usually seen in forests perching motionless on low branches or treetrunks watching for their prey (invertebrates and small lizards) which is taken on the ground. Songs and calls are simple but penetrating; dawn song usually begins before other birds, may reveal density of robins in forest as greater than one pair per hectare. Nests cup-shaped, built of bark lined with grass and leaves in upright forks, decorated with hanging strips of bark or lichen. Female incubates, fed on nest by male; both feed the young. White-breasted Robin is essentially a yellow robin that has lost yellow pigment.

Eastern Yellow Robin *Eopsaltria australis* 15 cm
Yellow-breasted robin with grey face, found in eastern and south-eastern Aust. *Juv:* dark brown streaked paler, moulting after several weeks into plumage like adult. Birds of north-east have rump bright yellow (race *chrysorrhoa*); south-eastern birds have rump olive (race *australis*). *Voice:* pre-dawn call whistle-like loud 'tewp tewp'; monotonous plaintive piping, particularly late afternoon. *Nesting:* bark cup with hanging strips of bark and/or lichen; 2–3 spotted light green eggs. *Range:* common forests, woodlands, well-planted gardens in eastern and south-eastern Aust from about Cooktown, Qld, to Naracoorte, SA.

Western Yellow Robin *Eopsaltria griseogularis* 15 cm
Yellow-bellied robin with grey breast and face, found in south-western and southern Aust (east to Eyre Peninsula SA). *Juv:* dark brown streaked paler, moulting shortly after leaving nest into plumage like adult. *Voice:* similar to Eastern. *Nesting:* bark cup lined with grass, leaves, decorated with long strips of bark; 2 pale green to bluish eggs. *Range:* common in open forest woodland, scrub and mallee in south and south-west from Eyre Peninsula SA, to Shark Bay, WA.

White-breasted Robin *Eopsaltria georgiana* 14 cm
Greyish robin with dark tail finely tipped white, found in south-west. *Juv:* similar to juvenile yellow robins, moulting shortly after leaving nest. *Voice:* double whistle; harsh double note when distressed. *Nesting:* bark cup sparsely lined with leaves, decorated with lichen and/or bark strips; 2 spotted olive to bluish eggs. *Range:* common in thickets in forest and coastal scrub in south-western Aust.

Pale-yellow Robin *Tregallasia capito* 12–14 cm
Yellow-breasted robin with white or pale buff face and brownish-grey head, found in eastern rainforests south to Port Stephens, NSW. Breast is greenish-yellow, greyish on flanks. Southern birds have face white (race *capito*); northern birds have face pale buff (race *nana*). *Voice:* high-pitched squeaking 'pit-it'; liquid churring; harsh scolding. *Nesting:* large cup of grass in lawyer vine; 2 blotched pale green eggs. *Range:* uncommon to common in rainforest in east, from Port Stephens, NSW, to Cooktown, Qld.

White-faced Robin *Tregallasia leucops* 12–13 cm
Yellow-breasted robin with white face and black head, found in north-eastern rainforests south to Rocky River, Qld. Large black eye in white face gives 'panda' look. Like Pale-yellow, takes food more often from leaves and tree trunks than other yellow robins. *Voice:* similar to Pale-yellow. *Nesting:* large neat cup of grass and rootlets decorated with lichen and moss; 2 spotted pale green eggs. *Range:* common in rainforest on Cape York south to Rocky River, Qld.

ROBINS

EASTERN

northern form

south-eastern form

juv

WESTERN

WHITE-BREASTED

southern form

PALE-YELLOW

northern form

WHITE-FACED

22 x 16   21 x 16   20 x 16   19 x 15

Western   White-breasted   Eastern 22 x 16   Pale-yellow   White-faced

ROBINS ('FLYCATCHERS' AND SCRUB-ROBINS)

'Flycatchers': similar to robins but taking much of their prey in the air; have broader bills to increase feeding efficiency. Nests are among smallest of any Aust bird, placed on horizontal fork. **Scrub-robins:** terrestrial robins similar to female red-breasted robin, but larger, with much longer legs and tails. Tail often raised to 45°, particularly while singing on low perch. Nest built on ground, surrounded by unusual pattern of radiating twigs.

Yellow-legged Flycatcher   *Microeca griseoceps* 11–13 cm
Yellow-breasted flycatcher with orange-yellow legs and yellow lower mandible to bill. Very active in outer canopy; one of least-known Aust birds, has been confused with yellow robins, other flycatchers and Grey Whistler, none of which have yellow legs. *Voice:* precise information required. *Nesting:* unknown. *Range:* rare in rainforest on tip of Cape York and Claudie R, Qld.

Brown Flycatcher (Jacky Winter)   *Microeca leucophaea* 13 cm
Brown-backed flycatcher with white outer tail feathers, more pronounced in western birds; underparts white, clouded grey on breast. *Juv:* mottled and streaked dark brown on breast, head and back (juvenile Lemon-breasted is similar). Often sits on upright dead limbs, wags tail in figure of eight. Kimberley form of Lemon-breasted is similar but has olive back and no white in tail except   extreme tip, confined to mangroves. *Voice:* rich 'peter-peter-peter. . .' *Nesting:* very small neat grass and cobweb cup in horizontal fork, often in dead timber; 2 blotched pale blue eggs. *Range:* common resident (partial migrant in south-east) in open forest and woodland throughout Aust.

Lemon-breasted Flycatcher   *Microeca flavigaster* 11–13 cm
Brown-tailed flycatcher with yellow or whitish underparts and dark legs. Varies in breast colour from east to west: on Cape York (race *terrareginae*), breast bright yellow; south of Cape York across to Ord R, WA, (race *flavigaster*), breast lemon yellow; around Kimberley coast (race *tormenti*), breast white. *Imm:* breast clouded grey, pale yellowish abdomen. *Juv:* mottled and streaked brown. *Voice:* musical 'chew-chew'; 'swee-so-wu-chew'. *Nesting:* tiny nest of bark and cobwebs in horizontal fork; 1 blotched pale greenish-blue egg. *Range:* locally common resident in woodland, paperbark swamp, riverine forest and mangrove in northern Aust from about Broome, WA, to about Bowen, Qld.

Southern Scrub-robin   *Drymodes brunneopygia* 23 cm
Brownish-grey scrub-robin of mallee heath. *Juv:* streaked on head, back and breast. Usually seen singly or in pairs, runs to cover when disturbed, scolding intruder. *Voice:* loud 'chip-por-wee'; harsh scolding. *Nesting:* concealed cup of twigs, bark and grass on ground with radiating twigs; 1 blotched green-grey egg. *Range:* uncommon resident in mallee with thick ground cover in south-eastern and south-western Aust.

Northern Scrub-robin   *Drymodes superciliaris* 22 cm
Reddish-brown scrub-robin of rainforest on Cape York (isolated population on Roper R, NT, may be extinct). Usually seen singly or in pairs on ground, flicking wings and raising tail; runs quickly, hops on to low perches; little known. *Voice:* more information required. *Nesting:* cup of roots and leaves on ground, surrounded by twigs; 2 blotched pale umber eggs. *Range:* rare resident in rainforest and vine scrub, particularly around edges on Cape York south to Coen, Qld; possibly extinct on Roper R, NT.

FLYCATCHERS

BROWN (JACKY WINTER)

YELLOW-LEGGED

ad

juv

LEMON-BREASTED

northern form

Kimberley form

imm

Cape York form

SCRUB-ROBINS

NORTHERN

SOUTHERN

15 x 12	20 x 15	19 x 14	22 x 18	25 x 19
Yellow-legged	Brown	Lemon-breasted	Northern	Southern

SHRIKE-TITS AND WHISTLERS

Shrike-tits are crested thickheads with powerful notched bills used for stripping bark from branches revealing insects and spiders, also for roughing up vertical twigs as anchor for deep cup-shaped nest. **Whistlers** are often colourful thickheads with melodious songs; many have black band on breast in male. Some are plainly coloured (p 225); Grey is often confused with Lemon-breasted Flycatcher.

Crested Shrike-tit *Falcunculus frontatus* 16–19 cm

Yellow-breasted thickhead with black and white crested head. *Male*: throat black. *Female*: throat mottled green. Eastern birds (race *frontatus*) have belly yellow and wing quills edged grey; northern birds (race *whitei*) have belly yellow and wing quills edged yellow; south-western birds (race *leucogaster*) have belly white. *Voice*: chuckling 'knock at the door jack'; whistling 'pit-you' in east; 'poo-wee-er' in west. *Nesting*: deep bark and cobweb cup in slender upright twigs near treetop; 2–3 blotched white eggs. *Range*: common (south-east), uncommon (south-west) to rare (north) resident in forest and woodland in south-west, east and north.

Golden Whistler *Pachycephala pectoralis* 17 cm

Male: yellow-breasted thickhead with black head, white throat and yellow edges to wing feathers (Mangrove has broader yellow collar). *Female*: plain grey-brown whistler with undertail coverts yellow (Qld), pale buff (south-east) or bright buff (west). *Imm*: similar but with rufous wings. *Juv*: rufous above and pale rufous below, moulted quickly. *Voice*: whistling 'wh-wh-wh-you wit'; single 'seeep'. *Nesting*: untidy flimsy grass cup in upright fork; 2–3 blotched cream eggs. *Range*: common resident or partial nomad in rainforest, forest and woodland in south-west, south-east, east and Tas.

Olive Whistler *Pachycephala olivacea* 21 cm

Olive-backed thickhead with white barred throat, grey breast and olive belly. *Female*: duller than male. Two forms occur: southern form (race *olivacea*) in Tas and south-east, has crown dark grey; and northern form (race *macphersonianus*) from Barrington Tops to McPherson Ra, with blue-grey crown, different song, confined to beech forest on highlands. Spends part of time on ground searching litter. *Voice*: in south, sweet 'chiff chiff chuff chuff'; 'peeee . . .'; 'te-wee-ee-chow'; in north, indrawn 'see-saw' ('see' octave higher than 'saw'); 'chor-wit' with upward inflection. *Nesting*: large cup of twigs, bark and leaves in bush; 2–3 blotched buff eggs. *Range*: common (south) to rare (north) in rainforest, dense gullies and woodland with thick undergrowth in Tas, south-east and in beech forest north from Barrington Tops, NSW, to McPherson Ra, Qld.

Mangrove Golden Whistler *Pachycephala melanura* 15 cm

Male: similar to Golden but smaller, with broader yellow collar, grey edges to wing feathers, black tail. *Female*: similar to female Golden but breast yellow in north and east (race *robusta*), with decreasing amount of yellow southwards on west coast (race *melanura*). Possibly subspecies of Golden, similar in most respects, but habitat preferences keep them apart. *Range*: common resident in coastal and riverine mangrove and rarely in nearby vine thickets in north-western, northern and north-eastern coast from Exmouth Gulf, WA, to about Bowen, Qld.

SHRIKE-TITS

CRESTED

eastern form

western form

northern form

juv

♀

WHISTLERS

GOLDEN

south-eastern

imm

juv

Qld ♀

♂

western ♀

MANGROVE GOLDEN

OLIVE

♀

northern ♀

♂

♂

23 x 17

28 x 19

Crested 24 x 15   Golden   Olive   Mangrove 21 x 16

WHISTLERS

These thickheads are rather plain in colour; voices often have indrawn pensive quality, but Rufous and White-breasted have vigorous songs – any loud noise such as slamming car door will cause them to call. Grey is often confused with other birds, enables lazy observer to tick off some rarely seen species without actually seeing them.

Grey Whistler   *Pachycephala simplex*   15–20 cm

Grey or brown thickhead with belly white or yellowish. Two forms occur: yellow-bellied form (race *peninsulae*) from north-east Qld, sometimes confused with Yellow-legged Flycatcher, but has dark legs and bill, is generally less active; brown form (race *simplex*) from northern NT, with whitish belly and buffish sides to breast, is sometimes confused with Kimberley form of Lemon-breasted Flycatcher (in some books called Brown-tailed Flycatcher), but ranges appear to be exclusive. *Voice:* whistling 'wh-wh-whee-w-whit'. *Nesting:* sparse cup of roots and grass; 2 spotted pale buff eggs. *Range:* uncommon resident in rainforest, mangrove, monsoon forest and riverine forest in coastal NT, from Port Keats to Gulf and on Cape York south to Cardwell.

White-breasted Whistler   *Pachycephala lanioides*   18–20 cm

Male: white-breasted mangrove thickhead with black breastband, white throat and chestnut collar. *Female:* stout black bill, greyish brown upperparts, underparts pale buff with darker streaks, similar to female Rufous but bigger, with heavier bill, less streaking on breast. *Voice:* vigorous song like Rufous; 'per-weet'; 'twit'. *Nesting:* flimsy cup of twigs and root fibres in mangrove fork; 2 spotted buff eggs. *Range:* uncommon resident in coastal and riverine mangroves around northern Aust,. from Carnarvon, WA, to Normanton, Qld.

Rufous Whistler   *Pachycephala rufiventris*   17 cm

Male: rufous-breasted thickhead with black breastband and white throat. *Female:* brownish-grey above, pale buff below with dark streaks, heaviest on breast. *Voice:* 'ee-chong'; 'joey-joey . . .'; loud vigorous song, often heard after loud noise; plaintive 'sweet' like Golden, but lower pitched. *Nesting:* untidy grass cup in fork, usually in clump of leaves; 2–4 freckled olive eggs. *Range:* common resident or partial migrant in open forest and woodland throughout Aust; vagrant Tas.

Gilbert's Whistler   *Pachycephala inornata*   19·5 cm

Greyish whistler with black lores and cinnamon throat (male) or grey throat (female). Feeds on ground and in trees, usually where cover is thick. Similar to Red-lored but has black lores, occupies wider range of habitat, more varied song; female looks like small shrike-thrush. *Voice:* whistling 'er-whit-er-whit . . .'; loud extended 'chop chop chop . . .'; slow 'pooo-eee'. *Nesting:* bark cup lined with grass, rootlets in low fork, or babblers' nest; 2–3 blotched cream eggs. *Range:* uncommon to rare resident in mallee and other woodland with ground cover in southern Aust.

Red-lored Whistler   *Pachycephala rufogularis*   22 cm

Buff-bellied, grey-breasted thickhead with cinnamon face and throat (including lores) found in south-east mallee. *Male:* face and throat bright cinnamon. *Female:* face and throat pale cinnamon. *Juv:* breast streaked. Shy, rather difficult to see well, flies some distance when disturbed. Similar to Gilbert's, which has black lores. If sighting suspected outside mallee, check juvenile Golden Whistler (p 222). *Voice:* sweet variable 'see saw', second note indrawn. *Nesting:* loose cup of twigs, lined with bark and grass, in low fork or spinifex clump; 2–3 blotched cream eggs. *Range:* rare in mallee thickets in eastern and north-eastern Vic, and in national parks at Round Hill and Cocoparra Ra, NSW.

WHISTLERS

GREY

yellow-bellied form

brown form

WHITE-BREASTED

♂   ♀

RUFOUS

inland   imm ♂   ♀

♂

GILBERT'S

♀

RED-LORED

♀

♂

♂

21 x 16   26 x 19   23 x 17   24 x 17   24 x 17

Grey   White-breasted   Rufous   Gilbert's   Red-lored

SHRIKE-THRUSHES

Large plainly-coloured thickheads with melodious voices; inhabit variety of habitats, from rainforests to wooded desert watercourses and tropical sandstone gorges. Food consists of large invertebrates and small vertebrates. Some feed mainly on ground, others in trees as well, usually on stout branches, trunks and fallen timber. Nests and eggs are similar, cup-shaped, mainly of bark, twigs and rootlets, placed in enclosed situations like hollow upright stump, leafy fork or sandstone crevice; 2–3 blotched creamy white eggs.

Little (Rufous) Shrike-thrush *Colluricincla megarhyncha* 17–19 cm
Small, short-tailed, robust-billed shrike-thrush with brown upperparts and rufous underparts. Rufous form (race *rufigaster*) from north-east is bright rufous with little white on throat and has pale brownish bill; black-billed form (race *parvula*) from NT and Kimberley is pale rufous below with whitish throat and has black bill. Rufous form is similar to Bower's, which has dark grey head and back, and diffuse grey streaks on throat and breast. *Voice:* variable mellow 'chip chip co-wee', lower than Grey; 'wee-too swit-you see-you', 'tu-whee-wet-wet'; 'ee-but eh-butch butcher'; rasping sneeze. *Nesting:* well-hidden cup of leaves, bark and twigs; 2–3 freckled white eggs. *Range:* uncommon resident in rainforest, riverine forest, paperbark forest and mangrove in north-eastern and northern Aust.

Bower's Shrike-thrush *Colluricincla boweri* 19–20 cm
Medium-sized, short-tailed thickhead with head and back dark grey and underparts rufous streaked grey on throat and upper breast. Similar to rufous form of Little, which is brown on head and back, has paler bill. *Voice:* quiet whistling, 'to whee to whit'; loud 'cluck'. *Nesting:* nest typical; 2–3 pearly-pinkish blotched eggs. *Range:* locally common in upland rainforests (above 400 m) in north-east from about Cooktown to Townsville, Qld.

Grey Shrike-thrush *Colluricincla harmonica* 22·5 cm
Large long-tailed grey thickhead with brown back. *Male:* bill black, white patch in front of eye. *Female:* bill pale at base, white eye-ring, faintly striped throat (heavier in younger birds). *Imm:* reddish eye-ring, stripes on throat and breast, reddish edges to wing feathers. *Juv:* heavily streaked on throat and breast. Eastern and northern birds (race *harmonica*) have back brown, paler in north, and underparts grey; western birds (race *rufiventris*) have back dark grey, buff undertail coverts. *Voice:* variable ringing calls, basically 'pip pip pip pip pip hoo-ee'; sharp 'yorick'; *juv:* 'klute'. *Nesting:* large cup of bark and grass in hollow, 2–4 freckled white eggs. *Range:* common resident in forest and woodlands throughout mainland Aust and Tas.

Crested Bellbird *Oreoica gutturalis* 20–23 cm
Large greyish-brown thickhead with orange eye. *Male:* black crest and breastband, face white. *Female:* greyish-brown head with black patch on crown, similar to Grey Shrike-thrush which lacks crown patch, has dark eye and over most of mutual range has buff undertail coverts. Feeds on ground, rarely perches high. *Voice:* haunting, ventriloquial: in east 'dick dick the devil' with either upward or downward inflection; in west more even 'pan-pan-pallela'; quiet chuckling while walking on ground. *Range:* common resident in drier woodlands particularly mulga throughout Aust, expanding into forests in south-west.

Sandstone Shrike-thrush *Colluricincla woodwardi* 25 cm
Large long-tailed brown thickhead usually seen on rocks in tropical sandstone gorges. Usually seen singly or in pairs, actively moving over sandstone or making short flights on rounded wings. *Voice:* most beautiful echoing notes; loud 'peter'. *Nesting:* typical nest and eggs, but placed in crevice in sandstone. *Range:* locally common in sandstone and other gorges in Kimberley, WA, and northern NT.

SHRIKE-THRUSHES

Cape York form

rufous form

LITTLE

BOWER'S

black-billed form

♀

GREY

♂

juv

western form

imm

CRESTED

♀

♂

SANDSTONE

25 x 19
Bower's

25 x 19
Little

28 x 20
Grey

30 x 21
Sandstone

27 x 20
Crested

MONARCHS

Monarchs: flycatchers generally with broad gape fringed with 'bristles' adapted for taking insects in flight and casque-like crowns of compact feathers that can be raised to give characteristic profile. Tails are long and often waggled. Calls are harsh and frog-like, but songs are often sweet. Many are migratory, arriving early Sep, leaving Mar–Apr. Nests are cup-shaped, placed on horizontal branch or slung in horizontal fork, often decorated with paperbark, spiders' eggsacs, lichen or moss.

Shining Flycatcher *Myiagra alecto* 17·5 cm
Male: shining blue-black flycatcher frequenting mangroves; in shade appears black; often calls with mouth wide open, revealing orange gape. *Female:* bright rufous with black head and white underparts. *Imm:* similar to female but duller. Very active, often flicking tail, raising flat crest, darting after prey on leaves, mangrove trunks or mud, less inclined to chase aerial insects. *Voice:* harsh frog-like 'cheeek', sweet triple whistle. *Nesting:* rather untidy decorated cup in sapling fork; 2–3 spotted pale blue or pale green eggs. *Range:* common in mangroves, dense vegetation near water and along tropical rivers from Broome, WA, to Noosa R, Qld.

Spectacled Monarch *Monarcha trivirgatus* 14–16 cm
Black-faced flycatcher with white belly and extensive white in black tail. *Female:* rather duller than male. *Imm:* lacks black face but has grey chin and whitish marks on face. Birds observed in most of eastern rainforests have variable amounts of rufous on breast and flanks; birds on Cape York have breast only rufous, sharply demarcated from white abdomen and flanks (race *albiventris*). *Voice:* loud 'swee-swee-swee', jingling chattering. *Nesting:* in fork of sapling usually at no great height, decorated with moss and scraps of lichens; 2 freckled pinkish eggs. *Range:* common in lower storey of rainforest and damp thick forest particularly in gullies from Cape York to central NSW. Migratory over southern range, arriving Sep and leaving early Mar.

Black-faced Monarch *Monarcha melanopsis* 15–20 cm
Black-faced flycatcher with rufous belly, grey wings and tail. *Imm:* lacks black face but has grey chin; buff tips on wing coverts diminish with wear; looks rather like a shrike-thrush but sharply demarcated rufous belly is diagnostic. *Voice:* distinctive whistler-like 'why-you-which-yee-ou'. *Nesting:* cup in fork of sapling, beautifully decorated with moss, rather like nest of Spectacled but more bulky and usually with no lichen decoration; 2–3 spotted and blotched white eggs. *Range:* common in rainforest, sclerophyll forests and woodland, usually in denser gullies, over eastern Aust from Dandenong, Vic, to Cape York Peninsula. Migratory over most of its range, with some sedentary populations in north; arrive Qld in Sep and leave Mar.

Black-winged Monarch *Monarcha frater* 18–19 cm
Black-faced flycatcher with rufous belly, black wings and tail. *Imm:* lacks black face and has buff edges to duller wing feathers. Feeds on insects and other invertebrates taken from leaves and branches, rarely in flight. *Nesting:* nest similar to Black-faced Monarch, decorated with moss lined with rootlets, strips of bark and other fibres, placed in fork of sapling up to 10 m; 2–3 spotted and blotched white eggs. *Range:* common in rainforest and nearby woodland on Cape York south to McIlwraith Ra; migrant from New Guinea, arriving Oct leaving Mar, some remain through winter.

MONARCHS

SHINING

♀

♂

SPECTACLED

BLACK-FACED

BLACK-WINGED

imm

ad

20 x 15	21 x 16	23 x 17	24 x 17
Shining	Spectacled	Black-winged	Black-faced

MONARCH FLYCATCHERS AND BOATBILL

Forest flycatchers with broad gapes fringed with bristles and loud whistling calls; most are boldly coloured and migratory in at least some of their range, breeding soon after arriving Sep–Oct. Nests are built in fork, usually in sapling 1–3 m high, beautifully decorated with lichen and moss or suspended on vine like small hammock. **Boatbills:** small forest flycatchers with outsized flat bills, brilliantly coloured males. One species only in Aust, found in north-eastern rainforest; builds hammock-like nest in outer branches.

Pied Monarch *Arses kaupi* 14–15 cm

Pied flycatcher with white collar, erectile frill on nape, black breastband and naked blue skin around eye. *Female:* less immaculate than male, has broader breastband, incomplete collar and greyish clouding on nape. *Imm:* brown, rather dirty-looking. Feeds on invertebrates often taken on treetrunks or in air after being disturbed from trunk. Very active, often raising frill. *Voice:* soft 'churr churr'; deep 'zreee zreee . . .'; harsh grating. *Nesting:* delicate decorated basket slung between two strands of vine, made from small twigs and tendrils. *Range:* uncommon in mangrove, rainforest and nearby woodland in north-eastern Qld from Townsville to Cooktown.

Frilled Monarch *Arses telescophthalmus* 14–17 cm

Pied flycatcher with white collar, erectile frill on nape, prominent naked blue skin around eye and no breastband. Female lacks male's black chin, has buff on sides of breast and black ticking on nape. *Imm:* brown where adults are black. Active feeder on treetrunks, often flushing and pursuing insects. *Voice:* similar to Pied. *Nesting:* similar to Pied; 2 blotched pinkish eggs. *Range:* uncommon in rainforest and nearby eucalypts north of range of Pied, on Cape York south to McIlwraith Ra.

White-eared Monarch *Monarcha leucotis* 13 cm

Pied flycatcher with white markings on head and white edges to wing feathers and outer tail feathers. Throat usually white but bases of feathers are black, and as feathers wear black patches appear. *Imm:* brown with reduced white patches on head, buff edges to wing feathers. Rather unobtrusive, active feeders in upper canopy, chasing insects in fantail fashion. *Voice:* mournful 'doo-dee-doo'; flutelike 'ee-choo'; harsh chattering 'chrrk chrrk chrrk'. *Nesting:* nest rarely found, usually high in rainforest tree, made from grass, bark and moss and decorated with spiders' eggsacs; 2 spotted and blotched white eggs. *Range:* uncommon resident or migrant in rainforest, sometimes mangroves and paperbark swamps, in eastern Aust from Tweed R, NSW, to Cape York.

Yellow-breasted Boatbill *Machaerirhynchus flaviventer* 11–12 cm

Small yellow-breasted flycatcher with enormous flat bill. *Male:* black above with yellow eyebrow and white edges to wing feathers and tail, bright yellow below. *Female:* similar but dull green above and paler yellow below. *Imm:* duller still and has boomerang-shaped bars on flanks (some breeding females retain these). Feeds like warbler, hovering in front of leaves. Sometimes adopts wren-like posture with tail cocked. *Voice:* soft but penetrating and persistent melodious trill; soft 'tizz tizz'. *Nesting:* flimsy basket rather like Pied but suspended in horizontal fork, 2–20 m, made from tendrils and pliable twigs; 2–3 spotted and blotched white eggs. *Range:* uncommon in rainforest in north-eastern Qld from Townsville to Cape York.

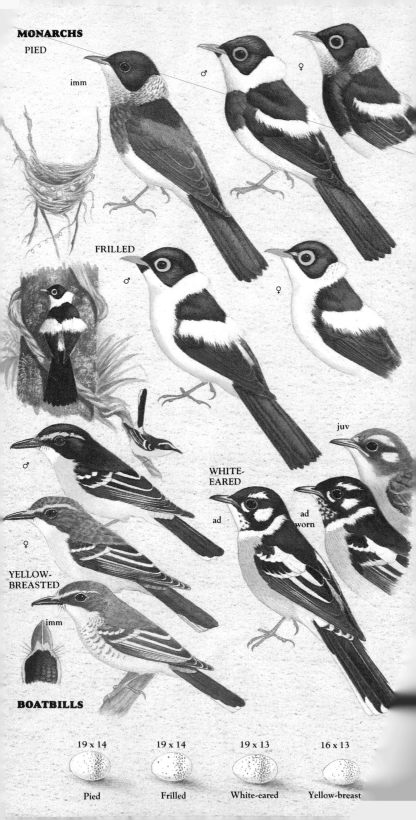

MONARCHS

PIED

imm

♂

♀

FRILLED

♂

♀

WHITE-
EARED

juv

ad

ad
worn

♂

♀

YELLOW-
BREASTED

imm

BOATBILLS

19 x 14	19 x 14	19 x 13	16 x 13
Pied	Frilled	White-eared	Yellow-breast

MONARCH FLYCATCHERS

These monarchs have flat casque-like crests which are raised when agitated; all have harsh frog-like calls as well as sweeter notes. Restless is rather like Willy Wagtail, but has white throat. The other three 'shiver' their tails; their females are very similar, presenting some identification problems; in particular Broad-billed requires care. Decorated cup nests generally on horizontal limbs in open, but usually sited just under another branch.

Restless Flycatcher *Myiagra inquieta* 19–21 cm

Shining black monarch with underparts white, faintly buff breast and paler lores in female. *Juv:* grey edges to wing feathers, buff breast. Eastern and south-western birds (races *inquieta* and *westralensis*) are large, rather greyish on back; northern birds (race *nana*, sometimes regarded as separate species, Paperbark Flycatcher), often associated with paperbark swamps, smaller, blacker on back. Active, often hovers, told from Willy Wagtail and Satin by white throat. *Voice:* several harsh notes like scissors being sharpened; musical 'tu-whee tu-whee'. *Nesting:* neat cup decorated with paperbark and spiders' eggsacs; 2–4 usually 3 blotched whitish eggs. *Range:* uncommon to common migrant or nomad (east) or resident (south-west and north) in woodlands, usually near water, in eastern, south-eastern, south-western and northern Aust.

Satin Flycatcher *Myiagra cyanoleuca* 16 cm

Male: shining blue-black with white lower breast and abdomen. *Female:* dark blue-grey above, glossed on head, throat and upper breast rufous-buff; larger and darker-looking than female Leaden and Broad-billed. *Juv:* mottled grey on breast, pale edges to wing feathers. *Voice:* more assertive than Leaden: hoarse 'queeark'; 'chu-wee, chu-wee chu-wee weer-to weer-to'. *Nesting:* neat cup of bark, grass and cobweb; 2–3 blotched whitish eggs. *Range:* uncommon migrant between south-east Aust/Tas and New Guinea, Sep–Mar in forest, particularly thick gullies.

Broad-billed Flycatcher *Myiagra ruficollis* 15 cm

Grey-brown flycatcher with throat and upper breast rusty buff with richer buff band or 'collar' on lower throat. *Male:* rather brighter and more glossy above than female; both similar to female Leaden, which is more brownish and has evenly-coloured throat patch. *Juv:* mottled greyish throat and breast, pale edges to wing feathers. *Voice:* harsh frog-like 'queeark'. *Nesting:* neat bark and cobweb cup in upright fork; 2–3 blotched whitish eggs. *Range:* common resident in mangroves, riverine forest and paperbark swamps in northern Aust from about La Grange, WA, to Cape Grenville, Qld.

Leaden Flycatcher *Myiagra rubecula* 15 cm

Male: dark blue-grey with glossy leaden head and upper breast, white lower breast and abdomen; not as dark as pied-looking Satin. *Female:* grey-brown above, bluer on head, throat rusty buff; much paler than female Satin; very similar to Broad-billed which has richer band of rufous on throat, lives in or near mangroves in north. Migratory form in south-east has grey lores; north of about Brisbane, lores are black. *Voice:* frog-like 'queeark'; loud 'peter peter . . .'. *Nesting:* neat pliable cup of bark and cobwebs; 2–3 spotted and blotched whitish eggs. *Range:* common migrant (Sep–Apr south-east) or resident in forest, woodland and mangrove in eastern and northern Aust, vagrant Tas, breeds occasionally.

FLYCATCHERS

RESTLESS

♀

♂

paperbark form

SATIN

♀

♂

BROAD-BILLED

LEADEN

♀

♂

21 x 16	18 x 14	17 x 14	17 x 14
Restless	Satin	Broad-billed	Leaden

FANTAILS

Active flycatchers characterised by habit of fanning tail, presumably to startle insects into flight. Most of their food taken on wing with aid of bristles around gape which act as scoops. Nest is neat cup, made from grasses, bark and cobwebs, on horizontal twig; most species append a tail to the nest, which may aid run-off in wet weather.

Grey Fantail   *Rhipidura fuliginosa*   16 cm
Small active grey fantail with white shafts to tail feathers. *Juv:* brownish-grey, moulted within several weeks, but never as reddish-brown as Rufous. Several easily recognisable forms occur, some of which may be separate species: southern forms (races *preissi* in south-west; *albiscapa* in Tas and southern Vic; and *alisteri* in south-east), mid-grey above, narrow breastband, many individuals migrate north in winter as far as Cape York; white-tailed form (race *albicauda*) from central and western desert easily told by extensive white in tail and different call; mountain form (race *keasti*) from rainforest on mountains above 700 m, from Cooktown to Townsville, Qld, much darker with broad black breastband; and mangrove form (race or species *phasiana*) from northern mangroves, pale brownish-grey, larger white bars in wing. *Voice:* vigorous sweet chatter; sharp 'jeck'. *Nesting:* neat cup with tail; 2–4 spotted pale buff eggs. *Range:* common migrant, nomad or resident in any habitat with trees throughout mainland and Tas.

Rufous Fantail   *Rhipidura rufifrons*   16 cm
Small active brown fantail with rufous forehead and rump. *Juv:* buff tips to wing feathers, lost quickly. Most often seen with tail spread in thickets and undergrowth close to ground (sits on ground in forest clearings), also seen in treetops, particularly on migration. *Voice:* high-pitched 'pseet' (among highest audible calls in bush); vigorous warble higher than Grey. *Nesting:* small neat cup with long tail; 2–3 spotted buff eggs. *Range:* common migrant (Sep–Apr) or resident in rainforest and forest in eastern and northern mainland; vagrant Tas and SA.

Northern Fantail   *Rhipidura rufiventris*   17 cm
Medium-sized inactive grey fantail with dark shafts to tail feathers. *Juv:* white tips to wing coverts, soon lost. Usually observed perched quietly in upright position on prominent twig, particularly in paperbark swamps. Easily told from Grey by larger size, less active habits, less white on eyebrow (may be covered by surrounding feathers), little white edging to wing and less in tail. *Voice:* sweet 'do-do-deed-a-day-do'; metallic 'chuck'. *Nesting:* neat cup with tail, often on slanting twig; 2–3 spotted white eggs. *Range:* uncommon resident or nomad in edges of rainforest, monsoon forest, riverine forest, paperbark swamps and mangroves in northern Aust from about Derby, WA, to Townsville, Qld.

Willy Wagtail   *Rhipidura leucophrys*   20 cm
Large black fantail with white belly and eyebrow. *Juv:* buff margins to wing feathers, buff eyebrow. One of commonest Aust birds, very much at home around human habitations; active, aggressive, fearless. *Voice:* 'sweet pretty creature' often at night in spring; harsh chatter when annoyed. *Nesting:* neat cup without tail; 2–3 spotted white eggs. *Range:* common resident in most habitats except treeless grassland and heavy forest throughout Aust, vagrant Tas.

FANTAILS

GREY

mountain form

southern forms

mangrove form

white-tailed form

RUFOUS

NORTHERN

WILLY WAGTAIL

ad

juv

16 x 13	18 x 14	18 x 14	21 x 16
Grey	Rufous	Northern	Willy Wagtail

WHIPBIRDS AND LOGRUNNERS

Whipbirds: long-tailed, crested birds with stout conical bills, spending most of their time on the ground or in low undergrowth where they feed on insects. Voices are loud and distinctive, some with ventriloquial effect. Nests are simple cups of grass, usually well hidden, eggs are beautiful blue with sparse spotting, look like rare Chinese porcelain. **Logrunners:** live in rainforest where they scratch in litter for food, tossing leaves sideways; tails have stout spines for added support. Bill is small, similar to that of unrelated Pilotbird (p 256) which eats similar food. Calls are loud and pervasive. Nests are very large for size of birds, heaped domes of sticks and moss each with approach ramp; eggs are unusual blunt oval shape.

Western Whipbird   *Psophodes nigrogularis*   20–25 cm
Elusive mallee whipbird with small crest and scratchy ventriloquial call. *Male:* head grey with white patch on each side of black throat, back olive, underparts grey in western mallee, grey with white belly in eastern mallee (race *leucogaster*). *Female:* duller, often lacks white cheek patches. *Juv:* dull grey-brown, more olive on back. *Voice:* scratchy 'its for teacher' from male, followed by 'pick it up' from female. Keeps silent if disturbed. *Nesting:* well-built cup of twigs, bark and grass in spinifex or swordgrass; 2 spotted blue eggs. *Range:* uncommon to rare resident in mallee in north-western Vic, south-eastern SA, Kangaroo I, southern Yorke Peninsula and Eyre Peninsula, SA; and south-west.

Eastern Whipbird   *Psophodes olivaceus*   25–30 cm
Eastern whipbird with prominent crest and loud whipcracking call. *Adult:* head and breast black with prominent white throat patch, back and tail olive-green; tail tipped white over most of range, pale brown in north Qld (race *lateralis*). *Juv:* rusty olive-brown, moults quickly after fledging. *Imm:* dull grey head with diffuse white sides to throat, back dull olive, underparts dull brown. *Voice:* loud whipcrack from male, followed by 'chew chew' from female; quiet chuckling while scratching for food. *Nesting:* sparse open cup of grass, twigs and fern fronds in low shrub, vine or lantana, sometimes conspicuous; 2 pale blue eggs with black hieroglyphics. *Range:* common in rainforest, dense undergrowth in woodland, in eastern Aust.

Chirruping Wedgebill   *Psophodes cristatus*   19–22 cm
Chiming Wedgebill   *Psophodes occidentalis*   19–22 cm
Two closely related almost identical arid country brown whipbirds with prominent brown crests and creaky incessant calls. Chirruping occupies eastern range, has faintly streaked breast, duets chirruping 'tootsie cheer' from male and 'tse-cheer' from female. Chiming occupies western range, has unstreaked breast and has chiming descending call 'did you get drunk', often duetting with female. *Nesting:* untidy cup of grass and twigs in low bush or fork in mulga or lignum bush, 2–3 spotted blue eggs. *Range:* common in mulga woodland, lignum and low shrubland in arid areas, Chirruping in east, Chiming in west.

Chiming

Chirruping

Logrunner   *Orthonyx temminckii*   17–20 cm
Spine-tailed ground bird with streaked back and rufous rump, barred wings. *Male:* throat white. *Female:* throat rufous. *Juv:* wings and tail as adult, remainder rufous-brown strongly barred black. *Voice:* loud resonant 'be-kweek-kweek-kweek . . .' sometimes in concert; soft 'tweet' when feeding. *Nesting:* large dome of sticks, ferns and moss with cap of moss, on or-near ground; 2 white eggs. *Range:* common to rare resident in rainforests from Blackall Ra, Qld, to Illawarra, NSW.

Chowchilla   *Orthonyx spaldingii*   26–28 cm
Spine-tailed dark brown ground bird with pale ring around eye, found in north-eastern rainforests. *Male:* throat and breast white. *Female:* throat cinnamon. *Juv:* similar to juvenile Logrunner but wings plain brown. *Voice:* loud 'chow-chilla', often followed by 'show-show-chowy-chook-chook' sometimes in concert early and late in day. *Nesting:* large dome of sticks and moss on or near ground; 1 white egg. *Range:* common resident in rainforest in north-eastern Qld, from Mt Spec to Cooktown.

WESTERN

western form

eastern form

WHIPBIRDS

EASTERN

ad

imm

juv

CHIRRUPING

CHIMING

CHOWCHILLA

♂

♀

LOGRUNNERS

♂

♀

LOGRUNNER

juv

26 x 19   28 x 20   24 x 17   24 x 17   28 x 20

Western   Eastern   Chiming   Chirruping   Chowchilla 37 x 26   Logrunner

QUAIL-THRUSHES

Quail-thrushes are plump, long-tailed, small-headed ground birds, rising with quail-like whirring of wings when close pressed. Usually avoid disturbance by running away, relying on soil-like coloration of upperparts to evade detection; usually keep back to observer. Most species live on stony ground, most often on ridges; exception is Chestnut which inhabits sandy areas. Calls are high-pitched but carry long distances. Feed on insects and seeds. Nests are similar, on ground usually under overhanging rock, fallen branch or among tree roots, cup-shaped lined with bark and grass; 2–3 spotted eggs. Several species inhabit arid areas; where species overlap in range hybrids may occur. *Note:* individual birds often puff out flank feathers, covering wing patterns.

Chestnut-breasted Quail-thrush   *Cinclosoma castaneothorax*
                                              21–25 cm

Long-tailed cinnamon-backed quail-thrush with cinnamon breast (male). *Female:* very difficult to tell from Cinnamon, but has darker breastband, longer tail and virtually exclusive range. Two widely separated forms occur: eastern form (race *castaneothorax*), darker with richer rufous breast and with feathers of brown upper back and tail coverts having darker centres; and western form (race *marginatum*), with more buff breast and cinnamon more extensive on back, with feathers lacking dark centres. *Voice:* far-carrying, high-pitched repeated whistle 'tee ti-ti tee tee', usually at dawn or evening; single high-pitched note. *Range:* uncommon on stony ridges at east and west extremes of mulga in central WA, western Qld and north-western NSW.

Nullarbor Quail-thrush   *Cinclosoma alisteri*   17·5–19 cm

Male: short-tailed quail-thrush with rich reddish-cinnamon back, black throat and breast, confined to Nullarbor Plain. *Female:* similar to female Cinnamon, with perhaps more restricted eyebrow. *Juv:* sandy cinnamon with faint speckling. Usually in pairs or small parties, difficult to approach. *Voice:* high-pitched piping similar to Cinnamon. *Range:* uncommon in saltbush and samphire on Nullarbor Plain, particularly about better vegetated dongas.

Cinnamon Quail-thrush   *Cinclosoma cinnamomeum*   19–22 cm
Male: short-tailed, cinnamon-backed quail-thrush with white breast. *Female:* difficult to tell from Chestnut-breasted but has paler grey breastband which extends up centre of throat, less dark centres to secondary feathers, shorter tail and virtually exclusive range. *Juv:* pale sandy-cinnamon with sparse dark speckling, unlike rich rufous profusely-speckled young of Chestnut-breasted. Some males (eg in Sturt's Stony Desert) have breast largely cinnamon. *Voice:* high-pitched far-carrying 'peet peet-peet eet eet', most often at dawn. *Range:* uncommon on stony plains, ridges and sandhills in southern central Aust.

Chestnut Quail-thrush   *Cinclosoma castanotum*   22–26 cm

Male: long-tailed, chestnut-backed quail-thrush with black breast; amount of chestnut on back varies with range, generally more extensive and bright in more arid habitat. *Female:* easily told from other quail-thrushes by grey flanks shading into pale umber; chestnut back less extensive than male's, virtually absent in south-eastern mallee and south-west. *Juv:* profusely speckled on head, back and breast. *Voice:* high-pitched almost inaudible 'seep'; penetrating repetitive trilling whistle. *Range:* uncommon in sandy areas in mallee, heath, mulga and woodland in drier areas of southern Aust.

Spotted Quail-thrush   *Cinclosoma punctatum*   26–28 cm

Male: profusely spotted quail-thrush with white ear patch, grey breast, black throat and patch on lower breast extending to spotted flanks. *Female:* throat white to pale buff, chestnut earpatch, grey breast extending to heavily spotted buff flanks. *Voice:* high-pitched 'tseep'; far-carrying repetitive notes similar to White-throated Treecreeper but slower, more plaintive. *Range:* common to uncommon on stony ridges in open forest in east from about Rockhampton, Qld, to Mt Lofty Range, SA.

QUAIL-THRUSHES

CHESTNUT-BREASTED ♀ eastern form

western form ♀

♂

CINNAMON ♀

♂

♂

NULLARBOR

SPOTTED ♀

♂

♀

♂

CHESTNUT

30 x 21	30 x 21	30 x 21	30 x 21	32 x 24
Chestnut-breasted	Nullarbor	Cinnamon	Chestnut	Spotted

BABBLERS

Long-tailed sociable birds with downcurved bills and powerful legs. Much of their food is taken on the ground, but they also prise bark from treetrunks looking for invertebrates. Usually observed in small noisy flocks of up to a dozen birds. Nests are bulky stick globes placed in upright forks or among outer branches of horizontal limbs, with entrances outwards. A number of nests may be built each year, some for roosting rather than for breeding. All babblers have white eyebrows and are basically similar in plumage pattern, but each is easily identifiable.

Chestnut-crowned Babbler   *Pomatostomus ruficeps*   21–23 cm
Chestnut-crowned babbler with white bars in wing. *Juv:* crown brown, eyebrown pale rufous, whitish patch behind eye; white bars in wing distinguish it from White-browed. *Voice:* 'witchee-chee-chee', 'chack-chack'. *Nesting:* large dome of sticks in upright form; 3–5 eggs. *Range:* uncommon in drier inland scrub – mulga, mallee, lignum, etc – in south-west interior from Channel Country and Mitchell Plains, Qld, to north-western Vic and eastern SA.

Hall's Babbler   *Pomatostomus halli*   23–26 cm
Dark-crowned babbler with broad white eyebrow and white breast, sharply demarcated from brown belly. Darker than other babblers, often appearing pied in colour. *Voice:* squeaky chattering. *Nesting:* large dome of sticks, smaller and neater than other babblers, in upright fork of mulga and casuarina, or horizontal eucalypt branch, 3–6 m above ground; 1–3 eggs. *Range:* locally common in mulga and eucalypt-lined watercourses in western Qld and north-western NSW.

White-browed Babbler   *Pomatostomus superciliosus*   20 cm
Dark-crowned babbler with narrow white eyebrow and white breast merging gradually into brown belly. Closest to Hall's but more diffusely patterned and not so dark in plumage, much more widely distributed. *Voice:* variety of calls from wild miaowing to sweet 'wit-wit'. *Nesting:* large untidy dome of sticks in upright fork, horizontal branch or prickly-leafed bush; 2–3 white or buff spotted and lined eggs. *Range:* common resident in open and scrubby woodland, prickly thickets (dryandra) in southern mainland west of Divide and avoiding wetter south-west.

Grey-crowned Babbler   *Pomatostomus temporalis*   25 cm
Large grey-crowned babbler. Two forms occur: grey-breasted form (race *temporalis*) in eastern Aust; and red-breasted form (race *rubeculus*) in northern and north-western Aust. *Voice:* loud 'yahoo', 'peeoo-peeoo . . .'. *Nesting:* large untidy dome of sticks, lined with grass, bark, wool, etc, 3–6 m above ground; 2–6 eggs. *Range:* common in open forest, woodland and scrubland, in eastern and northern Aust, becoming rarer in settled areas.

BABBLERS

CHESTNUT-CROWNED ad juv

HALL'S

WHITE-BROWED

GREY-CROWNED

grey-breasted form

red-breasted form

26 x 18 Chestnut-crowned

22 x 16 Hall's

24 x 17 White-browed

28 x 20 Grey-crowned

WARBLERS OF LONG GRASS AND REEDS

A number of similar warblers live in reeds. Some are notoriously difficult to identify, with probably only song reliable in field. In hand, relative lengths of primaries (wing formulae) are used. The common reed-warbler in Aust has a similar wing formula to Great, not like Clamorous of north-eastern Africa and south-eastern Asia, with which it has more recently been allied. Until this is investigated, it is preferable to retain old designation of Australian Reed-Warbler *Acrocephalus australis*. Grassbirds have streaked plumage on backs, can be identified easily by calls. Spinifexbird is arid-country grassbird, has unusually long tail coverts.

Gray's Grasshopper Warbler *Locustella fasciolata* 18 cm
Rarely-sighted vagrant warbler like a dark reed-warbler found in undergrowth. Upperparts greyish-brown with white eyebrow and rusty rump; underparts brownish-grey, more olive on flanks, paler on throat and centre of abdomen. Shy, usually keeps hidden in undergrowth. *Voice:* loud 'top-pin, kat-kat'. *Range:* migrant from central and eastern Asia to Philippines, Celebes and New Guinea; one record in Jan 1982, from Cumbungi reedbed at Harrison Dam, Humpty Doo, NT, likely in any tangled undergrowth in north.

Great Reed-Warbler *Acrocephalus arundinaceus* 19 cm
Similar to Australian Reed-Warbler with only very slight differences; some are slightly warmer in colour and more rusty buff on flanks with faint streaking on breast, but not consistently enough for reliable field identification; if direct comparison possible, Great has stouter shorter bill, looks larger; thus very difficult to identify unless voice is heard: harsh grating 'kret-kret shee-kaa-kaa-kaa' lower and more guttural than Australian, but unfortunately it does not sing often while in Aust. *Range:* rare summer migrant from Eurasia to reedbeds in northern Aust, mainly Kimberley and Top End.

Australian Reed-Warbler *Acrocephalus australis* 17 cm
Unstreaked reed bird with pale eyebrow, olive-brown back, pale buff underparts. *Adult:* lores pale. *Juv:* lores dark. Occurs in most large patches of reeds in Aust, usually well hidden but loud melodious voice betrays presence. *Voice:* rich 'twitchy twitchy twitchy quartz quartz quartz'; sharp 'cheet'. *Nesting:* deep cup of grass, dried reed sheaths, etc, supported by 3–4 reed stems (sometimes willows); 3–4 spotted and blotched bluish or brownish eggs. *Range:* common resident or migrant in reedbeds throughout mainland Aust and Tas.

Tawny Grassbird *Megalurus timoriensis* 19 cm
Unobtrusive grassbird with rufous streaked back, rich rufous rump and long drooping tail, unstreaked underparts. In summer male sings on prominent perch, gives display flight. *Voice:* rich 'ch-ch-ch-zzzzzzt lik lik'; loud 'see-lick'; high-pitched descending trill; metallic 'chuck chuck chuck . . .' often in flight. *Nesting:* well-hidden cup of grass in long grass; 3 freckled reddish eggs. *Range:* common resident or nomad in moist grassland and heath in eastern and northern Aust.

Little Grassbird *Megalurus gramineus* 14 cm
Unobtrusive grass and reed bird with streaked grey-brown back slightly richer on rump, small streaks on pale grey breast. Much harder to sight than Tawny, but reveals presence by mournful song. *Voice:* plaintive three-noted monotone, 'tee-ti-teee'. *Nesting:* deep lined cup of grass and reed scraps with one or two feathers over entrance; 3–5 freckled white or pinkish eggs. *Range:* common resident or nomad in reeds, long grass and dense undergrowth in eastern and south-western Aust and eastern and northern Tas.

Spinifexbird *Eremiornis carteri* 15 cm
Reddish-brown long-tailed grassbird confined to spinifex in arid areas of Aust. Usually seen perched above spinifex with drooping tail, sometimes cocked, or flying with tail-heavy look. *Voice:* pleasant warbling 'te tee te too'; grating 'chuk'; 'thrip thrip'; sharp 'tik'; high-pitched almost inaudible 'tsee'. *Nesting:* compact cup of grass and rootlets in spinifex clump; 2 speckled pinkish-white eggs. *Range:* common to uncommon resident in spinifex in arid Aust and islands off north-west coast.

WARBLERS

GOLDEN-HEADED

CISTICOLAS
(see overleaf)

GRAY'S

ZITTING

GREAT

dark-lored form

AUSTRALIAN

pale-lored form

TAWNY

LITTLE

SPINIFEX

16 x 12	15 x 12	20 x 14	20 x 15	19 x 14	18 x 12
Golden-headed	Streaked	Australian	Tawny	Little	Spinifex

CISTICOLAS AND EMU-WRENS

Cisticolas: small streaked warblers often called fantail-warblers because of their habit of flirting their tails, particularly when flushed. In summer they are very obvious with males singing on prominent perches or making spectacular display flights; in winter they rarely emerge from grasses. Male polygamous; female builds beautiful down nest in basket of leaves sewn together with cobwebs. **Emu-wrens:** tiny birds with unique long barbless tails of only six feathers, inhabiting heath or spinifex, usually in small parties of one or two blue-throated males and a number of females and juveniles. They keep well under cover, feeding on insects. The three forms are regarded variously as one, two or three species.

Zitting Cisticola   *Cisticola juncidis*   10 cm
Small streaked grassbird with greyish-white underparts and white tip to tail. In winter (Apr–Aug) tail longer, generally less obvious. *Male:* less streaking on crown than female. Similar to Golden-headed, which is more buff-looking and has buff tail-tip. *Voice:* loud metallic 'zit-zit' or 'lik-lik' (mainly Oct–Mar). *Nesting:* deep purselike cup of felted plant-down, spiderweb and grass, with entrance at top, placed in long grass; 3–4 spotted pale blue eggs. *Range:* locally common resident in swampy grasslands including canefield edges in north-eastern and northern Aust.

Golden-headed Cisticola   *Cisticola exilis*   10 cm
Small streaked grassbird with buff underparts and buff tip to tail; in winter (Apr–Aug) tail longer. *Male:* head golden in summer, streaked in winter like female. *Female:* crown streaked black. *Juv:* more heavily streaked above, pale yellow below. *Voice:* buzzing 'zzzt' often followed by double 'lik'; 'keet keet . . .'. *Nesting:* dome of grass, plant-down and spiderweb, stitched to leaves in long grass, entrance in side near top; 3–4 blotched blue eggs. *Range:* common resident in grasslands and crops in eastern, northern and north-western Aust, also King I.

Mallee Emu-wren   *Stipiturus mallee*   15–17 cm
Male: small, short-tailed emu-wren of mallee spinifex. *Female:* forehead rufous, nape streaked olive; younger females have brown to grey eyebrows and ear coverts which become blue with age. *Imm male:* pale blue or grey throat. Where range overlaps Southern in south-eastern SA, different habitats best pointer to species: Mallee has shorter, more densely filamentous tail but usually difficult to view clearly. Becoming extremely rare; observers should minimise disturbance during breeding period. *Voice:* high-pitched tinny reel; almost inaudible 'tseep'. *Nesting:* dome-shaped with side entrance, grass, rootlets and plant down, placed in spinifex clump. *Range:* uncommon resident in spinifex in mallee in south-east SA and north-west Vic.

Rufous-crowned Emu-wren   *Stipiturus ruficeps*   14–15 cm
Male: small, short-tailed emu-wren of desert spinifex. *Female:* forehead, crown and nape rufous, throat yellowish-buff. *Imm male:* similar to female but throat pale blue or grey. *Voice:* high-pitched tinny reel. *Nesting:* similar to Mallee. *Range:* common in desert spinifex from north-western Aust to central western Qld.

Southern Emu-wren   *Stipiturus malachurus*   15–18 cm
Male: comparatively large, long-tailed emu-wren of heath swamps and sandplain. *Female:* streaked olive crown. *Imm male:* pale blue or grey throat. Birds in eastern Aust have buff streaks on ear coverts; in WA streaks are white. Generally birds in damper heaths are richer in colour. Where range marginally overlaps Mallee Emu-wren in south-eastern SA, different habitat good pointer to species. *Voice:* high-pitched tinny reel; 'tseeep', inaudible to some observers. *Nesting:* dome-shaped nest with side entrance, grass, rootlets, moss, lined with fine grass, decorated with spiders' eggsacs, close to ground in tussock; 2–3 eggs. *Range:* uncommon in heaths, swampy vegetation and sandplain in south from near Gympie, Qld, to Shark Bay, WA.

SYLVIIDAE, MALURIDAE

CISTICOLAS

ZITTING

non-br

br

GOLDEN-HEADED

non-br

juv

♀

♂

EMU-WRENS

RUFOUS-CROWNED

♀

♂

MALLEE

♀

♂

SOUTHERN

imm ♂

♀

♂

western form

16 x 12   16 x 12   17 x 12

Mallee   Rufous-crowned   Southern

FAIRY-WRENS

Males are strikingly coloured, females and young males are plain greyish- or bluish-brown, although most have blue tails. Males moult completely after breeding season into brown eclipse plumage, then moult again into full plumage in one to six months depending on age; very old males may miss eclipse altogether. Usually found in small parties of up to eight birds, most in brown plumage, feeding on ground or in low vegetation. On this page are species whose males do *not* have a chestnut or red patch on shoulder.

Superb Blue Wren   *Malurus cyaneus*   14 cm
Male: only white-bellied blue fairy-wren without chestnut shoulders. *Female and juvenile male:* brown tail, pale bill and buff patch around eye. *Male eclipse:* tail blue, bill black, face grey-brown. *Voice:* soft sharp 'trrt'; trilling 'trrreee'; full-throated gushing song with downward inflection. *Nesting:* substantial dome of grass, bark, rootlets and spiderweb, well hidden in grass, low bush or reeds; 3–5 speckled pinkish eggs. *Range:* common resident in patchy undergrowth in forest and woodland giving both cover and access to open space for feeding, throughout south-east and Tas.

Purple-crowned Wren   *Malurus coronatus*   14–15 cm
Male: only fairy-wren with black mask and lilac crown. *Female:* only fairy-wren with chestnut earpatch. *Juv:* duller with longer white-edged tail. *Eclipse male:* most of lilac crown is lost but some remains on forehead; black on head confined to earpatch. Two forms occur: western form (race *coronatus*) with buff breast; and eastern form (race *macgillivrayi*) with creamy-white breast. *Voice:* soft high-pitched chirp; plaintive 'sweeet'; alarm call, loud 'churrr'; song unlike other fairy-wrens, extended bubbling. *Nesting:* bulky untidy dome of grass, bark and rootlets decorated with leaves in crown of pandanus or grass clump. *Range:* locally common in pandanus and canegrass-lined streams from Macarthur R, NT, to Leichhardt R, Qld (race *macgillivrayi*); rare and vanishing in similar habitat from Fitzroy R and Upper Drysdale R, WA, to Wattle Creek, NT (race *coronatus*).

Splendid Wren   *Malurus splendens*   14 cm
Male: mainly blue or purplish blue. *Female:* brown above, bright blue tail, chestnut patch around eye. *Juv:* tail browner, legs paler. *Imm male:* blue-green wash on wings. *Eclipse male:* like female but bill black, face dusky grey-brown, blue-green wash on wings. Three forms occur: western form (race *splendens*), underparts rich bluish-purple, no black patch on lower back; turquoise form (race *callainus*), throat bluish-purple, abdomen blue, black patch on lower back; and black-backed form (race *melanotus*), throat and abdomen blue, black patch on lower back. *Voice:* soft 'trrrp'; 'treeee'; high-pitched trilling alarm 'ts-s-s-s-s . . .'; rich gushing reel. *Nesting:* untidy dome of grass, leaves, bark and fur well hidden in low bush or among dead twigs; 2–4 speckled blotched pinkish-white eggs. *Range:* common resident in undergrowth from eucalypt forests of south-west to mulga woodland and mallee in east.

White-winged Wren   *Malurus leucopterus*   12–13 cm
Male: blue or black with white wings; blue birds vary in intensity of colour (perhaps due to age, or to fading). *Females and young males:* palest of all wrens with powder-blue tails. Each group has only one fully-plumaged male which does not have an eclipse plumage. Two forms occur: blue and white form (race *leuconotus*) on mainland; and black and white form (race *leucopterus*) on Barrow I and Dirk Hartog I, WA. *Voice:* very high-pitched 'tsit-tsit . . .'; trilling 'tsrrreeee'; alarm call 'trrit-trrit'; song is tinny reel like toy sewing-machine. *Nesting:* neat compact dome of grass, plant-down and spiderweb hidden close to ground in small bush; 2–4 sparsely freckled creamy-white eggs. *Range:* common on arid and semi-arid plains of Aust; also Barrow I and Dirk Hartog I, WA.

FAIRY-WRENS

PURPLE-CROWNED

♀

SUPERB

♂

♂

♀

SPLENDID

western form

turquoise form

black-backed form

eclipse ♂

fresh ♂

worn ♂

black and white form

WHITE-WINGED

♀

17 x 12	17 x 12	17 x 12	15 x 12	15 x 12
Superb	Purple-crowned	Splendid	White-winged (blue and white)	White-winged (black and white)

WRENS WITH CHESTNUT SHOULDER PATCHES

Lovely Wren   *Malurus amabilis*   14 cm

Tropical arboreal fairy-wren of rainforest verges. *Male:* similar to Variegated but with rounder ear covert same colour as crown, shorter broader tail edged white. *Female:* head, back and tail dull blue, paler on ear coverts; face white, bill black. Two forms occur: violet-headed form (race *amabilis*) of Cape York Peninsula; and blue-headed form (race *barroni*) from Cooktown to about Townsville. *Voice:* short rippling warble; high-pitched 'treeee'; churring alarm call. *Nesting:* untidy dome of grass and bark, bound with cobwebs; 2–4 sparingly freckled creamy-white eggs. *Range:* common in verges of coastal and subcoastal rainforest and vine scrub in north-eastern Qld from Edward R to just north of Townsville.

Blue-breasted Wren   *Malurus pulcherrimus*   15 cm

Male: uniform blue crown and ear coverts, dark blue breast, white flanks. *Female:* similar to female Variegated but greyer on breast, bill brown. *Male eclipse:* similar but with black bill. Male Variegated has black breast and ear coverts different blue to crown, generally in more open habitat. *Voice:* metallic reel, tinnier than Variegated; trilling 'trrrree'. *Nesting:* neat compact dome of grass and bark bound with cobweb; 2–3 blotched creamy-white eggs. *Range:* patchy distribution in heath and woodland thickets, particularly prickly poison peas, from Shark Bay to Eucla, avoiding forested areas; and on Eyre Peninsula, SA.

Red-winged Wren   *Malurus elegans*   15 cm

Largest and longest-tailed fairy-wren, confined to forested areas in south-west. *Male:* silvery-blue crown and ear coverts, deep-blue breast, buff flanks, duller chestnut wing patch than others. *Female:* blue-grey head, brown back, black bill. *Voice:* vigorous reeling warble; trilling 'treeee'. *Nesting:* large untidy dome of grass, leaves and bark close to ground in swordgrass or shrub; 2–4 blotched and spotted white eggs. *Range:* common resident in well-vegetated gullies and hillsides in forests, particularly tea-tree, swordgrass and bracken undergrowth in south-west.

Variegated Wren   *Malurus assimilis*   15 cm

Most widespread fairy-wren, found all over mainland except Cape York, lower south-east and south-west. *Male:* crown darker blue than ear coverts, blue patch where flank meets black breast. Two basic forms occur: (a) forms with grey-brown females (races *assimilis* west of Divide, and *lamberti* east of Divide) and rather variable males; and (b) rock-haunting forms with blue-grey females (race *dulcis* with white face in Top End, and race *rogersi* with chestnut face in Kimberley). Males are darkest blue in south-east mallee, looking much like Blue-breasted except for black breast and buffish flanks; brightest blue in Top End and Kimberley. *Voice:* mechanical reeling warble; trilling 'trrrt'. *Nesting:* untidy dome of grass and bark in tussock, shrub or dead fallen branch; 3–4 spotted blotched white eggs. *Range:* common resident in thickets of shrubs throughout Aust, avoiding Cape York, lower south-east and south-west.

Red-backed Wren   *Malurus melanocephalus*   12–13 cm

Smallest fairy-wren, found in tropical and subtropical woodland with grass understorey. *Male:* black with red back. *Female and eclipse male:* plainest coloured wren with no blue in tail and with no distinctive colour on face; bill brown. Two forms occur: scarlet-backed form (race *melanocephalus*) in eastern Aust from about Atherton, Qld, to Port Stephens, NSW; and crimson-backed form (race *cruentatus*) from about Pardoo, WA, to Atherton, Qld. *Voice:* tinny reel; soft 'tsst'. *Nesting:* neat dome of grass in tussock or small shrub; 3–4 spotted white eggs. *Range:* common resident or short-range nomad in grasslands throughout woodlands from Cape Keraudren, WA, to Port Stephens, NSW.

FAIRY-WRENS

violet-headed form ♂

LOVELY

blue-headed form ♂

♀

♀

BLUE-BREASTED

♂

♀

RED-WINGED

♂

assimilis

VARIEGATED

lamberti

lavender-flanked form

Kimberley form ♀

♂

♀ NT form

scarlet-backed form

♂

♂

crimson-backed form

RED-BACKED

♀

17 x 12	17 x 12	17 x 12	17 x 12	17 x 12
Lovely	Blue-breasted	Red-winged	Variegated	Red-backed

DARK GRASSWRENS

Large, coarsely-feathered, ground-dwelling wrens inhabiting spinifex, either on sandy plains or rocky hills, shrub steppe (saltbush and bluebush plains), lignum or canegrass and other vegetation on desert sandhills. Generally observed (with difficulty) in family parties, may form larger groups outside breeding season. Although not particularly shy, secretive by nature, keeping under cover when disturbed, darting rapidly from bush to bush like a bouncing ball if hard-pressed, losing an intruder with ease. Feed on insects and seeds. Nests are untidy domes of grasses, placed in bush or grass. Some forms at present regarded as subspecies could be full species, eg Horton's form of Dusky, and western form of Thick-billed.

Dusky Grasswren   *Amytornis purnelli*   16–17·5 cm

Dark cinnamon-brown grasswren with thin bill, found on rocky hills. *Male:* cinnamon-brown streaked white, darker on crown and paler below, rump tinged rufous. *Female:* similar but with rufous flanks. Two forms occur: Purnell's form (race *purnelli*) with fine white streaks on rufous-tinged breast, and greyish-brown flanks in male; and Horton's form (race *ballarae*) more reddish above and with dusky streaking on straw-coloured breast and grey flanks in male, confined to Mt Isa area. Similar to Thick-billed but darker and with thinner bill, found in different habitat, most of ranges exclusive. Told from reddish grasswrens (overleaf) by lack of prominent black 'moustache'. *Voice:* high-pitched silvery song 'sree-sree-sree teu teu'; high-pitched 'seet'. *Nesting:* loosely-built grass cup with half dome in spinifex clump or low bush; 2–4 spotted and blotched pinkish eggs. *Range:* locally common but often difficult to find on spinifex-covered rocky hills and gorges in central Aust and around Mt Isa, Qld.

Thick-billed Grasswren   *Amytornis textilis*   16–18 cm

Cinnamon-brown grasswren with thick bill, found on saltbush–bluebush plains. *Male:* faint dark moustache, lores tinged rufous, flanks pale cinnamon-brown, centre of abdomen white. *Female:* similar but with rufous flanks. Two forms occur: short-tailed central form (race *modestus*), underparts pale fawn only faintly streaked white; and long-tailed western form with underparts fawn prominently streaked (race *textilis* in WA and race *myall* on Eyre Peninsula, SA). Lives in more open habitat than other grasswrens, thus flies more often, easiest to see, but vanishing in range. *Voice:* high-pitched squeak; silvery recurrent song from top of bush, 'swit-swit-swit-swee'. *Nesting:* loose grass cup with wispy dome in bush or grass clump; 2–3 blotched white or pink eggs. *Range:* locally common but vanishing on plains of saltbush, bluebush, cottonbush, canegrass and flood debris on creeks in central and mid-western Aust and northern Eyre Peninsula, SA.

White-throated Grasswren   *Amytornis woodwardi*   20–22 cm

Male: large black grasswren with white throat and breast, flanks and abdomen tawny, found among tumbled rocks of western Arnhem Land. *Female:* flanks and abdomen rufous. A busy bird with self-important air, often cocking tail, usually in small parties. *Voice:* wren-like trill, deeper and slower; sharp 'tsst'. *Nesting:* loose dome shape, of grass, bark and leaves placed in spinifex; 2 speckled and blotched pinkish eggs. *Range:* sandstone escarpment of Arnhem Land, NT.

Black Grasswren   *Amytornis housei*   21 cm

Male: large black grasswren, found among tumbled rocks on western escarpment of Kimberley Plateau. *Female:* flank and abdomen rufous. Cocks its tail less than other grasswrens, often looks like black rat scurrying among rocks and spinifex. Moves in small groups, 4–8. *Voice:* wren-like reel lower and harsher; loud ticking. *Range:* western escarpment of Kimberley Plateau, WA, from Mitchell R to Mt Barnet Station.

GRASSWRENS

DUSKY
Purnell's form
♀
♂
Horton's form
♀
♂

THICK-BILLED
central form ♀
♂
western form ♀
♂

WHITE-THROATED ♀
♂

BLACK
♀
♂

21 x 16
Dusky

21 x 16
Thick-billed

23 x 16
White-throated

REDDISH GRASSWRENS

These grasswrens have reddish or greyish-cinnamon plumage on back, with pale underparts and black 'moustache'. Favoured habitat is spinifex or canegrass, but Grey is only grasswren occupying swampy country.

Grey Grasswren *Amytornis barbatus* 18–20 cm
Long-tailed greyish-cinnamon grasswren with distinctive facial pattern; appears grey at a distance, confined to lignum and canegrass swamps on Goyder's Lagoon and Bulloo R Overflow (Bullerine). *Voice:* soft, high-pitched 'trip-ip-ip' or 'tsi-tsi-tist'; alarm note 'seep'. *Nesting:* loose, bulky dome of grass in clump of lignum or canegrass; 2 freckled or blotched white or pale pink eggs. *Range:* rare resident or local nomad in lignum and canegrass swamps at Goyder's Lagoon, SA, and the Bullerine on border of NSW and Qld; other suitable swamps occur between these localities.

Eyrean Grasswren *Amytornis goyderi* 14 cm
Reddish grasswren with stout finch-like bill and white underparts, found in canegrass and spinifex on dunes of Simpson Desert. *Female:* similar but with rufous flanks. *Juv:* browner above with faint streaking, flanks and abdomen washed rufous. *Voice:* cricket-like trilling 'tsee-ee tsee-ee'; high-pitched 'zeet'; silvery gushing song. *Nesting:* cup of rootlets, grass and leaves with sparse dome in tussock; 2–3 spotted and blotched white eggs. *Range:* locally common to rare, depending on season, in canegrass, spinifex and dune pea hummocks on sand dunes in Simpson Desert.

Carpentarian Grasswren *Amytornis dorotheae* 17 cm
Reddish grasswren with dark head contrasting with white throat, found on spinifex-covered rocks on Carpentarian sandstones. *Male:* flanks, belly and undertail coverts rich buff. *Female:* flanks, belly and undertail coverts rufous. *Voice:* cricket-like 'ssstz' or 'tzzt'; extended wren-like song 'ssst ssstz, seeze-ooo chrr chree, trrr trrr, chrrrp chrrrp chuck, chrrrp chrrrp chick will'. *Nesting:* bulky dome of grass, leaves and spinifex in top of spinifex hummock; 2–3 spotted and blotched pinkish eggs. *Range:* rare resident in spinifex-covered sandstone hills and gorges from Buckalara Ra, NT, to China Wall, Qld.

Striated Grasswren *Amytornis striatus* 15–17·5 cm
Reddish grasswren with thin bill, white throat and buff or reddish abdomen, found in spinifex in arid areas. *Female:* rufous flanks. Several forms occur: mallee form (race *striatus*) with greyish-brown back, found in spinifex in mallee from central NSW and north-western Vic to northern edge of Nullarbor; short-tailed form (race *merrotsyi*) with reddish-brown back, found in spinifex on rocky hills in northern Flinders Ra; sandhill form (race *oweni*) with reddish back having white streaks faintly margined darker, found in spinifex on sandhills in WA and NT; and Pilbara form (race *whitei*) with bright rufous back having black margins to white streaks, rusty abdomen and undertail coverts, found in spinifex on rocky hills in north-west. *Voice:* high-pitched trilling 'tseeee'; rich extended wren-like warble from perch during breeding. *Nesting:* substantial dome of grass, bark and spinifex in spinifex clump; 2–3 sparsely spotted and blotched white or pinkish eggs. *Range:* locally common to rare according to seasons in spinifex, either in mallee, on sandhills or on rocky hillslopes in arid areas.

GRASSWRENS

GREY

EYREAN

♀

♂

CARPENTARIAN

♂

♀

STRIATED

♀ mallee form

short-tailed form

♂

sandhill form

Pilbara form

19 x 15

19 x 14

23 x 16

21 x 16

Grey

Eyrean

Carpentarian

Striated

SCRUBWRENS WITH PROMINENT 'EYEBROWS'

Scrubwrens are thicket-frequenting warblers which spend much of their time on the ground where they take most of their insect food. They tend to be noisy, revealing their presence in dense undergrowth, and can generally be attracted into view by 'squeaking'. Nests are bulky dome-shaped structures lined with feathers sited in a variety of situations; most are well hidden under leaves, in grass or in small bushes, but two species use obvious hanging nests on lawyer-vine tendrils and one builds in sandstone caves.

White-browed Scrubwren *Sericornis frontalis* 11–14 cm
A variable white-browed scrubwren with dark lores and pale yellow eye. Several forms occur: spotted form (race *maculatus*) with breast spotted or streaked, occurs west of Spencer Gulf, SA; south-eastern form (race *frontalis*), has black lores in male, brown in female, underparts pale cream; Tasmanian form (race *humilis*), is more evenly coloured brown with less obvious eyebrows; buff-breasted form (race *laevigaster*), has black lores and ear coverts in male, black lores and brown ear coverts in female, underparts buff, occurs in Queensland and north-eastern NSW; McPherson Ra form like buff-breasted form but has dark underparts. *Voice:* rich, varied, including mimicry: 'ts-cheer . . .', 'sit you sit you . . .' (second syllable higher). *Nesting:* dome of grass, leaves and bark well hidden under leaves, in grass or thick bushes; 2–3 pale buff spotted eggs. *Range:* common resident in thick undergrowth in eastern and southern Australia from about Cairns to Shark Bay, WA; and in Tas.

Scrub-tit *Sericornis magnus* 11 cm
Tasmanian scrubwren with grey lores and obscure whitish eyebrow, light brown eye, tail with dark subterminal band and whitish tip. White-browed Scrubwren in Tas has darker lores, paler eye, more uniformly brown plumage, lacks black bar in tail; Tasmanian thornbills lack prominent white throat. *Voice:* 'to-wee-too' with whistling variations: experience required to differentiate from thornbills and White-browed. *Nesting:* dome of grass, bark, fern fibre, hidden on or near ground in undergrowth; 3–4 white speckled and blotched eggs. *Range:* common resident in undergrowth, fern gullies and forest in Tas.

Yellow-throated Scrubwren *Sericornis citreogularis* 13–15 cm
Yellow-throated scrubwren with dark eye. *Male:* facial mask black. *Female:* facial mask brown. Quiet and tame, usually seen hopping on rainforest floor or along forest tracks. *Voice:* sharp 'tick'; beautiful varied song including mimicry. *Nesting:* obvious bulky football-shaped dome suspended on vine tendril; 3 pink or brown eggs. *Range:* common resident in upland rainforest, forest and dense gullies in eastern Aust between Cooktown and Townsville, Qld; and between Bunya Mts, Qld, and Bega, NSW.

Fernwren *Crateroscelis gutturalis* 12–14 cm
Dark tropical scrubwren with white throat and eyebrow and black upper breast. *Juv:* eyebrow, throat and upper breast brown. Usually seen on forest floor, often burrows noisily among leaves looking for food. *Voice:* repeated high 'peep'; single whistle; staccato scolding. *Nesting:* well-hidden dome of moss, leaves and ferns, often under creek bank; 2 white eggs, sometimes speckled. *Range:* common to uncommon in upland rainforest in north-eastern Qld between Cooktown and Paluma Ra.

SCRUBWRENS

buff-breasted form ♂

♀

WHITE-BROWED

♂

south-eastern form

♀

spotted form

McPherson Ra form

Tas form

SCRUB-TIT

YELLOW-THROATED

♀

♂

FERNWREN

juv

♀

♂

19 x 14	26 x 18	18 x 14	23 x 17
White-browed	Yellow-throated	Scrub-tit	Fernwren

SCRUBWRENS WITHOUT PROMINENT 'EYEBROWS'

These are generally plainly coloured scrubwrens. Large-billed and Atherton are easily confused where they co-exist, best told by relative prominence of eye in facial plumage. Rock Warbler and Pilotbird are aberrant scrubwrens generally classified as 'sandstone warblers' as they are confined to sandstone areas of south-east. Redthroat is an arid-country scrubwren.

Tropical Scrubwren *Sericornis beccarii* 11 cm

Small red-eyed scrubwren confined to Cape York. Two forms occur: northern form (race *minimus*) has yellowish underparts, more pronounced facial pattern, white tips on wing coverts; southern form (race *dubius*) is more cinnamon-looking with obscure facial pattern and buff tips on wing coverts, rather similar to the more arboreal Large-billed, which lacks buff tips on wing coverts. *Voice:* similar to White-browed. *Nesting:* dome of vine tendrils, rootlets and leaves well hidden on or near ground or suspended on vine; 2–4 pinkish freckled eggs. *Range:* common resident in rainforest and vine scrub on Cape York, Qld, south to about Cooktown.

Atherton Scrubwren *Sericornis keri* 13·5 cm

Plain-coloured scrubwren with dark face and relatively obscure eye, reddish-brown tail, centre of underparts paler than sides of breast and flanks, confined to Atherton Tablelands. Usually seen on or near rainforest floor. Arboreal Large-billed has paler face, more prominent eye. *Nesting:* large dome of grass and leaves hidden in vegetated bank; 2 speckled white or brownish eggs. *Range:* uncommon resident in upland rainforest on Atherton Tableland, Qld.

Large-billed Scrubwren *Sericornis magnirostris* 12–13 cm

Plain-coloured scrubwren with prominent eye in pale buff face, most often seen in trees and shrubs fluttering after insects like warbler or thornbill. Bill has unusual shape, with straight upper edge and curved lower edge, giving a slightly 'Donald Duck' look. Atherton has darker face with less obvious eye, more reddish tail, bill usual shape with curved upper edge, usually seen low in forest or on ground. *Voice:* musical sibilant twitter 'tsip tsip . . .'; harsher scolding 'si scheer s-cheer . . .'. *Nesting:* either uses Yellow-throated nest or builds own dome of bark, grass and moss in swordgrass, bushes, etc; 3–4 spotted white or brownish eggs. *Range:* common resident in rainforest and wet sclerophyll forest in east from Cooktown, Qld, to Dandenong Ra, Vic.

Redthroat *Sericornis brunneus* 12 cm

Greyish scrubwren with white-tipped black tail, found in dry-country undergrowth. *Male:* throat pale rufous; *female:* throat grey. Usually keeps under cover but when disturbed flutters loudly away revealing white-tipped tail. *Voice:* rich varied song based on full-throated 'wheet wheet widda widda wheet whee'. *Nesting:* bulky dome of coarse strips of bark, grass and feathers placed in low bush; 3–4 reddish-brown eggs. *Range:* uncommon in inland undergrowth particularly in mulga, mallee and bluebush in southern and central Aust.

Pilotbird *Pycnoptilus floccosus* 17 cm

Large reddish-brown scrubwren with forehead and breast scaly rufous found in south-eastern temperate wet forest. Usually seen hopping on forest floor, flicking tail up and down; runs quickly when disturbed. *Voice:* soft 'tu-eewit'; male, loud 'guinea-a-week'; female answers 'whit a wit wee'. *Nesting:* bulky dome of twigs, bark and leaves, hidden in vegetation on or near ground. *Range:* common resident in rainforest and wet sclerophyll forests in south-east from about Port Hacking, NSW, to Wilson's Promontory, Vic.

Rock Warbler *Origma solitaria* 14 cm

Rock-haunting scrubwren with grey throat and rufous underparts, confined to Hawkesbury sandstone. *Juv:* throat buff. Usually seen hopping actively over rocks, flicking tail, investigating crevices. *Voice:* loud, sad 'goodbye-goodbye . . .'; harsh scolding. *Nesting:* bulky dome of rootlets, bark, moss and grass hung in cave or under overhanging rock; 3 white eggs. *Range:* common to uncommon resident in suitable Hawkesbury sandstone and nearby limestone in central eastern NSW from about Lake Macquarie to Bermagui.

SCRUBWRENS

southern form

TROPICAL

♀

northern form

♂

LARGE-BILLED

ATHERTON

♂

♀

REDTHROAT

PILOTBIRD

ROCK WARBLER

19 x 14   19 x 14   19 x 14

Tropical   Atherton   Large-billed

Redthroat   Rock   Pilotbird
19 x 14   20 x 15   28 x 20

SCRUBWRENS WITH STREAKED BREASTS

All of these except Speckled carry tail cocked. All have white eyebrow. They have been variously known as heathwrens, fieldwrens, calamanthus, and hylacolas – it is probably best to call them all scrubwrens. Some prefer to regard Chestnut-rumped and Shy as one species and also combine Striated and Rufous.

Speckled Warbler   *Chthonicola sagittata*   12 cm
Heavily streaked scrubwren with whitish face, usually found in open woodland. *Male*: black streak above white eyebrow; *female*: reddish-brown streak above white eyebrow. *Juv*: not so heavily streaked. Usually observed hopping among grass, tussocks, litter and rocks in open forest, often with other small ground-feeding birds. *Voice*: pleasant warble with falling cadence. *Nesting*: dome of grass, leaves, bark and moss concealed under tussock, bush or fallen twigs; 3–4 reddish chocolate eggs. *Range*: locally common in open woodland in east and south-east from about Rockhampton, Qld, to western Vic.

Chestnut-rumped Heathwren   *Sericornis pyrrhopygius*   13–14 cm
Cock-tailed scrubwren with unstreaked back, chestnut rump, faint streaking on underparts and no white patch in wing. Similar to Shy which is brighter and bolder in colour, more heavily streaked and has white patch in wing. Usually keeps under cover, feeding on ground; song is often first indication of presence. *Voice*: beautiful swelling canary-like song incorporating mimicry. *Nesting*: well-hidden dome of bark, rootlets and grass on or near ground; 3 freckled pink eggs. *Range*: common to uncommon resident in undergrowth in dry sclerophyll forest, woodland and heath in south-east from Cunningham's Gap, Qld, to south-east SA.

Shy Heathwren   *Sericornis cautus*   12–14 cm
Cock-tailed scrubwren with unstreaked back, chestnut rump, bold streaking on underparts and white patch in wing. Chestnut-rumped is less boldly marked, lacks white patch in wing; Rufous has streaked back, lacks chestnut rump. Usually keeps under cover but will respond to 'squeaking'. *Voice*: rich 'shee-shee-chick a dee' with variations and mimicry. *Nesting*: concealed dome of grass, bark and twigs on or near ground; 2–3 freckled pinkish eggs. *Range*: uncommon resident in mallee undergrowth in southern Aust from about West Wyalong, NSW, to Murchison R, WA.

Striated Fieldwren   *Sericornis (Calamanthus) fuliginosus*   13 cm
Cock-tailed scrubwren with streaked greenish-olive back and brown eye, confined to heath and moors in south-east and Tas. Rufous has streaked fawn or rusty back and cream eye, occupies range more westerly and inland. Usually keeps to cover but ascends into open to sing, often with tail depressed; when flushed shows white-tipped tail with dark subterminal bar. *Voice*: rich musical phrases less sepulchral than Rufous. *Nesting*: concealed dome of grass under grass tussock or shrub; 3–4 freckled buff to pale chocolate eggs. *Range*: uncommon to common resident in heaths, moors and buttongrass plains in south-east and Tas.

Rufous Fieldwren   *Sericornis (Calamanthus) campestris*   12–14 cm
Cock-tailed scrubwren with streaked sandy-buff to rufous back and cream eye. Coloration varies with aridity; dry-country birds tend to be more rufous; south-western birds are greyer, almost pipit-like in colour. Striated has more easterly coastal range, has streaked greenish-olive back, darker eye. Often flushed along outback roads, showing white-tipped tail with dark subterminal band. (Redthroat shows wholly black tail with broad white tip.) *Voice*: musical 'Whirr-whirr-chick-chick-whirr-ree-ree', with sepulchral quality. *Nesting*: similar to Striated; 3–4 clouded light brown eggs. *Range*: uncommon to common resident in heath, mallee undergrowth, saltbush and samphire in drier areas west of a line from south-west Qld to north-west Vic.

SCRUBWRENS

SPECKLED

♂

♀

SHY

CHESTNUT-RUMPED

STRIATED

♂

♀

RUFOUS

southern

mid-
western

north-
western

19 x 15	19 x 14	19 x 14	22 x 16	22 x 16
Speckled	Chestnut-rumped	Shy	Striated	Rufous

GERYGONES (AUSTRALIAN WARBLERS)

Warblers are relatives of scrubwrens and thornbills with a similarity in appearance and behaviour to Old World warblers. However, their nests are quite different, being dome-shaped with a side entrance usually half-concealed by an overhanging 'verandah', suspended from a slender twig. Most live in or near mangroves or rainforest; only Western and White-throated are regularly seen in more open habitat. They feed mainly by fluttering near clumps of outer foliage to flush out insects. Some thornbills have similar habits, but told by streaked or freckled forehead and ear coverts. Some observers prefer name 'Gerygone' (pronounced 'grig-on-ee') to 'Warbler'.

Brown Warbler *Gerygone mouki* 10 cm
Brown-backed warbler with white eyebrow, red eye, white in tail tip, incessant call 'what-is-it', found in eastern rainforests. Northern form (race *mouki*) greyer above than southern form (race *richmondi*), but lacking grey 'face', and with eyebrow smaller. Very active and vocal in rainforest, often most obvious bird. *Voice:* in south, incessant busy 'what-is-it, what-is-it'; in north, 'having-a-good-time'. *Nesting:* pendant dome of bark, twigs, moss, rootlets and lichen; 2–3 freckled white or pinkish eggs. *Range:* common resident in eastern rainforest, occasional wet sclerophyll and mangrove, between eastern Vic to about Cooktown, Qld.

Mangrove Warbler *Gerygone levigaster* 10–11 cm
Brown-backed warbler with white eyebrow, red eye, white tip in tail, found in mangroves on east and north coast. *Juv:* pale lemon throat, no eyebrow, lemon eye-ring. Similar to Brown, which sometimes enters mangroves, but has very different song, finer bill, more active feeder. Dusky is greyer, has white eye, no white in tail, lacks prominent eyebrow. Large- billed lacks eyebrow, inhabits taller mangroves like Dusky rather than smaller ones close to shore. *Voice:* sweet cadence, midway between Western and White-throated. *Nesting:* pendant dome of grass and bark, decorated with spiders' eggsacs; 2–3 speckled pinkish eggs. *Range:* common to uncommon resident in mangroves from about Broome, WA, to about Lake Macquarie, NSW, but expanding southwards.

Western Warbler *Gerygone fusca* 10–11 cm
Brownish-grey warbler with red eye, white eyebrow, white tail tip and white in base of tail, found in open forest and woodland mainly west of Divide. Only warbler over most of Aust, replaced east of Divide and in north by White-throated. *Voice:* beautiful falling cadence with unfinished quality. *Nesting:* pendant dome of grass and bark; 2–3 blotched and speckled pinkish eggs. *Range:* common resident in west, uncommon in east, in open forest and woodland in broken distribution; isolated population on Eyre Peninsula, SA.

Green-backed Warbler *Gerygone chloronota* 10 cm
Green-backed warbler with red eye, no white in tail and no distinct eyebrow, found in northern and north-western mangroves and thickets. *Voice:* repeated three-noted falling cadence. *Nesting:* pendant dome of grass, rootlets, and bark decorated with spiders' eggsacs and long 'tail'. *Range:* locally common resident in north-west and northern mangroves, riverine forest, paperbark swamps, monsoon forest, bamboo, fig and wattle thickets, from about Derby, WA, to Groote Eylandt, NT.

Dusky Warbler *Gerygone tenebrosa* 11·5 cm
Grey-brown warbler (paler in south) with faint white eyebrow, white eye and no white in tail, found in north-western mangroves. *Juv:* yellowish wash on underparts, darker eye. Mangrove is browner, has red eye and white tip to tail, tends to inhabit smaller mangroves closer to shore. Western has much white in tail. *Voice:* sweet plaintive notes, midway between Western and Large-billed. *Nesting:* pendant dome of bark decorated with spiders' eggsacs; 2 blotched white eggs. *Range:* common to uncommon in north-western mangroves from about Shark Bay to Kunmunya, WA.

WARBLERS

BROWN

southern form

northern form

MANGROVE

eastern form

northern form

juv

WESTERN

GREEN-BACKED

ad

DUSKY

juv

17 x 12	17 x 12	18 x 13	18 x 13	17 x 12
Brown	Mangrove	Western	Green-backed	Dusky

GERYGONES (AUSTRALIAN WARBLERS)

Included here are two birds rarely recorded in Aust: Rusty-tailed Warbler, very similar to Large-billed, is known from two specimens collected long ago and only recently recognised as different; field observations are required. Arctic Warbler, from northern hemisphere, has been recorded only once but migrates to Indonesia so may regularly overshoot.

Large-billed Warbler   *Gerygone magnirostris*   11 cm
Brown-backed warbler with no white in tail and 'eyebrow' confined to divided white eye-ring, found in north-eastern and northern mangroves and thickets. *Voice:* continuous descending reel of 3–4 notes. *Nesting:* untidy pendant dome with long 'tail' usually near water, looks like flood debris; 2–3 freckled whitish eggs. *Range:* common resident in rainforest, mangroves, paperbark swamps, riverine forest, in northern Aust, from north-west Kimberley, WA, to about Mackay, Qld.

Rusty-tailed Warbler   *Gerygone ruficauda*
Virtually unknown brown-backed warbler similar to Large-billed but with rusty-brown rump and tail, short white eyebrow and white underparts with faint grey clouding on breast. When spread, tail shows no white tips nor any subterminal dark band, common features of all similar birds. *Range:* some doubts about locality; specimens were labelled Rockingham Bay and Wide Bay; presumably the place to look is in mangroves between Tully and Fraser I, Qld.

Fairy Warbler   *Gerygone palpebrosa*   10–11·5 cm
A yellow-breasted warbler with little or no white in tail found in or near north-eastern rainforest. Two forms occur, with intergrades between differing in males only: southern form (race *flavida*), throat white with dusky chin, similar to White-throated but with white eye-ring and with little white in tail; and northern form (race *personata*), throat blackish-brown with white 'moustache'. Females of both have throat greyish-white, differ from White-throated in white eye-ring, little white in tail; Green-backed Honeyeater (p 298) very similar, but has pale eye (not red), blue legs (not slate-grey), more 'puffy' appearance, confined to small area at Iron Ra; juveniles of both have throat lemon-yellow, lack white spot in front of eye. Often moves through forest in feeding association with other small birds. *Voice:* reeling 'whit ee whit you'. *Nesting:* pendant dome of grass, bark and rootlets, often near wasp nest; 2 freckled white eggs. *Range:* common resident in mangroves, rainforest or contiguous woodland in north-east from about Rockhampton, to Cape York, Qld.

White-throated Warbler   *Gerygone olivacea*   10–11 cm
A yellow-breasted warbler with much white in tail tip. *Female:* paler yellow than male. *Juv:* throat lemon-yellow, pale eye-ring. Similar to Fairy Warbler but no white around eye (except juv), much more white in tail. Yellow form of Weebill has much shorter bill, different song, thornbills with yellowish breasts have ear coverts or breast streaked. *Voice:* lovely descending melody, evocative of falling leaf, followed by more explosive 'phee-ee-ew'. *Nesting:* pendant dome of bark and grass with flimsy tail, decorated in south with spiders' eggsacs, in north with bag-nest caterpillar droppings; 2–3 freckled and blotched pink eggs. *Range:* common to uncommon resident or migrant (south-east) in forest and woodland in eastern and northern Aust from about Broome, WA, to south-eastern SA.

Arctic Warbler   *Phylloscopus borealis*   13 cm
Olive-backed migratory warbler with prominent white eyebrow, pale underparts tinged yellow on breast and pale pinkish-orange legs; some show faint wing bar. *Voice:* monotonus buzzing trill; buzzing 'tzz-yeet'. *Range:* known in Aust only from one picked up dead on Scott Reef off Kimberley coast, WA; some local warblers similar but none has such prominent eyebrow nor has pale orange legs; most have some white in tail. (Watch for similar Eastern Crowned Warbler *Phylloscopus coronatus*, which also migrates from northern hemisphere to Indonesia, is greener olive above, has yellowish eyebrow and centre of crown pale olive.)

WARBLERS

LARGE-BILLED

RUSTY-TAILED

FAIRY

northern ♂

intermediate ♂

southern ♂

♀

juv

WHITE-THROATED

ad

juv

ARCTIC

17 x 12
Large-billed

18 x 13
Fairy

18 x 13
White-throated

WHITEFACES AND SMALL THORNBILLS

Whitefaces inhabit drier areas of southern and central Aust. They are rather like thornbills but have stout almost finch-like bills and a distinctive facial pattern. They feed on the ground on invertebrates and seeds, flying up into trees or bushes when disturbed, showing white tip to tail. **Thornbills** have slender bills and freckled or streaked foreheads; those on this page are rather like warblers, feeding in similar fashion and building similar nests. The Weebill has a stubbier bill than thornbills and warblers, but is similar in other respects.

Striated Thornbill   *Acanthiza lineata*   10 cm
Small active thornbill with streaked breast, yellowish flanks, brown forehead with paler streaks, rump same colour as back, constant 'zit-zit'. Brown (p 268) is similar but less active, has reddish rump, red eye, generally feeds lower in forest, rather sepulchral extended call. *Voice:* constant 'zit-zit' like Yellow but softer, several fill the bush with sound. *Nesting:* neat pendant dome of bark, grass and rootlets; 2–4 freckled pinkish eggs. *Range:* common resident in forest and woodlands in southeast from about Maryborough, Qld, to Kangaroo I, SA.

Yellow Thornbill   *Acanthiza nana*   10 cm
Active thornbill with pale yellowish underparts, grey eye, streaked ear coverts. Several forms occur: coastal form (race *nana*) with yellowish underparts; inland form (race *modesta*) with almost no yellow on underparts; and northern form (race *flava*) with bright yellow underparts. *Voice:* 'tsit-tsit', harsher than Striated, two or three make the bush sound 'full of birds'. *Nesting:* pendant dome of bark, grass, moss and lichen; 2–4 freckled white eggs. *Range:* patchily common resident or nomad in dry forest and woodland, particularly wattle and she-oak, in south-eastern and eastern Aust, with isolated population (race *flava*) on Atherton Tableland, Qld.

Weebill   *Smicrornis brevirostris*   8–9 cm
Smallest Australian bird, with whitish eye, blunt bill, pale eyebrow, olive brown in south grading to yellowish-olive in north; underparts pale yellowish-buff in south grading to clear pale yellow in north. Yellow Thornbill has longer bill, more sibilant call, darker eye, streaked ear coverts. *Voice:* vigorous 'wee-willey-weet weet'; throaty 'tchick'. *Nesting:* compact dome of grass, leaves and plant-down in outer clump of leaves; 2–3 freckled white to pale buff eggs. *Range:* common resident or nomad throughout Aust in most wooded habitats except wettest forests.

Southern Whiteface   *Aphelocephala leucopsis*   10 cm
Rather plain whiteface with pale grey underparts and flanks grading from buffy-grey in east (race *leucopsis*) to chestnut in west (race *castaneiventris*). *Voice:* repeated twitter; soft 'tik-tik-tik' contact call. *Nesting:* untidy dome of grass, bark, etc, in hollow limb or fencepost; 2–5 freckled or blotched white-pink-purplish eggs. *Range:* locally common resident in woodland and grassland, favouring dead trees, in southern Aust west of Divide and avoiding south-west.

Chestnut-breasted Whiteface   *Aphelocephala pectoralis*   10 cm
A chestnut-breasted whiteface, confined to rocky desert in central SA. Rarely observed and little known due to isolated habitat. *Voice:* weak chatter. *Nesting:* untidy dome of twigs in bluebush; 3 blotched pink eggs. *Range:* uncommon resident or nomad in mulga and bluebush on gibber plains and desert tablelands in central SA.

Banded Whiteface   *Aphelocephala nigricincta*   10 cm
A whiteface with black breastband, in central Aust. Feeds on ground, often with other birds. *Voice:* repetitive twitter, more musical than Southern; contact notes 'pee-pee-pee'. *Nesting:* in west, untidy dome of sticks in low bush; in east, grass and bark nest in hollow limb; 3–4 freckled or blotched pinkish or purple eggs. *Range:* fairly common resident in saltbush, bluebush, mulga, sandhills and gibber plains in central Aust from south-west Qld to mid-western WA.

THORNBILLS

STRIATED

YELLOW

WEEBILL

northern form

southern form

south-western form

SOUTHERN

western form

eastern form

WHITEFACES

CHESTNUT-BREASTED

BANDED

17 x 12	18 x 12	15 x 11	19 x 14	18 x 13	Banded 18 x 13
Straited	Yellow	Weebill	Southern	Chestnut-breasted	

WHITE-EYED THORNBILLS

These thornbills have coloured rumps, contrasting in flight with dark, pale-tipped tail and feed mostly on ground (except Mountain). All have forehead scalloped. Usually in small groups. Calls are basically similar but easily learnt with experience.

Buff-rumped Thornbill *Acanthiza reguloides* 11 cm

White-eyed thornbill with rump and base of tail yellowish-buff. Northern birds (race *squamata*) are more yellow, rather like Yellow Thornbill but have pale yellow rump. In south-eastern SA very similar to Samphire but has more buff on rump. Feeds mainly on ground and lower treetrunks in small groups. *Voice:* tinkling reel, similar to Yellow-rumped but harder; 'chip chip' in flight. *Nesting:* bulky dome of grass and bark in hollow limb, behind loose bark or in grass tussock; 4 freckled white eggs. *Range:* common resident in open forest in east and south-east, from Mt Lofty Ra, SA, to Atherton Tableland, Qld.

Western Thornbill *Acanthiza inornata* 10 cm

Western representative of Buff-rumped, with rump more buff-olive not contrasting so much with back. Usually in small groups in understorey of forest. *Voice:* similar to Buff-rumped including mimicry; loud whistling 'wh wh wh wh-whit'; contact call 'tsip tsip'. *Nesting:* small dome of grass and bark hidden behind loose bark or under skirts of grass tree; 3 freckled whitish eggs. *Range:* common resident in forest and woodland in south-west.

Slender-billed (Samphire) Thornbill *Acanthiza iredalei* 9 cm

White-eyed thornbill with pale buff rump. Three forms occur: Iredale's form (race *iredalei*) with back pale grey-brown, large rump patch, found in samphire around salt lakes and on Nullarbor Plain; Rosina's form (race *rosinae*) with darker olive-grey back, smaller rump patch, found in samphire on saltmarshes and lakes about Gulf St Vincent, SA; and Hedley's form (race *hedleyi*) with pale olive-grey back, small rump patch, found in heath in south-eastern SA and north-western Vic. Similar to Buff-rumped but buff rump more restricted. *Voice:* warbling song rather like Buff-rumped; contact call 'chip chip'. *Nesting:* dome of grass and plant down in low bush; 3 freckled white eggs.

Yellow-rumped Thornbill *Acanthiza chrysorrhoa* 10–12 cm

White-eyed thornbill with bright yellow rump, crown black with white spots. Inland and northern birds are paler and more yellowish below. Usually in small groups, move jerkily on ground while feeding; when perched often fall backwards on perch. *Voice:* pleasant reeling warble; contact call 'chip chip' in flight. *Nesting:* large untidy dome with rudimentary open 'nest' on top slung in outer branch among leaves; 3–4 spotted dull white eggs. *Range:* common resident in open woodland throughout southern two-thirds of Aust and Tas.

Chestnut-rumped Thornbill *Acanthiza uropygialis* 10 cm

White-eyed thornbill with chestnut rump. Usually feeds on ground in small groups, favours dead timber. Other dry-country thornbills with chestnut rumps (Brown, Slate-backed p 268) have red eyes, seldom feed on ground. *Voice:* 'see-ti-ti-ti-ti-seeee'; harsh 'teu'; pleasant warble. *Nesting:* dome of grass and bark hidden in hollow limb with small knothole; 2–4 spotted white eggs. *Range:* common in dry country woodland and timbered grassland, saltbush and bluebush.

Mountain Thornbill *Acanthiza katherina* 10 cm

White-eyed thornbill with chestnut rump, found in upland rainforest in north-east Qld. Usually feeds actively in small groups well up in rainforest trees. No other thornbills share its habitat; local warblers and scrubwrens are either browner or yellow underneath. *Voice:* downward trill preceded by single note; descending whistles rather like Little Bronze Cuckoo but five instead of four notes. *Nesting:* large pendant dome of grass, tendrils, rootlets and moss, usually high among outer leaves; eggs undescribed. *Range:* common resident in highland rainforest from about Cooktown to Mt Spec.

THORNBILLS

BUFF-RUMPED

northern form

southern form

WESTERN

SLENDER-BILLED

Iredale's form

Rosina's form

YELLOW-RUMPED

CHESTNUT-RUMPED

MOUNTAIN

16 x 12	16 x 12	15 x11	18 x 13	16 x 12	16 x 12
Buff-rumped	Western	Slender-billed	Yellow-rumped	Chestnut-rumped	Mountain

RED-EYED THORNBILLS

These are more arboreal than white-eyed thornbills, seldom if ever feeding on ground; they usually move in the lower and middle storeys of woodlands and forest, often in association with warblers, silvereyes and honeyeaters. They are less active than warblers and generally take insects from among leafy clumps rather than by hovering outside them. Nests are small domes usually placed in low bushes, in grass clumps or in grasstree skirts, and sometimes quite high in outer clumps of hanging leaves (south-western form).

Brown Thornbill
Acanthiza pusilla 10 cm

Variable red-eyed thornbill with chestnut rump, streaked breast and scalloped forehead. Usually seen in pairs or small groups low in forest or woodland, often with tail half-cocked in dry-country forms. Several forms occur, some of which have been regarded as species in past: (a) *Birds with olive backs, rufous foreheads and narrow dark band in tail* (mainly in humid areas): south-eastern form (race *pusilla*) forehead pale rufous, flanks buff-olive; Tasmanian form (race *diemenensis*) darker above, flanks darker, whitish edges to wing feathers, very similar to Tasmanian Thornbill which has little scalloping on rufous forehead, has whitish flanks and rufous edges to wing feathers, generally in denser habitat; yellow-bellied form (race *mcgilli*) from Clarke Range near Mackay to Proserpine, Qld; rare long-billed form (race *archibaldi*) from King I; grey-backed form (race *zietzi*) from Kangaroo I; south-western form (race *leeuwinensis*) greyer above, scalloped rufous forehead, clearer buff flanks. (b) *Birds with grey backs, greyish-brown scalloped foreheads and broad dark band in tail, often carry tail cocked in wren fashion* (mainly inland): broad-tailed form (race *apicalis*) in south-west inland, with dark chestnut rump, pale buff flanks, olive-grey above; white-bellied form (race *albiventris*) paler grey above with reddish rump, white below; Tanami Desert form (race *tanami*) palest above with cinnamon-rufous rump. *Voice:* rather sepulchral song, quiet but far-carrying; mimicry; loud 'pee-orr'; contact call 'tsip-tsip'. *Nesting:* small dome of grass and bark placed in low bush, grass tussock or pendant foliage; 3 speckled and blotched white eggs. *Range:* common in forest and woodland throughout southern two-thirds of mainland Aust and Tas.

Tasmanian Thornbill
Acanthiza ewingii 10 cm

Red-eyed thornbill with chestnut rump, unscalloped rufous forehead, throat and breast mottled, whitish flanks, rufous edges to wing feathers. Very similar to Brown, which has scalloped forehead, streaked rather than mottled breast, buff-olive flanks and whitish edges to wing feathers. *Voice:* variable, rather like Brown; 'zit zit zit whoorl'. *Nesting:* neat small dome of grass, bark and moss, in outer foliage of bush or low tree; 3–4 freckled pinkish-white eggs. *Range:* common resident in forest and woodland, particularly moist gullies, in Tas.

Slate-backed Thornbill
Acanthiza robustirostris 9 cm

Red-eyed thornbill with chestnut rump, grey forehead with fine black streaks and slaty-grey back, found in small groups mainly in mulga. Similar to less-vocal Chestnut-rumped, which has white eye, scalloped forehead; female Redthroat also similar but lacks chestnut rump. *Voice:* loud 'whippy-chew'; single soft 'pseet'; harsh 'thrip thrip', 'pitsuid-piteet', 'pitsuid-grarrr-sweet-eet'. *Nesting:* untidy loose dome of grass; 3 freckled white or pale pink eggs. *Range:* uncommon resident in mulga woodland in central Aust from about Shark Bay, WA, to Eromanga, Qld.

THORNBILLS

BROWN

broad-tailed form

Tanami form

yellow-bellied form

white-bellied form

south-western form

south-eastern form

Tas form

King I form

SLATE-BACKED

TASMANIAN

16 x 12
Brown

16 x 12
Slate-backed

18 x 13
Tasmanian

SITTELLAS AND TREECREEPERS

Sittellas are unmistakable small dumpy birds with short wings and tails, usually seen in flocks hopping along and under branches, pecking in bark crevices for invertebrate prey, very noisy, attracting attention; when flying, contact is kept with the flock by loud call notes. In flight, rump shows white and wings show pale bar; white in northern birds and orange in southern forms. Nest closely resembles a knot in branch on which it is built. Members of flock attend nest, but one female usually does most of the building and incubating. Various forms of sittella haven't quite achieved specific status. The map opposite shows approximately where one would expect to find various forms and where hybrids are likely to occur. Females have more extensive dark feathering on head; juveniles of all forms are similar, with dark streaks on basically white head. **Treecreepers** are strong-clawed birds most often seen climbing up treetrunks, looking for invertebrate prey in bark crevices. Some species feed on ground as well as on treetrunks; Rufous is very fond of fallen timber; White-throated pecks into soft wood after boring grubs. In flight, show a whitish or buffish 'window' or wingstripe. An untidy cup-shaped nest of bark lined with animal fur or plant-down, placed in hollow limb, usually upright with entrance at top. Rufous more likely to use horizontal limbs. All except White-throated and Little may have extra helpers at the nest, probably progeny from previous broods. Colour of back tends to reflect colour of dominant trees in each species' preferred habitat, much darker in tropics than farther south.

Varied Sittella   *Daphoenositta chrysoptera*   10–12 cm
Variably-plumaged small dumpy bird most often seen hopping up or down outer branches of tall trees. Patterns vary according to locality: Orange-winged form (race *chrysoptera*), greyish-brown streaked head, back and breast; orange bar in wing, south-east. White-headed form (race *leucocephala*), white head, streaked back and breast; orange bar in wing, south-eastern Qld with eastern NSW. Pied form (race *albata*), white head, streaked back and breast; white bar in wing, small range in central eastern Qld. Striated form (race *striata*), darker race with dark head, more extensive in female, streaked back and breast; white bar in wing, north-eastern Qld. White-winged form (race *leucoptera*), black cap, more extensive in female, streaked back; white bar in wing; white underparts except barred undertail coverts; northern Aust except north-eastern Qld. Black-capped form (race *pileata*), black cap, more extensive in female, unstreaked back, white underparts except barred undertail coverts; orange bar in wing, southern Aust except south-east. *Voice:* high-pitched, twittering 'chip chip . . .'. *Nesting:* beautiful deep camouflaged cup in thin upright fork, of bark and lichen bound with cobwebs, 5–20 m above ground; 3 eggs. *Range:* common throughout Aust wherever there are trees, except dense rainforest.

Black-tailed Treecreeper   *Climacteris melanura*   16–20 cm
Very dark tropical treecreeper without pale eyebrow. *Male:* throat streaked black and white. *Female:* throat streaked white with reddish patch on upper breast. Two forms occur: northern form (race *melanura*), underparts dark brown, undertail coverts black with minute white spots; north-western form (race *wellsi*), paler back, underparts reddish brown, streaks on throat more extensive. Ranges do not meet. *Voice:* loud piping 'peet peet peet . . .'. *Nesting:* cup of grass lined with feathers in upright spout or hollow limb of tree; 2–3 spotted pinkish eggs. *Range:* common to scarce in open woodland in northern Aust from Broome, WA, to Leichhardt R, Qld (race *melanura*); and in river gums in north-western Aust between De Grey R and Fortescue R, WA (race *wellsi*).

Rufous Treecreeper   *Climacteris rufa*   15–17 cm
Reddish-brown south-western treecreeper with rusty underparts and rich buff 'window' in wing. *Male:* centre of breast streaked black and pale buff. *Female:* centre of breast streaked buff. *Juv:* greyish tone to plumage except rump which is rufous. Often feeds on ground or on fallen timber. *Voice:* loud 'peet'. *Nesting:* loose pad of grass lined with plant-down and feathers in hollow limb, often horizontal; 2–3 spotted white eggs. *Range:* common in open eucalypt forest and woodland in south-western corner from Shark Bay, WA, to Eyre Peninsula, SA.

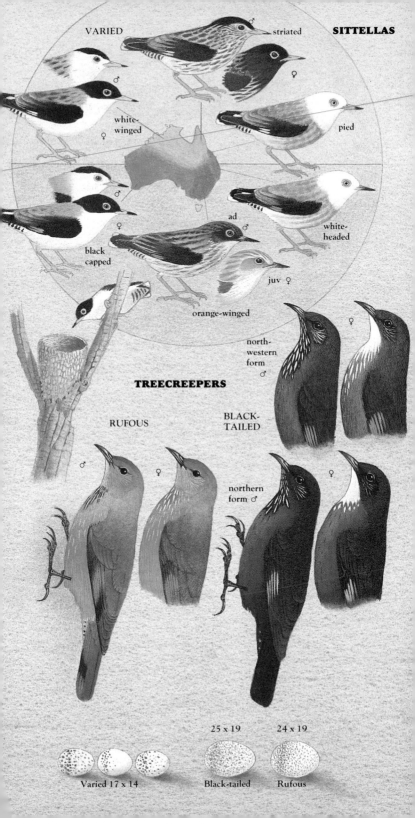

SITTELLAS

VARIED

striated

♂ white-winged ♀

pied

♀ black-capped ♂

ad ♂ white-headed

juv ♀

orange-winged

TREECREEPERS

north-western form ♀ ♂

RUFOUS

BLACK-TAILED

♂ ♀

northern form ♂ ♀

25 x 19   24 x 19

Varied 17 x 14   Black-tailed   Rufous

TREECREEPERS

White-throated Treecreeper   *Cormobates leucophaea*   13–15 cm
Grey-brown treecreeper without eyebrow. *Male:* throat white. *Female:*
throat white with orange patch below ear. *Juv:* rump rich rufous in
females. Two forms occur: large southern form (race *leucophaea*), with
boldly streaked flanks; and small northern form (race *minor*), from
Eungella Ra north mostly in highland rainforest, with less prominent
streaks on flanks. Feeds singly or in pairs on treetrunks, seldom on
ground; longer bill than other species, often pecks into decayed wood.
Voice: loud 'peet peet peet . . .'; also 'clicks' tail loudly. *Nesting:* loose
cup of bark lined with fur and feathers in hollow branch, usually
upright; 2–3 spotted white eggs. *Range:* common resident in rainforest,
forest, woodland and riverine forest in eastern and south-eastern Aust
from Mt Lofty Ra, SA, to Cooktown, Qld.

Red-browed Treecreeper   *Climacteris erythrops*   15 cm
Olive-grey treecreeper with rusty eyebrow confined to south-eastern
forests. *Male:* upper breast grey, lower breast and abdomen heavily
streaked black and white. *Female:* upper breast streaked chestnut and
white, lower breast and flanks streaked black and white. Very similar to
White-browed, which inhabits mulga woodland; looks very like White-
throated against trees, when reddish eyebrow and more heavily streaked
underparts difficult to see. *Voice:* 'peet peet peet . . .' softer than
White-throated. *Nesting:* lined loose cup of bark in hollow limb; 2–3
spotted pinkish eggs. *Range:* uncommon resident in sclerophyll forest,
mainly on steep country in south-eastern Aust from near Melbourne,
Vic, to about Tewantin, Qld.

White-browed Treecreeper   *Climacteris affinis*   14 cm
Grey-brown treecreeper with prominent white eyebrow and streaked
ear coverts. *Male:* throat and upper breast grey; breast and abdomen
heavily streaked black and white. *Female:* rusty streak in white eyebrow
(visible only at close range), breast pale grey with faint rusty and white
streaks merging into heavily streaked underparts. Feeds extensively on
ground, often on ants; usually occur in pairs or small groups. *Voice:* loud
'peet peet peet . . .'. *Nesting:* tree hollow lined with grass and bark; 2–3
spotted pink eggs. *Range:* uncommon resident in arid woodlands
particularly mulga and she-oak in central Aust from about Cobar,
NSW, and Cunnamulla, Qld, to Upper Gascoyne R, WA.

Brown Treecreeper   *Climacteris picumnus*   16–18 cm
Brown or blackish-brown treecreeper with pale buff eyebrow, buff or
grey breast and lightly streaked underparts. *Male:* small patch of black
and white streaks on upper breast. *Female:* small patch of rufous streaks
on upper breast. Two forms occur: southern brown-backed form (race
picumnus); and northern black-backed form (race *melanota*) on Cape
York Peninsula. Often occurs in same areas as White-throated, easily
told by large whitish-buff eyebrow. Feeds often on ground in small
groups as well as pairs. *Voice:* 'pink-pink-pink . . .'. *Nesting:* grass-lined
hollow often on base of porous material; 2–3 freckled and streaked
pinkish eggs. *Range:* common resident in drier forests and woodlands
particularly among fallen timber in eastern Aust from Spencer Gulf,
SA, to Cape York, Qld.

TREECREEPERS

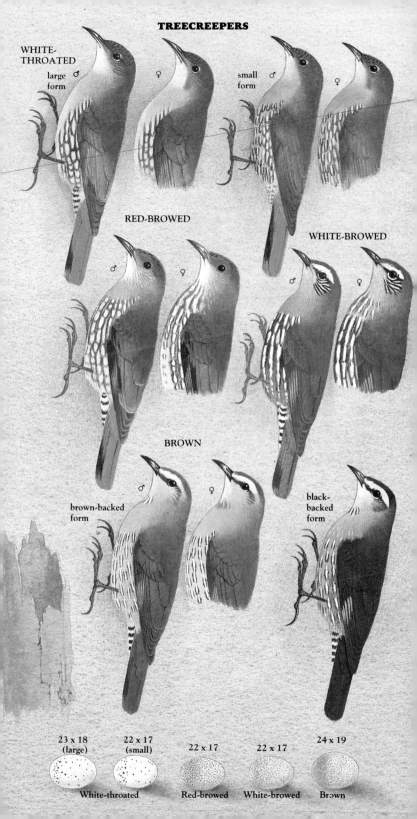

WHITE-THROATED

large form ♂   ♀   small form ♂   ♀

RED-BROWED

♂   ♀

WHITE-BROWED

♂   ♀

BROWN

brown-backed form ♂   ♀   black-backed form

23 x 18 (large)	22 x 17 (small)	22 x 17	22 x 17	24 x 19
White-throated		Red-browed	White-browed	Brown

WATTLEBIRDS

Honeyeaters are nectar-feeding birds with long brush-tipped tongues; bills are curved, often long, reflecting to some extent the sorts of flowers they frequent. They vary in size from tiny (Scarlet Honeyeater) to large (Yellow Wattlebird), thus are capable of exploiting (and pollinating) the full range of nectar-producing plants. As well as nectar they feed extensively on insects and other invertebrates and some eat fruit as well. Nests are generally cup-shaped, suspended by the rim, but two build domed, warbler-like nests. Many are migratory or nomadic within Aust. **Wattlebirds** are among the largest honeyeaters, named for the pendant wattle of skin on the ear of some species. They are noisy obvious birds with raucous calls, favouring flowers of banksia, dryandra and introduced coral tree.

Spiny-cheeked Honeyeater   *Acanthagenys rufogularis*   22–27 cm

Small wattlebird with buff throat and bicoloured bill; wattle is confined to fleshy extension of gape. Adults have a patch of whitish or pale yellow stiff feathers behind gape; in juveniles these are yellow. In flight, pale rump and white-tipped tail very obvious; among commonest birds flushed from roadside vegetation in arid areas. *Voice:* plaintive whistle followed by ascending notes 'whee-you-weer, wh wh wh . . .'; explosive 'quock'; querulous bubbling. *Nesting:* flimsy deep cup of grass and leaves suspended in horizontal fork in creeper or among outer leaves; 2–3 spotted and blotched white–buff eggs. *Range:* common resident or nomad (perhaps migratory in south-east) in wide variety of dry habitats usually with understorey of tall shrubs throughout southern two-thirds of Aust, avoiding wetter forests and woodlands in east and south-west.

Yellow Wattlebird   *Anthochaera paradoxa*   38–48 cm

Large wattlebird with long pendant yellowish wattles, confined to Tas and King I. *Juv:* wattle undeveloped. Vigorous, acrobatic, showy bird, often in small flocks, sometimes with Little, attracted to blossoming trees. *Voice:* raucous gurgling. *Nesting:* large untidy saucer of twigs, grass, bark and wool placed in high fork; 2–3 spotted and blotched salmon eggs. *Range:* common to uncommon resident or nomad (more common in east) in forest, woodland and scrub in Tas and King I.

Red Wattlebird   *Anthochaera carunculata*   31–39 cm

Large wattlebird with yellow abdomen, small red pendant wattle below ear. Young birds lack wattles, have less yellow on abdomen. White tips on tail and wingtips very obvious in flight. Noisy active aggressive and vigorous, often in groups, occasionally large flocks. *Voice:* loud sneezing or 'chock a lock'; barking 'cheeokk'; grating chuckle. *Nesting:* flimsy saucer of grass twigs and bark, placed in fork in tree or bush; 2–3 spotted pink or pinkish-buff eggs. *Range:* common in forest, woodland and heath from about Shark Bay, WA, to south-eastern Qld; movements are complex, associated with blossom, also altitudinal and north/south migration in east and small-scale east/west migration in south-west.

Little Wattlebird   *Anthochaera lunulata*   26–30 cm

Small brown wattlebird with silvery ear patch and red eye. In flight shows chestnut wing patch, narrow white tips to primaries and tail. Lacks visible wattles. Raucous, aggressive and vigorous, favouring banksia and dryandra thickets. *Voice:* bubbling, twittering, cackling and chuckling; bill snapping; loud 'kwock'; duetting, female with higher-pitched calls. *Nesting:* untidy cup of twigs, bark and grass in concealed fork; one pinkish-buff spotted egg. *Range:* common resident or nomad in heath, woodland and forest in south-west.

Brush Wattlebird   *Anthochaera chrysoptera*   27–33 cm

Small brown wattlebird with silvery ear-patch and grey to blue-grey eyes. In flight shows chestnut wing patch, broad white tips to primaries and tail. Lacks visible wattles. Noisy and aggressive, favouring banksias. *Voice:* incessant chuckling, bill clicking and metallic shrieks; loud 'chock'; duetting, female with higher-pitched calls. *Nesting:* untidy cup of twigs, grass and down in concealed fork, 2–3 pinkish-buff spotted eggs. *Range:* common resident or nomad in coastal heath and woodland in south-east and Tas.

WATTLEBIRDS

SPINY-CHEEKED

YELLOW

RED

LITTLE

BRUSH

25 x 18
Spiny-cheeked

36 x 24
Yellow

33 x 22
Red

29 x 21
Little

29 x 21
Brush

FRIARBIRDS

Large unmistakable honeyeaters with dark naked skin on head and, in most species, a knob on top of the bill. Similar to wattlebirds in many ways, noisy, aggressive, and active, often chasing other birds from flowering trees. Nests are neater than wattlebirds', deep cups suspended by rim in outer foliage, often covered with wool.

Little Friarbird
Philemon citreogularis 25–29 cm

Small friarbird with naked bluish or leaden skin on face and no knob on bill. *Juv:* yellow patch on throat, pale edges to feathers on back. *Voice:* in east, pleasant 'quook' like Singing Honeyeater; raucous 'rrach-coo'; in Kimberley, raucous 'rackety crookshank'. *Nesting:* deep cup of grass, bark and wool suspended in outer foliage, favouring paperbark; 2–3 spotted pink, salmon or purplish-red eggs. *Range:* common resident or nomad in north, migrant (Sep–Apr) in south-east, in open forest and woodland mainly in eucalypt, grevillea and paperbark blossoms in eastern and northern Aust from Eyre Peninsula, SA, to about Port Hedland, WA.

Helmeted Friarbird
Philemon buceroides 30–37 cm

Large tropical friarbird with backward-sloping knob on bill, face black, crown, nape and throat greyish-brown, tail without white tip. *Juv:* knob vestigial, face dark grey, feathers on back edged paler. Several forms occur: North Queensland form (race *yorki*); Melville Island form (race *gordoni*) found in mangroves and monsoon forest in coastal NT; sandstone form (race *ammitophila*) in sandstone gorges in Arnhem Land escarpment. *Voice:* varied raucous calls: 'watch out . . .'; 'chilane chilane'; 'chank chank . . .'; 'wack a where'; 'poor devil poor devil'. *Nesting:* large deep cup of bark, twigs and rootlets in leafy horizontal fork; 3–4 spotted pink eggs. *Range:* common nomad or resident in forest and woodland (Qld), mangroves, monsoon forest and sandstone gorges (NT).

Silver-crowned Friarbird
Philemon argenticeps 27–32 cm

Large tropical friarbird with prominent sloping knob on bill; crown, nape and throat silver, tail with faint white tip. Similar to adult Noisy, which lacks feathers on crown and has white patch of feathers on black throat; juvenile Noisy is much more similar, but has pale edges to back feathers, yellowish breast. *Voice:* very variable, harsh nasal cackling or grating sounds; a common call is repeated 'more tobacco, uh'. *Nesting:* deep suspended cup of bark and grass in leafy fork; 2–3 spotted pinkish eggs. *Range:* common nomad following blossoms, mainly eucalypt and paperbark, and fruiting figs in woodlands in northern Aust from north of Derby, WA , to Townsville, Qld.

Noisy Friarbird
Philemon corniculatus 25–29 cm

Large eastern friarbird with upright knob on bill, entire head naked black skin except tufts of feathers on eyebrow and chin, tail with prominent white tip. *Juv:* vestigial knob, feathers of frilly nape extending to crown, patch of silver feathers on chin, pale edges to back feathers, yellowish upper breast; Silver-crowned similar to young northern birds but has throat silver. *Voice:* loud cackling, often harsh yet pleasant based on 'four o'clock . . .'; 'chock . . .'; 'chewlip'. *Nesting:* large deep pendant cup of bark, rootlets, leaves and wool in horizontal leafy fork; 2–4 blotched pink eggs. *Range:* common nomad or migrant (Sep–Apr in south) in forest and woodland, favouring eucalypt and paperbark as well as fruiting trees, in eastern Aust from Murray R, SA, to Cape York Peninsula, Qld.

FRIARBIRDS

LITTLE

ad

juv

HELMETED

SILVER-
CROWNED

NOISY

| 28 x 20 | 34 x 23 | 30 x 21 | 33 x 22 |
| Little | Helmeted | Silver-crowned | Noisy |

LARGE WATTLED HONEYEATERS

A group of honeyeaters characterised on a wattle of coloured skin behind or over the eye (see overleaf for smaller species). All are sociable in behaviour, usually observed in small to large flocks depending on species. All are very vocal, reflecting need of flocks to keep in contact; calls range from harsh (Blue-faced) to charming (Bell Miner) and melodious (Black-chinned). Most have a hawk-alarm call, a continuous piping whistle which allows observers the opportunity of spotting raptors that would otherwise be missed. Food is predominantly insects, spiders and nectar. Most species have extended breeding seasons – as many as ten birds may attend one nest. Blue-face usually lays in abandoned nest of Grey-crowned Babbler, but occasionally makes its own miner-like nest.

Blue-faced Honeyeater   *Entomyzon cyanotis*   25–30 cm

Large noisy honeyeater with extensive area of skin around eye blue (adult), yellowish-green (imm) or brownish (juv). Two forms occur: northern form (race *albipennis*) has large white patch in wing, looks rather like butcherbird with khaki back; eastern form (race *cyanotis*) has small buff patch in wing. *Voice:* loud metallic 'keet' or 'kwok'. *Nesting:* lines old nest of Grey-crowned Babbler with bark, grass, etc; rarely builds own nest (rather like miner's) in dense clump of leaves; may also use friarbird, miner or Apostlebird nests; 2–3 eggs. *Range:* common in small flocks in open woodland to edges of rainforest, mangrove and paperbark swamps in northern and eastern Aust from Broome, WA, to south-eastern SA.

Yellow-throated Miner   *Manorina flavigula*   25–28 cm

Grey-brown, sociable miner with forehead and sides of neck yellowish; rump noticeably paler than back. Several forms occur: white rumped form (race *flavigula*) over most of Aust; black-eared form (race *melanotis*) confined to mallee in north-west Vic and eastern SA, being hybridised out of existence by white-rumped form as clearing allows contact, has darker rump, more extended black on face and less white in tail tip; dusky form (race *obscura*) in south-west WA, similar to black-eared form but less black on face. *Voice:* very variable; commonest call querulous 'teee teee teee'; hawk-alarm call, high-pitched 'ti-ti-ti-ti-ti . . .', is helpful indication to observers of raptor approaching. *Nesting:* untidy cup of grass, twigs, lined with wool and hair, more substantial than Noisy Miner, usually placed in fork or among drooping outer leaves, mistletoe clump favourite site; communal; 3–4 eggs. *Range:* common in flocks up to 30–50 in open woodland, mostly west of Great Dividing range.

Noisy Miner   *Manorina melanocephala*   25–28 cm

Grey-brown, sociable miner with white forehead and black crown, rump same colour as back, *Voice:* very variable: commonest are loud 'zwit', 'teu teu teu' in flight; nasal 'weet weet weet . . .' and lovely pre-dawn song. *Nesting:* untidy sparse cup of grasses lined with hair and fur placed in vertical fork of sapling or horizontal fork among branches of larger tree; communal; 3–4 sometimes 5 eggs. *Range:* among commonest birds of eucalypt woodland in flocks up to 30 in eastern Aust from about Townsville, Qld, to Adelaide, SA.

Bell Miner   *Manorina melanophrys*   18–20 cm

Olive-green, sociable miner with loud bell-like call. *Juv:* beak darker, lacks yellow patch in front of eye and bare skin behind eye greenish-yellow. *Voice:* incessant bell-like 'tink'; harsh 'tok' in alarm; nasal 'peet'. *Nesting:* sparse cup of twigs, grass and bark decorated with spiders' eggsacs, leaves or lichen, usually in horizontal fork ranging from shrubs near ground level to treetops; communal; usually 2, sometimes 1 or 3 eggs. *Range:* common in colonies of up to 300 birds in sclerophyll gullies usually with tall trees and dense undergrowth in south-eastern Aust from about Fraser I, Qld, to Otway Ra, Vic.

northern form

eastern form

BLUE-FACED

MINERS

ad

juv

YELLOW-THROATED

white-rumped form

black-eared form

NOISY

BELL

32 x 22	26 x 18	26 x 18	27 x 18	24 x 16
Blue-faced	Yellow-throated	Black-eared	Noisy	Bell

SMALL WATTLED HONEYEATERS

White-throated Honeyeater *Melithreptus albogularis* 13–15 cm
Small black-headed honeyeater with white nape extending to eye, white chin, pale blue wattle over eye, bright olive-yellow back. *Juv:* head brown, darker on ear coverts, chin white. Very similar to White-naped which has red wattle, white nape not extending to eye, black chin, less piercing more sibilant call, duller back. Black-chinned much larger, with prominent black chin, orange legs, beautiful rich song. *Voice:* rather peevish, high pitched 'tserp-tserp . . .'; 'tee tee, tee tee . . .'; loud alarm 'ti ti . . .'. *Nesting:* suspended deep cup of bark, grass, plant down in outer foliage; 2 blotched and freckled pink eggs. *Range:* common resident in forest, woodland and riverine vegetation in north and north-east Aust from Broome, WA, to Macleay R, NSW.

White-naped Honeyeater *Melithreptus lunatus* 13–15 cm
Small black-headed honeyeater with white nape not meeting eye, black chin, red wattle over eye (in east; white in south-west), dull olive-green back. *Juv:* head brown, more gingery on crown, very similar to Juvenile white naped with duller, dusky chin, orange wattle over eye. Two forms occur: eastern form (race *lunatus*) with red wattle; south-western form (race *whitlocki*) white wattle. *Voice:* sibilant 'sherp sherp'; mellow 'tew tew tew . . .'; loud alarm 'ti ti ti . . .'. *Nesting:* suspended deep cup of bark, grass and plant-down in outer foliage; 2 blotched and freckled pink eggs. *Range:* common resident, nomad or migrant in eastern Aust, many moving out of south-east Mar–Apr/Aug–Oct, wintering in northern NSW and Qld; common resident or nomad in south-west, mainly in forest and woodland.

Brown-headed Honeyeater *Melithreptus brevirostris* 13–15 cm
Small brown-headed honeyeater with white nape reaching eye, white chin, pink wattle, dull olive-grey back. Similar to juveniles of White-throated and White-naped but lacks green tone on back, has underparts dirty white and characteristic 'chip chip . . .' call. *Voice:* 'chip chip . . .' mainly in flight; alarm call 'ti ti ti . . .'. *Nesting:* deep cup of bark, grass and fur either suspended in outer foliage or in slender upright fork; 2 spotted pink eggs. *Range:* common resident or nomad in dry woodland and open forest, mostly in eucalypts, in southern Aust from south-west to central Qld.

Black-headed Honeyeater *Melithreptus affinis* 13–15 cm
Small black-headed honeyeater without white nape, found in Tas and Bass Strait islands. *Juv:* head brown, darker on ear coverts, yellowish tinge on upper breast. *Voice:* sharp whistle; loud 'ti ti ti . . .'. *Nesting:* suspended deep cup of bark, grass and down in outer foliage; 2–3 spotted pinkish eggs. *Range:* common nomad in forest, woodland and heath in Tas, Flinders I and King I.

Strong-billed Honeyeater *Melithreptus validirostris* 15–17 cm
Large black-headed honeyeater with white nape, blue wattle, dusky breast and abdomen, white sides to throat, found in Tas and Bass Strait islands. Very similar to Black-chinned of mainland, but juvenile very different from Black-chinned juv with black head, orange wattle, yellowish sides to throat. Often seen with Black-headed. *Voice:* loud, single, double or repeated 'cheep'. *Nesting:* suspended deep cup of bark, grass and fur in outer foliage; 2–3 spotted pinkish eggs. *Range:* common nomad in forest, woodland and heath in Tas, Flinders I and King I.

Black-chinned Honeyeater *Melithreptus gularis* 15–17 cm
Large black-headed honeyeater with broad white nape, blue or green wattle, olive or golden back, dark chin extending to breast, leaving white 'moustaches'. *Juv:* head brown, darker on ear-coverts, wattle pale blue. Two forms occur: olive-backed form (race *gularis*) in mainly interior, eastern and south-east; and golden-backed form (race *laetior*) in central and northern Aust. *Voice:* beautiful rich 'prrp prrp scorp scorp . . .'. *Nesting:* deep cup of bark, grass, down and fur suspended in outer foliage; 2 spotted pink eggs. *Range:* uncommon nomad or resident in dry forest and woodlands particularly along rivers in northern, central and eastern Aust.

HONEYEATERS

WHITE-THROATED

ad

juv

WHITE-NAPED

eastern form

south-western form

juv

BROWN-HEADED

BLACK-HEADED

ad

juv

STRONG-BILLED

ad

juv

BLACK-CHINNED

juv

ad

ad

northern form

eastern form

18 x 14	18 x 14	19 x 14	16 x 13	22 x 17	22 x 16
White-throated	White-naped	Black-headed	Brown-headed	Strong-billed	Black-chinned

RB

WHITE-GAPED HONEYEATERS

Lewin's Honeyeater *Meliphaga lewinii* 19–21 cm
Large yellow-gaped honeyeater with short bill and 'half-moon' earpatch. *Voice:* pervasive staccato machinegun-like call. Large size, vigorous call and 'half-moon' earspot distinguish it from Yellow-spotted and Graceful where ranges overlap. *Nesting:* deep decorated cup of plant-down, bark and grass, 1–5 m; 2–3 eggs. *Range:* common in any dense vegetation and nearby eucalypts in east from about Cooktown, Qld, to Dandenong, Vic, inland through brigalow to Carnarvon Ra, Qld, Nandawar Ra, NSW and ACT.

Yellow-spotted Honeyeater *Meliphaga notata* 17–19 cm
Medium-sized, yellow-gaped honeyeater with short bill, rounded earpatch. *Voice:* melodious 'chip', sharp 'quick quick quick', rattling machinegun notes less vigorous than Lewin. Told from Graceful by short bill, and from Lewin by round earpatch. *Nesting:* flimsy cup of shredded bark and grass, in horizontal fork, 1–4 m; 2 eggs. *Range:* common in rainforest, mangrove and nearby eucalypts, gardens, from Cape York to about Townsville, Qld.

Graceful Honeyeater *Meliphaga gracilis* 14–16 cm
Small yellow-gaped honeyeater with long fine bill, diamond-shaped yellow earpatch, sharp single call 'tuck'. Lewin has short stout bill and half-moon earpatch, loud machinegun-like call; Yellow-spotted has shorter bill, round earpatch, calls pleasant 'chip' and machinegun-like call less rich than Lewin. *Nesting:* deep decorated cup of grass and moss in horizontal fork; 2 eggs. *Range:* uncommon in rainforest and nearby eucalypts, from Cape York to Ingham, Qld.

White-gaped Honeyeater *Meliphaga unicolor* 18–20 cm
Very plain grey-brown honeyeater with white gape, almost invisible red ring around eye. Very active and aggressive. *Voice:* 'ch ch ch-aweya'; 'quock'. *Nesting:* deep cup of grass, tendrils, rootlets and hair lined with soft grass, 1–3 m; 2 eggs. *Range:* common to uncommon in riverine swamp and mangrove vegetation from below Broome, WA, to Burdekin R, Qld, occasionally farther south.

White-lined Honeyeater *Meliphaga albilineata* 18–20 cm
White-gaped honeyeater with white line below pale blue eye, confined to scrub remnants on tropical escarpments. Brown has yellow gape, small yellow tuft behind eye, more olive plumage, lacking faint mottling on breast, different song. *Voice:* melodious, vigorous, flutelike 'wheee-wheee-wheeooo'. *Nesting:* deep cup of vine strands, spider web and plant down, 1–5 m; 1–2 eggs. *Range:* locally common in rainforest remnants on escarpments in Arnhem Land, NT, and north Kimberley, WA.

Bridled Honeyeater *Lichenostomus frenatus* 18–22 cm
Large bustling white-gaped honeyeater with bicoloured bill, yellow line below, and white line behind, blue eye; dark throat and ear edged with yellow tuft and pale buff crescent. *Voice:* loud melodious 'we-are'; rather metallic 'pretty creature'. *Nesting:* deep cup of fine twigs and tendrils lined with plant-down, 1–3 m; 2 eggs. *Range:* common in upland rainforest and contiguous vegetation, lower in winter, from Cooktown to Townsville, Qld.

Eungella Honeyeater *Lichenostomus hindwoodi* 18 cm
Recently recognised, white-gaped honeyeater with black bill, white line below eye ending in yellow tuft, white tuft behind eye and another below ear; underparts brownish-grey streaked paler, back with fine buff streaks. Formerly considered conspecific with Bridled Honeyeater. *Range:* common in upland rainforest and contiguous vegetation in Eungella Ra, near Mackay, Qld.

HONEYEATERS

GRACEFUL

YELLOW-SPOTTED

LEWIN'S

WHITE-LINED

WHITE-GAPED

BRIDLED

EUNGELLA

25 x 15
Bridled

21 x 14
Eungella

26 x 18
Lewin's

23 x 16
Yellow-spotted

20 x 15
Graceful

22 x 17
White-gaped

26 x 14
White-lined

MASKED HONEYEATERS

These honeyeaters have a dark mask passing through eye, usually with a contrasting tuft of feathers behind ear, varying from bold black mask in Yellow-tufted to faint dusky mask in Yellow. Generally noisy with loud distinctive calls.

Yellow-tufted Honeyeater   *Lichenostomus melanops*   17–23 cm
Masked honeyeater with broad black mask, small yellow ear tuft, golden or greenish-yellow crown, yellow throat with narrow dusky chin, greyish-yellow breast and dark olive back. Several forms occur: helmeted form (race *cassidex*) has conspicuous golden crest, confined to small area east of Melbourne (Yellingbo Reserve); olive-crowned forms varying from pale (race *meltoni*) to dark (race *gippslandica*). Usually found in small parties, sometimes quite large colonies, in tall eucalypts with understorey of shrubs. *Voice:* harsh deep 'chop chop'; sharp 'yip'; melodious 'tooey-ti-tooey ti tooey . . .'. *Nesting:* deep cup of grass and bark decorated with spiders' eggsacs in slender fork; 2–3 spotted and blotched pinkish-buff eggs. *Range:* locally common to rare resident or nomad in sclerophyll forest and woodland in eastern Aust from about Naracoorte, SA, to Caloundra, Qld.

Yellow-throated Honeyeater   *Lichenostomus flavicollis*   18–22 cm
Masked honeyeater with dark head and yellow throat , found in Tas and Bass Strait islands. *Voice:* loud 'tonk-tonk'; 'chur-ock churock'; 'pick-em-up'. *Nesting:* small cup of bark and grass in slender fork; 2–3 spotted pinkish eggs. *Range:* common resident in forest woodland and heath in Tas, Flinders I and King I.

Yellow Honeyeater   *Lichenostomus flavus*   16–18 cm
Greenish-yellow honeyeater with faint dusky mask found in tropical Qld. *Juv:* similar but more greenish, particularly on back. Noisy flutter of wings in flight. *Voice:* loud 'cheweer cheweer . . .'; 'tut-tut-tut'. *Nesting:* shallow cup of bark and grass in horizontal fork; 2 spotted and blotched white to pink eggs. *Range:* common resident in mangroves, rainforest verges, woodland, gardens, orchards in northen Qld from Mt Isa to Cape York, south to St Lawrence.

White-eared Honeyeater   *Lichenostomus leucotis*   18–22 cm
Masked honeyeater with dark head and prominent white eyepatch. Two forms occur: eastern form (race *leucotis*) with yellowish abdomen; and south-western form (race *novaenorciae*) smaller with more olive abdomen. Eastern race fearless particularly at nesting time; western race shy. *Voice:* varied: loud 'chock'; 'cheery-bob'; falling 'chung chung chung'; five-noted metallic chime (inland Qld); in west, 'tew-wh-wh' first note higher. *Nesting:* deep cup of bark and grass in slender horizontal form; 2–3 spotted and blotched whitish eggs. *Range:* locally common resident (altitudinal migrant in south-east) in dry forest, woodland, mallee and heath in eastern and southern Aust from about Dawson R, Qld, to Murchison R, WA.

Yellow-faced Honeyeater   *Lichenostomus chrysops*   15–18 cm
Masked honeyeater with black mask divided by narrow yellow stripe below eye, and a narrow white comma behind eye. Breast faintly streaked grey. *Voice:* loud cheery 'chick-up'; 'pirrup pirrup pirrup . . .'. *Nesting:* flimsy deep cup of bark, grass and moss in thin fork; 2–3 spotted and blotched pinkish to reddish-buff eggs. *Range:* common resident (in north); or migrant (leaves south-east Apr–May, returns July–Oct) in forest and woodland, particularly in dense understorey, in eastern Aust from Cooktown, Qld, to Mt Lofty Ra, SA; also King I.

HONEYEATERS

Helmeted race

Other races

YELLOW-TUFTED

YELLOW-THROATED

WHITE-EARED

YELLOW-FACED

YELLOW

ad

juv

3 x 16	23 x 15	21 x 15	21 x 14	22 x 16
...fted	Yellow-throated	White-eared	Yellow-faced	Yellow

RB

...e honeyeaters have a narrow black mask passing through the eye, below which is a yellow streak ...alar plume') ending in a white or yellow tuft behind the ear. They are basically very similar but ...nerally live in different habitats.

Varied Honeyeater   *Lichenostomus versicolor*   18–21 cm

Mangrove-dwelling masked honeyeater with yellow malar plume ending in extended white tuft. Two forms occur: southern form (Mangrove Honeyeater, race *fasciogularis*) with yellow, faintly barred throat merging into dusky breast and streaked abdomen, found in or near mangroves from south-eastern Qld (and lower Clarence R and Macleay R in NSW) to Townsville, Qld; and northern form (Varied Honeyeater, race *versicolor*) with yellow underparts faintly streaked brown, found in or near mangroves from Cardwell to Cape York, Qld. Enters well-planted gardens in coastal towns, particularly in north Qld. *Voice:* rich and varied: 'wood a woow'; 'whit-a-we-u'; 'ch-ch-weeyo'. *Nesting:* deep cup of grass and rootlets in horizontal fork; 2 spotted pink eggs. *Range:* rare in NSW common resident in Qld in mangroves and nearby woodland, parks and gardens from Macleay, NSW, to Cape York, Qld.

Singing Honeyeater   *Lichenostomus virescens*   17–22 cm

Masked honeyeater with yellow malar plume ending in white tuft, underparts grey faintly streaked darker, found in woodland and shrubland over most of Aust west of Divide. Grey-headed similar but paler, yellowish underparts, has yellow tuft below ear, grey head; Purple-gaped is darker, inhabits mallee, has small yellow tuft below ear, unstreaked underparts, purple stripe below mask visible at close range only. *Voice:* (misnamed!) loud single, double or repeated 'prrp'; persistent 'kitch-ee-wok' dawn song; loud staccato alarm call. *Nesting:* deep flimsy cup of grass suspended in horizontal fork; 2–3 spotted pinkish or buffish eggs, sometimes unspotted. *Range:* locally common resident in shrublands and woodlands usually with dense understorey throughout Aust west of Divide, also on many inshore islands.

Purple-gaped Honeyeater   *Lichenostomus cratitius*   16–20 cm

Rather shy mallee-dwelling masked honeyeater with two malar stripes, the upper being a purple featherless extension of gape, the lower yellow feathers ending in small yellow plume; underparts plain brownish-grey. *Juv:* malar skin yellow. Singing Honeyeater is paler, has streaked underparts, white tuft below ear, less wary; Grey-headed similar but paler, occurs north of mallee. *Voice:* variable: harsh chattering; 'twit twit' in flight; whistling 'too-whip'. *Nesting:* flimsy cup of bark, grass and down slung in horizontal fork usually low; 2 spotted white eggs. *Range:* uncommon nomad in mallee woodland, sometimes gathering in number where blossom thick, in mallee areas of southern Aust, including Kangaroo I.

Grey-headed Honeyeater   *Lichenostomus keartlandi*   15–17 cm

Grey-headed masked honeyeater with yellow malar plume ending in small yellow tuft; underparts pale yellow faintly streaked grey; found in desert woodland particularly in gorges. Singing Honeyeater lacks yellowish tone, is more brownish-grey with large white tuft below ear. *Voice:* loud 'kwot', 'chee-toyt', 'chip chip chip . . .' in flight. *Nesting:* small cup of bark, grass and down suspended in slender fork in low bush; 2 spotted and blotched white to pale pink eggs. *Range:* locally common in arid woodland particularly along creeks and in gorges, in central and central western Aust.

HONEYEATERS

VARIED

northern
form

southern
'mangrove'
form

SINGING

PURPLE-
GAPED

GREY-
HEADED

21 x 16	24 x 18	22 x 16	21 x 15	20 x 14
(uthern)	(northern)	Singing	Purple-gaped	Grey-headed
Varied				

...han masked honeyeaters, these birds are similar, differing mainly in lack of facial mask;
the eye is a small white or yellow plume made more obvious by a narrow black crescent.

...ey-fronted Honeyeater   *Lichenostomus plumulus*   15–17 cm
...ellow-plumed honeyeater with dark grey plume, large yellow plume,
...aint streaking on underparts. Two forms occur; southern form has
narrow grey strip across forehead (barely visible in field); northern form
lacks this grey 'front'. Similar to Yellow-plumed, which has more tuft-
like plume and heavily streaked underparts; Yellow-tinted has yellow
lores, lacks streaking on underparts. *Voice:* Loud 'it-wirt wirt . . .';
Nesting: small neat cup of bark, grass and down, mostly in vertical
foliage, sometimes in horizontal fork; 2 spotted pink eggs. *Range:*
common resident or nomad in woodland and riverine vegetation in
drier areas of Aust.

Yellow-plumed Honeyeater   *Lichenostomus ornatus*   15–18 cm
Yellow-plumed honeyeater with dark grey and olive lores, tuft-like
plume, prominent streaking on underparts. Similar to Grey-fronted,
which has only faintly streaked underparts. *Voice:* harsh 'chick-a-dee';
rapid alarm call 'ti-ti-ti . . .'; *Nesting:* small cup of grass and down
suspended in hanging foliage; 2 freckled white to pale reddish-buff eggs.
Range: common resident to nomad in dry forest and woodland,
particularly mallee, wandoo and tuart, in drier areas of southern Aust
from western NSW and Vic to Murchison R, WA.

White-plumed Honeyeater   *Lichenostomus penicillatus*   15–19 cm
White-plumed honeyeater with yellowish lores, unstreaked underparts
(but faintly mottled on breast). Birds in south-east are greenish,
merging into yellowish birds in arid regions where they frequent river
gums, being among commonest and most obvious birds along inland
watercourses. *Voice:* variable: 'chee-uck-oo-wee'; 'chick-wist chick-wist
. . .' in flight; loud persistent alarm 'ti ti ti . . .'. *Nesting:* deep frail cup
of grass and down in vertical or horizontal fork; 2–3 spotted white, pink
or buff eggs. *Range:* common resident in woodland, usually near water,
in south-eastern, central and midwestern Aust and inland Qld.

Fuscous Honeyeater   *Lichenostomus fuscus*   14–17 cm
Yellow-plumed honeyeater with dull olive unstreaked underparts, small
plume, among plainest-looking honeyeaters. *Adult breeding:* bill and
eye-ring black. *Non-breeding:* base of bill, gape and eye-ring yellow. In
north (ie Atherton Tableland) plumage is yellower, approaching
Yellow-tinted, which generally inhabits riverine vegetation rather than
forest. *Voice:* rich 'arig arig a taw taw', reminiscent of Black-chinned;
'tew tew . . .'; alarm call 'ti ti ti . . .'. *Nesting:* deep cup of grass and
down in horizontal or vertical fork; 2 spotted pink or buffish eggs.
Range: common resident or nomad in north, partial migrant in south,
in forest and woodland in south-eastern and eastern Aust, from Flinders
Ra, SA, to Cooktown, Qld.

Yellow-tinted Honeyeater   *Lichenostomus flavescens*   14–17 cm
Yellow-plumed honeyeater with yellow lores, unstreaked underparts
with yellowish throat and upper breast. Similar to Grey-fronted, which
is not as yellow-looking and has dark lores; sometimes considered a race
of Fuscous Honeyeater, which is much duller looking and also has dark
lores. *Voice:* 'porra-chew, porra-chew, porra-chew-chew-chew'; 'tew-
tew'; alarm call 'ti ti ti . . .'. *Nesting:* flimsy deep cup of grass and down
in vertical fork in eucalypt or horizontal fork in bauhinea; 2 spotted pink
eggs. *Range:* common resident in woodland and riverine vegetation in
northern Aust from Herberton, Qld, to Derby, WA.

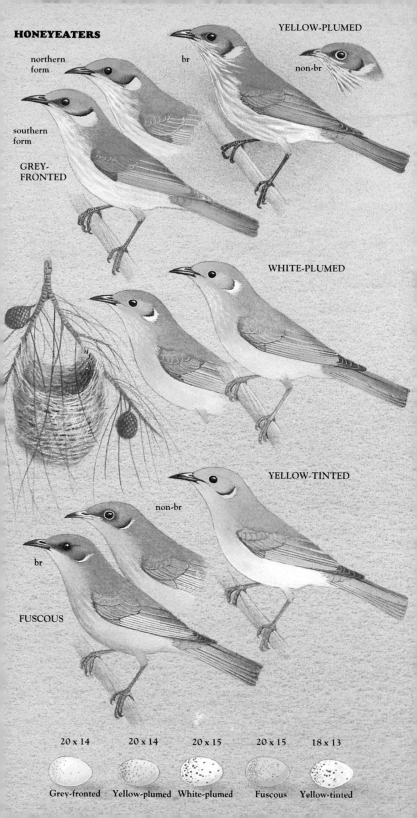

HONEYEATERS

YELLOW-PLUMED

northern form

br

non-br

southern form

GREY-FRONTED

WHITE-PLUMED

YELLOW-TINTED

non-br

br

FUSCOUS

20 x 14	20 x 14	20 x 15	20 x 15	18 x 13
Grey-fronted	Yellow-plumed	White-plumed	Fuscous	Yellow-tinted

YELLOW-WINGED HONEYEATERS

Many honeyeaters have yellow edges to wing and tail feathers, but in these the yellow is accentuated into bold splashes of colour – grouping them here is for convenience only; neither Regent nor Painted is related to the others, which are more closely allied and occur commonly throughout their ranges. Regent and Painted tend to be encountered sporadically.

Crescent Honeyeater *Phylidonyris pyrrhoptera* 15 cm
Yellow-winged honeyeater with crescent on side of breast. *Male:* back dark grey, crescent black. *Female:* back brownish-grey, crescent brown. *Voice:* loud cheery 'egypt'; simple 'jip' in winter. *Nesting:* deep cup of bark, dry grass, spiderwebs and plant-down, in fork in low bush or shrub; 3 spotted pink to reddish eggs. *Range:* common resident or migrant in woodland, forest, wetter heaths, gullies and gardens in south-eastern mainland and Tas.

White-fronted Honeyeater *Phylidonyris albifrons* 16–18 cm
Yellow-winged streaked honeyeater with white face and small red wattle behind eye. *Juv:* no red wattle. Western birds have more abrupt demarcation between black breast and white abdomen. *Voice:* musical 'tsooee'; metallic nasal 'trreep'; 'pert-pertoo-weet'; harsh 'truk truk'. *Nesting:* deep cup of grass and bark decorated with spiders' eggsacs in fork of bush or small tree; 2–3 spotted pinkish or buff eggs. *Range:* uncommon nomad in heaths, mallee and mulga in arid areas.

New Holland Honeyeater *Phylidonyris novaehollandiae* 16–19 cm
Yellow-winged boldly streaked honeyeater with white eye and wispy black beard on black breast and white tip to tail. *Juv:* browner with grey eye. *Voice:* loud 'tchlik'; weak whistling 'pseet'; shrill harsh chattering. *Nesting:* untidy cup of grass and twigs in fork of low bush usually prickly; 2–3 spotted and blotched pinkish eggs. *Range:* common resident in heath and undergrowth in forest and woodland, particularly where banksias are common in south-west, south-east and Tas.

White-cheeked Honeyeater *Phylidonyris nigra* 16–19 cm
Yellow-winged boldly streaked honeyeater with dark eye, large white cheek patch and no white tip to tail. *Juv:* duller. Two forms occur: western form (race *gouldi*) with longer bill and white cheek patch compact and pointed; and eastern form (race *nigra*) with shorter bill and fan-shaped white cheek patch. *Voice:* vigorous 'chip-choo-chippy-choo'; 'twee-ee-twee-ee'; rapid alarm call 'tee-tee-tee . . .' often in concert with other honeyeaters; harsh 'chak-a-chak'. *Nesting:* untidy cup of grass, rootlets and twigs in fork of low bush or clump of grass; 2–3 spotted and blotched pinkish eggs. *Range:* common resident in heath and undergrowth in woodland in south-west; eastern Aust between Atherton Tableland, Qld, and Bermagui, NSW.

Painted Honeyeater *Grantiella picta* 15–17 cm
Yellow-winged honeyeater with pink bill, black upperparts and white underparts. *Male:* narrow black streaks on flanks. *Female:* brown above, no streaks on flanks. *Voice:* loud 'george-ee'; 'pretty-pretty . . .'; 'kow-kow-kow'. *Nesting:* flimsy cup of rootlets and grass bound with spiderweb suspended in leafy branch; 2–3 spotted salmon-red eggs. *Range:* nomad or irregular migrant in mistletoes in woodland and open forest in eastern Aust.

Regent Honeyeater *Xanthomyza phrygia* 20–23 cm
Yellow-winged honeyeater with bare warty face, yellow edges to feathers on back and black edges to feathers on belly. *Juv:* browner with yellowish bill. *Voice:* metallic 'chink-chink-chink'; musical 'quipper-quip'; 'cloop-cloop-cloop'. *Nesting:* substantial cup of bark and grass in upright fork of sapling or in mistletoe; 2–3 spotted salmon eggs. *Range:* uncommon nomad in woodlands and open forest in south-east from Kangaroo I to Rockhampton, Qld.

HONEYEATERS

WHITE-CHEEKED

NEW HOLLAND

western form

eastern form

PAINTED

WHITE-FRONTED

REGENT

CRESCENT

♂

♀

21 x 15   21 x 15   20 x 15   20 x 14   19 x 15   24 x 18

ew Holland   White-cheeked   Painted   White-fronted   Crescent   Regent

HONEYEATERS

Fasciated honeyeaters: two species with barring on breast, bold in one, faint in other, differing from all other Aust honeyeaters in construction of nest, which is bulky and *domed* with entrance near top.
Conopophila honeyeaters: three small short-billed honeyeaters less dependent on nectar than most species. Rufous-throated and Rufous-banded breed during tropical wet season, build purse-like nests.
Tawny-crowned honeyeater is a 'yellow-winged honeyeater' (previous page) without any yellow in wing.

Bar-breasted Honeyeater   *Ramsayornis fasciatus*   12–15 cm
Small brown honeyeater with boldly barred breast, streaked flanks, white face with fine black 'moustache', dull reddish bill. *Juv:* breast and flanks streaked with 'teardrop' markings. *Voice:* rapid piping; soft 'phew'. *Nesting:* large dome of paperbark and rootlets bound with spiderweb suspended in outer leaves of paperbark tree usually over water; 2–3 freckled white eggs. *Range:* nomad or migrant in paperbarks on northern rivers and swamps from Fitzroy R, WA, to Burdekin R, Qld.

Brown-backed Honeyeater   *Ramsayornis modestus*   11–12 cm
Small brown honeyeater with obscurely barred breast, dark face with white 'moustache', reddish bill. *Juv:* duller with faint streaking on breast. *Voice:* rapid 'mick, mick . . .'; chattering 'shee-shee . . .'. *Nesting:* large dome of paperbark strips bound with spiderweb, decorated with spiders' eggsacs, in paperbark branch often over water, often in loose colonies; 2 freckled white eggs. *Range:* common migrant in paperbarks, mangroves and swampy woodlands in north-east.

Rufous-throated Honeyeater   *Conopophila rufogularis*   11–13 cm
Small brown honeyeater with short bill, rufous throat and yellow edges to wing feathers. *Juv:* lacks rufous throat; similar to juv Rufous-banded but has brown head. *Voice:* rasping sparrow-like chatter; sharp 'zit-zit'. *Nesting:* deep felted purse-like cup of fine grass, paperbark and plant-down bound with cobweb in hanging branch; 2–3 spotted white eggs. *Range:* common resident in riverine vegetation and gardens in north; migrant to drier areas and down to about Ingham, Qld.

Rufous-banded Honeyeater   *Conopophila albogularis*   11–13 cm
Small brown honeyeater with short bill, rufous breastband and yellow edges to wing-feathers. *Juv:* lacks breastband; similar to juv Rufous-throated but has grey head. *Voice:* harsh 'zwee' like young cuckoo; 'shwee-whit-chi-ti'. *Nesting:* deep cup of paperbark strips bound by cobweb and tendrils in hanging branch often over water; 2–3 spotted white eggs. *Range:* common nomad in riverine forest, mangroves, swampy woodland and gardens Top End and north-east coast of Gulf of Carpentaria.

Tawny-crowned Honeyeater   *Phylidonyris melanops*   15–18 cm
Long-billed brown honeyeater with pale fawn crown, black crescent on side. *Juv:* duller with yellow throat and pale yellow edges to wing feathers. *Voice:* nasal sneezing 'kneep'; melodious rising 'peer-peer-pee-pee . . .'. *Nesting:* untidy cup of grass and bark bound with spiderweb in fork in low bush; 2–3 blotched white eggs. *Range:* common nomad or resident in heaths, mallee and woodland in southern Aust; in heath and buttongrass plains in Tas.

Grey Honeyeater   *Conopophila whitei*   11–12 cm
Small grey honeyeater with white tip to tail, found mainly in mulga woodland. *Adult:* grey above with faint yellow wash on wing quills, white below with greyish clouding on breast, faint white eye-ring. *Juv:* obvious white or pale lemon eye-ring, pale base to bill, looks rather like a grey Silvereye. Usually found in small parties, sometimes with thornbills and warblers, Can be confused with Western Warbler, but white in tail confined to tip, eye brown not red, and no white eyebrow. *Voice:* silvereye-like 'te-te-dee' or 'tsee-youee'. *Nesting:* sketchy hammock of bark, grass and cobwebs slung on horizontal fork; 2 spotted white eggs. *Range:* uncommon to rare nomad in mulga woodland and acacia thicket in spinifex throughout arid western and central Aust.

HONEYEATERS

BROWN-BACKED

BAR-BREASTED

ad

juv

RUFOUS-THROATED

RUFOUS-BANDED

TAWNY-CROWNED

GREY

ad

juv

20 x 13	21 x 14	18 x 13	Rufous-throated	21 x 14	17 x 12
Brown-backed	Bar-breasted	Rufous-banded	18 x 13	Tawny-crowned	Grey

LONG-BILLED HONEYEATERS

Many tubular flowers such as grevilleas have developed pollination mechanisms involving these small long-billed honeyeaters as well as those shown overleaf. Pollen is dabbed on bird's forehead and carried from flower to flower, often showing as yellow or orange 'forehead'. Some birds are strongly nomadic, following flowering seasons of various trees and shrubs, others have regular local movements to coincide with blossoming. Often hover in front of flowers while feeding.

Dusky Honeyeater *Myzomela obscura* 12–14 cm

Plainest honeyeater, uniformly dusky brown with darker chin. *Juv:* base of bill yellow. Usually in parties or small groups, aggressive and vivacious, often seen with other honeyeaters. Similar to females of Scarlet and Red-headed but darker below and lacks reddish chin. *Voice:* soft whistle followed by chirping; rapid 'see-see-see'. *Nesting:* flimsy cup of rootlets and grass bound by cobweb in outer foliage; 2 freckled pinkish eggs. *Range:* common in mangrove, rainforest edges, riverine and swampy forests, gardens and woodland in northern and eastern Aust from Top End, NT, to Brisbane, Qld.

Scarlet Honeyeater *Myzomela sanguinolenta* 10–11 cm

Male: red honeyeater with black wings and tail and white abdomen. *Female:* brown honeyeater paler below with reddish chin. *Juv:* similar to female but yellow base to bill. *Immature male:* varying amounts of red on head and breast; at one stage rather similar to Red-headed. *Voice:* high-pitched 'to-see . . .'; spirited falling song 'plid-o-willy-erk'. *Nesting:* flimsy cup of bark and grass bound with cobwebs in foliage, often thick such as wild passionfruit vine; 2–3 spotted and blotched white eggs. *Range:* common nomad in forest and woodland in eastern Aust, less common south of Sydney, NSW.

Red-headed Honeyeater *Myzomela erythrocephala* 10–12 cm

Male: dark brown honeyeater with red head and rump. *Female:* brown honeyeater with reddish chin and forehead, paler below. *Voice:* harsh 'chiew-chiew-chiew'; sibilant 'chirp'. *Nesting:* flimsy cup of bark and rootlets bound with cobwebs slung in outer clump of mangrove leaves; 2 spotted and blotched white eggs. *Range:* common resident in mangrove and contiguous vegetation, gardens, riverine vegetation and woodland when flowering in northern Aust, from Derby, WA, to Cape York Peninsula, Qld.

Eastern Spinebill *Acanthorhynchus tenuirostris* 13–16 cm

Fine-billed eastern honeyeater with extensive white in outer tail, loud fluttering wings in flight. *Male:* head and crescent on sides of breast, black; abdomen cinnamon. *Female:* head dull olive-grey, underparts duller. *Juv:* upperparts olive-grey, underparts pale cinnamon with throat paler. *Voice:* staccato piping, softer 'chee-chee-chee . . .'. *Nesting:* small, well-hidden cup of grass, moss and hair bound with cobweb; 2–3 spotted pinkish eggs. *Range:* common to uncommon resident or short-range migrant in forest, woodlands, gardens, thickets and heaths in eastern Aust and Tas.

Western Spinebill *Acanthorhynchus superciliosus* 13–16 cm

Fine-billed south-western honeyeater with rufous nape. *Male:* throat and nape rich rufous, black and white breast band. *Female:* throat and underparts soft grey, nape pale rufous. *Juv:* similar to female but base of bill yellow. *Voice:* shrill 'kleat-kleat'. *Nesting:* neat cup of bark and grass bound with cobweb in low bush or tree, sometimes conspicuous; 1–2 spotted and blotched pinkish or bluish eggs. *Range:* common in undergrowth in forest, woodland and heath in south-western Aust.

HONEYEATERS

SCARLET

imm

♂

♀

DUSKY

RED-HEADED

♀

♂

SPINEBILLS

WESTERN

♀

EASTERN

♂

imm

♂

16 x 12	16 x 12	17 x 13	17 x 13	18 x 13
Scarlet	Red-headed	Dusky	Eastern	Western

LONG-BILLED HONEYEATERS

Certhionyx honeyeaters: long-billed honeyeaters with pied males. Black and Pied have brown females, occupy arid zone, primarily mulga and similar woodlands; Banded has female similar to male, occupies tropical woodland, principally bauhinea/eucalypt associations. Black and Banded have flimsy cup-shaped nests, Pied has substantial miner-like nest, may be more closely allied to miner than to Black and Banded. Displays by males are spectacular: Pied dives with wings closed and tail spread like jet fighter; Black has lovely stiff-winged aerobatic ballet. **Bare-eyed honeyeaters:** two tropical forest honeyeaters with long stout bills, representatives of similar New Guinea species; bills may be adaptation to probing into bark crevices and clusters of dead leaves for insects and into fruit rather than for flowers. Nests are typical of honeyeaters.

Black Honeyeater   *Certhionyx niger* 10–12 cm
Male: black with abdomen and sides of breast white. *Female:* brown above, with pale line behind eye; below pale buff with brown delicately scalloped breastband. *Voice:* plaintive 'peee'; 'pee-p-pee'; pretty almost inaudible chattering song. *Nesting:* flimsy hammock of grass, twigs and rootlets in low bush, sometimes among dead fallen branches; 2 spotted pale buff eggs. *Range:* locally common nomad mainly in mulga but irrupting into other woodland throughout drier areas of Aust.

Banded Honeyeater   *Certhionyx pectoralis* 11–14 cm
Only pied honeyeater with black breastband. *Female:* faint buff tinge to abdomen. *Juv:* brown where adults are black, yellowish earpatch. *Imm:* like adult but back feathers edged buff. *Voice:* jaunty 'tseep'; husky 'didrap'; tinkling twittering song. *Nesting:* flimsy cup of bark and grass bound with cobweb, usually in small bauhinea, often in loose colonies; 2 speckled pale buff eggs. *Range:* locally common nomad usually in loose flocks in tropical woodland and mangroves in northern Aust.

Pied Honeyeater   *Certhionyx variegatus* 15–18 cm
Male: pied honeyeater with white patch on wing and only tip of tail black, patch of pale blue skin below eye. *Female:* brown above with complete eye-stripe, pale fawn below with faint streaking on breast, patch in wing pale grey with black centres to feathers; small blue wattle below eye. *Voice:* plaintive 'tee-titee-tee-tee'. *Nesting:* locally common nomad in arid woodland, mostly in mulga following rain, throughout drier parts of Aust.

Tawny-breasted Honeyeater   *Xanthotis flaviventer* 18–22 cm
Bare-eyed, olive-brown honeyeater with small bare eye-patch, streaked nape, yellow breaks below eye and buff breast. *Voice:* cheerful 'which witch is which'. *Nesting:* neat cup of bark and rootlets suspended in foliage, often quite high; 2 spotted pinkish eggs. *Range:* uncommon resident in mangroves, rainforest and woodland on Cape York Peninsula, Qld, to Archer R in west and McIlwraith Ra in east.

Macleay's Honeyeater   *Xanthotis macleayana* 18–21 cm
Boldly streaked bare-eyed honeyeater with large orange bare patch around eye. *Voice:* loud 'chew-it-che-wew' or 'tweet your juice'. *Nesting:* neat cup of leaves, bark and fibre bound with cobweb in bush or low in tree; 2 spotted pinkish-buff eggs. *Range:* common resident in mangroves, rainforest and adjacent woodland and gardens in north-eastern Qld, from Cooktown to Cardwell.

HONEYEATERS

BANDED

imm

juv

♂

BLACK

♀

♂

PIED

♀

♂

MACLEAY'S

TAWNY-BREASTED

17 x 12

Banded

15 x 12   22 x 16

Black   Pied

25 x 17   32 x 17

Tawny-breasted   Macleay's

HONEYEATERS AND SUNBIRDS

Green-backed Honeyeater is little known in Aust, could be confused with warbler or silvereye, has most restricted range of any Aust bird except Noisy Scrub-bird. Brown Honeyeater, delightful songster, is widely spread throughout western and northern Aust. Striped and White-streaked have lanceolate feathers on breast, forked tails, but differ in habitat and habits; Striped has short bill, feeds much on leaf-rolling insects and spiders. **Sunbirds:** superficially like honeyeaters but have iridescent plumage in males. Feed on nectar, insects and spiders; build long pendulous nests, often in gardens.

Green-backed Honeyeater   *Glycichaera fallax*   11–12 cm
Warbler-like honeyeater with puffy feathers particularly on lower back and rump, upperparts greyish-green, throat white, breast and abdomen pale yellow, narrow white ring around pale grey eye; similar in colour to female Fairy Warbler, which is slightly smaller, has blunt bill, grey not blue legs and red eye. *Voice:* chicken-like 'peep'; abrupt 'twit'; rapid 'twee-twee-twit-twit'. *Nesting:* unknown. *Range:* locally common resident in small area of rainforest and adjacent woodland on Cape York Peninsula in vicinity of Claudie R.

Brown Honeyeater   *Lichmera indistincta*   12–16 cm
Plain olive-brown honeyeater with small yellow patch behind eye, and usually with yellow gape (except male while breeding, when it is black); edges of wing feathers washed greenish-yellow. *Voice:* loud rich 'plik'; varied song, rich and musical. *Nesting:* neat cup of grass, bark and leaves decorated with eggsacs, suspended in fork in foliage in bush or low tree; 2–3 spotted white eggs. *Range:* common resident and nomad in forest, woodland, gardens, mangroves, scrubland and wooded watercourses in western, northern and north-eastern Aust.

White-streaked Honeyeater   *Trichodere cockerelli*   15–19 cm
Untidy-looking dark brown honeyeater with yellow edges to wing and tail feathers, yellow streaks below and behind eye, streaked whitish underparts with lanceolate feathers on breast. *Juv:* duller with yellowish chin. *Voice:* persistent scolding; sweet four-noted song like Brown. *Nesting:* flimsy cup of grass and rootlets slung in fork of low bush; 2 spotted and blotched pink eggs. *Range:* common resident or short-range nomad in paperbark swamps, woodland, heath and vine thickets on Cape York Peninsula south to Edward R on west coast and Coen in east; isolated population at Shipton's Flat on east coast.

Striped Honeyeater   *Plectorhyncha lanceolata*   20–23 cm
Short-billed honeyeater with boldly striped head and neck, grey back, buffy-white underparts with lanceolate feathers on breast. Usually in pairs. *Voice:* rich cheerful 'cher-cher-cherry-cherry . . .', often as duet. *Nesting:* deep, thickly-felted purse of grass, plant-down, feathers and cobwebs, suspended in hanging foliage; 2–4 freckled pink eggs. *Range:* common resident in drier woodland, mulga and mallee in drier areas of eastern Aust.

Yellow-bellied Sunbird   *Nectorinia jugularis*   10–12 cm
Long-billed sunbird with yellow eyebrow and bright yellow underparts. *Male:* throat and breast black with blue and purple iridescence. *Female:* entire underparts yellow. *Voice:* shrill 'tsee-tsee-tsee tss-ss-ss'; high-pitched 'dzit-dzit'. *Nesting:* long pendant with hooded entrance in side, made of bark, grass, leaves and feathers bound with cobwebs, suspended from twig (or hanging rope or wire); 2–3 mottled greenish eggs. *Range:* common resident in rainforest margins, mangroves, riverine vegetation and gardens in north-eastern Qld, from Cape York to Gladstone.

HONEYEATERS

BROWN

GREEN-
BACKED

STRIPED

WHITE-
STREAKED

SUNBIRDS

YELLOW-
BELLIED

17 x 13	23 x 17	16 x 14	17 x 12
Brown	Striped	White-streaked	Yellow-bellied

WHITE-EYES AND FLOWERPECKERS

White-eyes: small yellow-green birds with short pointed bills, brush-tipped tongues like honeyeaters and a ring of white feathers around each eye. Feed on nectar, insects and fruit; usually move in flocks except when breeding. Nest is flimsy cup, usually slung on low fork; eggs are beautiful unmarked pale blue or blue-green. **Flowerpeckers:** tiny birds with brightly-coloured males. Single Aust species feeds mainly on mistletoe berries, has reduced gut; wings have only nine primaries (as in pardalotes), flies swiftly and erratically, looks swallowlike in flight. Nest is delicate, felted dome 'like baby's bootee'.

Pale White-eye
Zosterops citrinella 12 cm

White-breasted white-eye found on islands off Cape York Peninsula, closest in appearance to more southerly Barrier Reef form of Silvereye, which lives on islands of Capricorn–Bunker Group, has grey back. *Voice:* like other white-eyes. *Nesting:* neat cup of grass bound by cobweb in horizontal fork of shrub; 2–4 pale blue-green eggs. *Range:* common in thickets on islands of Cape York Peninsula from about mid Torres Strait to Eagle I, near Lizard I.

Yellow White-eye
Zosterops lutea 11 cm

Yellow-breasted white-eye found in or near tropical mangroves. *Voice:* similar to other white-eyes but harsher and louder, presumably to complete with sound of waves. *Nesting:* deep soft cup of grass bound with cobweb, suspended from mangrove fork; 3–4 pale blue eggs. *Range:* common in or near mangroves on coast and along rivers in northern Aust from Shark Bay, WA, to Edward R on western Cape York Peninsula; and in mangroves on north-east coast in vicinity of Ayr, Qld.

Silvereye
Zosterops lateralis 12 cm

Grey-breasted white-eye found in forest, woodland and heaths in south-western, southern and eastern mainland Aust and Tas; all eastern forms have grey backs, western form has olive back. Several identifiable forms occur: Tasmanian form (race *lateralis*) with rufous flanks, greyish throat and undertail coverts, migrates to south-east north to about Caloundra, Qld, in autumn; eastern forms (race *halmaturina, familiaris* and *ramsayi*) vary from buff-flanked with greyish throat and undertail coverts in SA to grey-flanked with yellow throat and undertail coverts in Qld – as there is autumn migration, different variants can be seen together in winter; Barrier Reef form (race *chlorocephala*) is larger, has white undertail coverts, confined to islands of Capricorn–Bunker Group of southern Barrier Reef; and western form (race *gouldii*) with olive-green back, from North-West Cape, WA, to eastern coast of Bight in SA and isolated population at Kalgoorlie, WA; intergrading with eastern birds on Eyre Peninsula and Kangaroo I. *Voice:* loud 'tsee'; pleasant warbling including mimicry; rapid 'giggle'. *Nesting:* compact cup of grass, plant down and hair bound with cobweb, suspended in low fork or in vine; 2–4 pale blue eggs. *Range:* common resident, nomad or migrant in south-western, southern and eastern Aust from North-West Cape, WA, to Cape York, Qld; and Tas.

Mistletoebird
Dicaeum hirundinaceum 10–11 cm

Red-breasted or grey-breasted flowerpecker with pinkish-red undertail coverts. *Male:* red breast with black stripe in centre of belly, iridescent dark blue upperparts. *Female:* pale grey breast with some faint speckles, dusky grey streak on centre of belly, upperparts dark grey. Flight rapid and erratic, rather like startled swallow. *Voice:* high-pitched 'swee-swit' or 'swit'; sharp 'wit' in flight; high-pitched warbling including mimicry. *Nesting:* neat suspended 'baby's bootee' of plant-down and spiderweb, sometimes decorated with caterpillar droppings; 3–4 white eggs. *Range:* common nomad throughout Aust wherever mistletoe occurs; also feeds in gardens on introduced plants such as Japanese pepper.

WHITE-EYES

YELLOW

PALE

Barrier Reef form

SILVEREYE

south-eastern form

Tas form

western form

♀

MISTLETOEBIRD

♂ FLOWERPECKERS

17 x 13
Yellow

17 x 13
Pale

17 x 13
Silvereye

17 x 11
Mistletoebird

CHATS

Small colourful birds similar to honeyeaters but much less dependent on nectar. Most of their food is insects taken on ground where they walk with characteristically nodding heads. They usually occur in small groups and sometimes nest in loose colonies. Nests are open shallow cups of grass built in low bushes such as samphire, bluebush or Ptilotis. Some species are highly nomadic in the interior, exploiting temporary lush growth after rain.

White-fronted Chat   *Ephthianura albifrons* 11–12 cm

White-headed or grey-headed chat with dark breastband. *Male:* head white with black hood, breastband broad and black. *Female and juv:* head grey, breastband narrow and dusky. Usually in small groups, often seen perched on low bushes or samphire. *Voice:* metallic 'tang', rather like Zebra Finch. *Nesting:* neat cup of grass in low bush, grass tussock or clump of weeds; 3–4 spotted eggs. *Range:* common resident or nomad in samphire flats, damp paddocks, low shrublands and grasslands in southern mainland and Tas.

Gibber Chat   *Ashbyia lovensis* 12–13 cm

Yellow-breasted chat with rump same colour as back (cinnamon-brown) usually found on stony plains ('gibbers') in central Aust. *Male:* breast yellow with buffish tinge. *Female:* breast duller yellow. Usually singly or in pairs, most often seen on ground or perching on rocks. *Voice:* 'wheat-wheat wheat'; 'chip-chip' in flight. *Nesting:* sparse cup of grass in depression on ground usually under grass clump; 3 spotted white eggs. *Range:* uncommon resident or nomad on arid treeless stony plains in south-west Qld, north-west NSW, south-east NT and north-east SA.

Orange Chat   *Ephthianura aurifrons* 10–12 cm

Orange-rumped chat with breast either orange or yellowish-grey. *Male:* bright orange with black face, rich orange rump. *Female:* pale yellowish-grey, browner on wings, pale orange rump similar to female Yellow which has white eye. *Juv:* ashy-brown with pale orange rump. Usually in small groups in samphire, bluebush or saltbush; runs quicker than Crimson, does not nod head. *Voice:* 'check-check' like woodswallow; 'cheel-cheel' in flight. *Nesting:* untidy cup of grass and rootlets in low bush; 3 spotted white eggs. *Range:* common resident on many inland samphire flats, but mostly highly nomadic in low shrublands in drier areas of Aust.

Yellow Chat   *Ephthianura crocea* 11–12 cm

Yellowish chat with bright yellow rump, found on river flats of northern coastal rivers and on bore drains and grassy depressions in south-west Qld. *Male:* bright yellow on head and underparts with black band on chest, broad in north, narrow in central Aust. *Female:* paler yellow below, lacks breastband. *Juv:* yellowish-grey below. Often shy, flying high in large semi-circle when disturbed. Gibber Chat occupies same range in western Qld, lacks yellow rump, white edges to wing feathers; female Orange Chat has more orange rump, less yellowish below, has dark red eye. *Voice:* nasal 'nang'; pleasant 'pee pee peep' with downward inflection. *Nesting:* shallow cup of grass in long grass or small bush; 3 spotted white eggs. *Range:* rare (locally common in subcoastal NT) resident in grassy swamps in north, from Fitzroy R, WA, to Fitzroy R, Qld; and on bore drains in south-west Qld from Coorooboolka Station to Windorah.

Crimson Chat   *Ephthianura tricolor* 10–12 cm

Brownish or reddish chat with crimson rump. *Male:* unmistakable with crimson crown, breast and rump. *Female:* more brownish-looking with reddish mottling on breast. *Juv:* more reddish-brown above and fawn below. After good rains inland large numbers may appear and nest in loose colonies. *Voice:* high-pitched single or multiple 'tseee'; pleasant warble. *Nesting:* shallow cup of grass in low bush; 3 spotted white eggs. *Range:* common nomad with periodic irruptions in plain, grassland, mulga and open woodland in drier areas of Aust.

CHATS

WHITE-FRONTED

juv

♀

♂

juv

ORANGE

♂

♀

GIBBER

♀

YELLOW

imm

♂

CRIMSON

juv

♂

♀

18 x 14	20 x 15	17 x 13		17 x 13	17 x 14
White-fronted	Gibber	Orange		Yellow	Crimson

PARDALOTES

'Pardalote' derives from a Greek word meaning 'spotted'. Striated has small red, yellow or orange spot in wing, Red-browed has spotted crown and Spotted is profusely marked; Forty-spotted actually has more than 40 spots, but few observers get opportunity to count. They are pudgy little birds with short tails, strong legs, stout blunt bills (adapted for taking scale insects from leaves – compare with Weebill, p 264, which takes similar food). They feed mostly in outer canopy, mainly high in eucalypts, but most nest in holes dug in earth banks, including roadside cuttings; others build in hollow limbs. Flight is undulating and rapid, showing pale stripe in wings which have only nine primaries, unique among perching birds to pardalotes and flowerpeckers. Often chased by magpies, wattlebirds and drongos.

Spotted Pardalote   *Pardalotus punctatus*   8–9·5 cm
Profusely spotted pardalote with red or yellow and red rump. *Male:* crown black with white spots, yellow throat and undertail coverts. *Female:* crown brown with buffish spots, throat whitish, undertail pale yellow. *Juv:* crown greyish with green tinge, only faintly spotted. Two forms occur: red-rumped form (race *punctatus*) in sclerophyll forest and woodland in east, south-east, Tas and south-west; and yellow-rumped form (race *xanthopygus*) with red and yellow rump, found in mallee from south-east to south-west. *Voice:* monotonous 'sleep baby'. *Nesting:* bark globe in hole dug in earthen bank or sloping ground; 3–5 white eggs. *Range:* common nomad or resident in sclerophyll forest, woodland and mallee from Atherton Tableland, Qld, to Jurien Bay, WA; and Tas.

Forty-spotted Pardalote   *Pardalotus quadragintus*   9–10 cm
Greenish Tasmanian pardalote with freckled crown and dully spotted back. Never widely spread in Tas, now restricted to few areas in south-east, but may be on increase. Similar to juvenile Spotted which is more brownish rather than greenish, has red rump. *Voice:* 'whee-wit', first note louder. *Nesting:* dome or cup of bark lined with grass, either in hole in ground or more commonly on hollow limb; 4 white eggs. *Range:* rare resident or partial nomad in dry sclerophyll forest in coastal south-eastern Tas.

Red-browed Pardalote   *Pardalotus rubricatus*   10–12 cm
Yellow-rumped pardalote with spotted crown, red eyebrow and yellow edges to wing feathers, found in drier areas of central and northern Aust. *Voice:* loud five-noted call 'whee whee wh-wh wit', similar in cadence to western Crested Bellbird. *Nesting:* dome or cup of bark lined with grass in hole in earthern bank along creek or roadside cuttings; 3 white eggs. *Range:* common resident in dry woodlands particularly along watercourses in central and northern Aust, generally avoiding higher-rainfall areas.

Striated Pardalote   *Pardalotus striatus*   9–11·5 cm
Variable pardalote with crown black, with or without stripes. Many identifiable forms occur, distinguished by colour of rump, number of feathers edged white in wing, whether wing spot is red or yellow and whether crown is black or striped; interbreeding occurs between forms.
(a) Black-crowned forms: yellow-rumped form (race *uropygialis*) in north; and cinnamon-rumped form (race *melanocephalus*) in north-east, south to northern NSW.
(b) Stripe-crowned forms: south-eastern form (race *ornatus*) with red spot and narrow white stripe in wing; western form (race *substriatus*) with red spot and broad white stripe on wing; yellow-spotted form (race *striatus*) with yellow spot and narrow stripe in wing, breeding in Tas and migrating to south-east. *Voice:* local variations: 'wit wit', 'witta witta', 'rigby dick', 'pick-it-up'. *Nesting:* dome or cup of bark lined with grass in hole in bank or tree hollow (usually in small knothole); 2–5 white eggs. *Range:* common in forest and woodland throughout mainland Aust and Tas.

PARDALOTES

SPOTTED

red-rumped form

yellow-rumped form

juv

♀

RED-BROWED

juv

typical form

Gulf form

FORTY-SPOTTED

yellow-rumped form   STRIATED   cinnamon-rumped form

western form

south-eastern form

yellow-spotted form

juv

19 x 15

Striated

16 x 13   17 x 13   Red-browed

Spotted   Forty-spotted   18 x 15

FIRETAILS

Finches are small seed-eating birds with stout conical bills adapted for crushing seeds, often occurring in flocks, sometimes in large numbers. **Firetails** are colourful finches with bright red rumps, feeding either on ground or in seeding grasses. Most live in areas of good rainfall; and dry-country species are limited in range by access to surface water. Nests are large globes of grass with a spout-like entrance.

Star Finch   *Neochmia ruficauda*   10–12 cm
Profusely spotted finch with red face and olive back. *Female:* duller than male. *Juv:* head and breast greyish, back dull olive-brown. Two forms occur: northern form (race *clarescens*), still locally common, has back greenish-olive, centre of breast yellow; and eastern form (race *ruficauda*), becoming rare, with back olive-brown, no yellow on underparts. *Voice:* high-pitched 'seet'; rapid sibilant 'polit'. *Nesting:* globe of grass lined with feathers; 3–7 white eggs. *Range:* locally common in wooded grasslands near water in northern Aust from Ashburton R, WA, to western Cape York, Qld (race *clarescens*); and rare in wooded grasslands in upper Dawson R and Mackenzie R, Qld (race *ruficauda*).

Red-browed Firetail   *Aegintha temporalis*   11–12 cm
Olive-backed firetail with red eyebrow, grey underparts and golden patch on side of neck. Birds in Cape York are more yellowish-olive on back and have more extensive grey on head. Usually in small flocks feeding in seeding grasses. *Voice:* high-pitched 'pseet'. *Nesting:* globe of grass placed in clump of leaves, in creeper or in crown of sapling; 4–6 white eggs. *Range:* common resident in forest woodland and mangrove in eastern Aust from Cape York, Qld, to Kangaroo I, SA; introduced population in Darling Ra, WA, near Perth.

Painted Firetail   *Emblema pictum*   10 cm
Brown desert firetail with red face and black underparts spotted with white and mottled with red on belly. *Male:* extensive red face, less heavily spotted on black breast. *Female:* restricted red on face, breast heavily spotted. *Juv:* black bill, no red on face or rump. Usually in pairs or flocks up to 30 among spinifex. *Voice:* harsh 'trut', 'check' or 'check-it-up'; wheezy 'che-che-che-che-che-che-che werreee-ooeee'. *Nesting:* untidy globe of spinifex, twigs and rootlets, usually in spinifex clump; 3–5 white eggs. *Range:* common but patchy in spinifex-covered hills, resident near water, nomadic elsewhere (possibly extending range) in arid Aust from western Qld to north-west coast, south to northern Flinders Ra, SA.

Diamond Firetail   *Emblema guttatum*   12 cm
Brown-backed firetail with pale grey head, black breastband and black flanks spotted white. *Juv:* browner-looking with olive-brown flanks and indistinct breastband. Usually in small flocks feeding on ground. *Voice:* penetrating 'twooo-hee' or 'pairrr'. *Nesting:* globe of grass lined with feathers, with long entrance spout placed in dense bush or high in tree, mistletoe or babbler's nest; 4–9 white eggs. *Range:* uncommon resident in open forest woodland and wooded grassland in eastern Aust generally west of Divide from about Kirrama, Qld, to Eyre Peninsula and Kangaroo I, SA.

Red-eared Firetail   *Emblema oculatum*   11–12 cm
Olive-backed firetail with red earpatch, barred breast and spotted abdomen. *Juv:* duller with black bill, lacking red earpatch, abdomen barred. Usually in pairs or small groups feeding on seeding grasses among low bushes. *Voice:* mournful 'oowee'. *Nesting:* largest nest of Aust finches, made from long grass stems, usually in clump of eucalypt or paperbark leaves; 4–6 white eggs. *Range:* uncommon resident in heath, forest undergrowth and dense vegetation along creeks in south-western Aust from about Perth to Esperance.

Beautiful Firetail   *Emblema bellum*   11–12 cm
Olive-backed firetail with finely-barred underparts and black face with blue eye-ring. *Male:* centre of abdomen and undertail black. *Female:* similar but with centre of abdomen barred. *Juv:* duller with black bill. Usually in pairs or small flocks, feeds on ground in clearings. *Voice:* simple 'pee' or 'pee-you'. *Nesting:* substantial globe of grass with long spout; 5–8 white eggs. *Range:* uncommon, becoming rare, in heaths and dense riverine vegetation in south-eastern Aust from about Muswellbrook, NSW, to Kangaroo I, SA, and Tas.

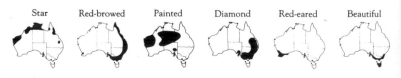

Star   Red-browed   Painted   Diamond   Red-eared   Beautiful

FIRETAILS

STAR

eastern form

northern form

RED-BROWED

juv

ad

juv

ad

PAINTED

♀

DIAMOND

ad

♂

juv

RED-EARED

juv

BEAUTIFUL

ad

| 6 x 11 | 14 x 12 | 18 x 13 | 14 x 11 | 18 x 12 | 14 x 11 |

d-browed   Star   Diamond   Painted   Beautiful   Red-eared

FINCHES

Double-barred Finch *Poephila bichenovii* 10–11 cm

Brown-backed finch with two black bars on breast. Two forms occur: black-rumped form (race *annulosa*) in northern Aust west of Burketown, Qld; and white-rumped form (race *bichenovii*) in north-eastern Aust. Usually in small flocks feeding on ground or among seeding grasses. *Voice:* nasal 'tiaaat, tiaaat' and low 'tat-tat'. *Nesting:* small globe of grass lined with feathers placed in bush or low tree; 4–5 white eggs. *Range:* common resident in open woodland in north and north-east from Broome, WA, to south-eastern NSW, expanding southwards.

Zebra Finch *Poephila guttata* 10 cm

Red-billed finch with barred black and white tail, vertical black and white streak on face. *Male:* chestnut earpatch, finely barred breast, flanks chestnut spotted white. *Female:* fawn-grey breast. *Juv:* bill black. Usually in pairs or small groups but sometimes large flocks near inland waterholes. *Voice:* nasal 'tang'; aggressive 'woot'; trilling song by male. *Nesting:* untidy dome of grass and rootlets lined with feathers placed in shrubs, trees, hollows, under hawk nests etc; 3–6 white eggs. *Range:* common resident or nomad in woodland and grassland throughout Aust except wetter areas.

Plum-headed Finch *Aidemosyne modesta* 11–12 cm

Brown-backed finch with broadly barred underparts and dark purple crown. *Male:* eyebrow dark. *Female:* eyebrow white, barring less distinct. *Juv:* plain greyish-brown bird with white markings on wings, paler on underparts, often in flocks without adults, distinguished from other small brownish birds by flocking habit, upright stance, slender build. *Voice:* soft 'tlip'; loud 'pyiit' soft song. *Nesting:* small globe, without spout, of grass, placed in tall grass or low bush; 4–7 white eggs. *Range:* uncommon to common nomad in grassy woodland from upper Burdekin R, Qld, to about Canberra, ACT; vagrant west Vic.

Blue-faced Finch *Erythrura trichroa* 12 cm

Green finch with blue face and dark red tail. Confined to eastern Cape York, Qld. *Female:* duller than male with less blue on face. *Juv:* lacks blue face. Usually in pairs or small flocks, in rainforest clearings or mangroves. *Voice:* high-pitched trill. *Nesting:* pear-shaped globe of grass, moss and vine tendrils in sapling; 3 white eggs. *Range:* rare in rainforest clearings and edges and mangroves in north-eastern Qld south to Cairns.

Crimson Finch *Neochmia phaeton* 12–14 cm

Red-tailed finch with red face. *Male:* crown blue-grey, underparts red with white spots on side, belly black or white. *Female:* greyish-brown with reddish edges to wing feathers. Several forms occur: black-bellied forms (races *phaeton* and *iredalei*) in north-eastern and northern Aust excluding Cape York; and white-bellied form (race *evangelinae*) on Cape York, mainly on western side. *Voice:* loud 'chee-chee-chee'; sharp 'chip'. *Nesting:* bulky dome of flattened grass stems lined with feathers, in pandanus and paperbarks, even in eaves of houses; 5–8 white eggs. *Range:* common resident in riverside vegetation from about Broome, WA, to Proserpine, Qld, disappearing where cattle disturb habitat.

Gouldian Finch *Erythrura gouldiae* 14 cm

Colourful finch with green back, purple breast and yellow abdomen. *Female:* breast paler purple. *Juv:* greyish to olive-green above, grey-brown below. Several phases occur: black-headed phase outnumbers red-headed phase 3:1; rare golden-headed phase occurs once in several hundred. *Voice:* sibilant 'ssit'. *Nesting:* scanty grass dome usually in hollow spout, sometimes in low bush; sometimes in small colonies; 4–8 white eggs. *Range:* scarce nomad (common where undisturbed) in open woodland in north from Derby, WA, to about Atherton Tableland, Qld.

FINCHES

ZEBRA

southern form

northern form

DOUBLE-BARRED

BLUE-FACED

PLUM-HEADED

juv

ad

juv

black-bellied form

CRIMSON

GOULDIAN

white-bellied form

16 x 11	15 x 11	17 x 12	18 x 13	14 x 12	17 x 13
Double-barred	Zebra	Plum-headed	Blue-faced	Crimson	Gouldian

R.B.

FINCHES

Masked Finch   *Poephila personata*   12–13 cm
Yellow-billed finch with black mask extending only to chin, brown head, white rump and long tail. *Juv:* duller brown, lacks black face, bill black at first. Two forms occur: brown-eared form (race *personata*) in northern Aust except Cape York; and white-eared form (race *leucotis*) on Cape York. *Voice:* loud 'tsit'; quiet 'tat-tat'. *Nesting:* bulky globe of grass lined with soft material and charcoal, on ground, in long grass or bush; 4–6 white eggs (sometimes stained with charcoal). *Range:* common resident in grassy woodland in northern Aust from Derby, WA, to Chillagoe, Qld.

Long-tailed Finch   *Poephila acuticauda*   15–16 cm
Red- to yellow-billed finch with grey head, black throat, white rump and long tail. *Juv:* bill black, tail shorter. Bill colour of adults grades from yellow in Kimberley (race *hecki*) through orange to red in NT and Qld (race *acuticauda*). *Voice:* loud 'teeweet'; soft 'tet'; pretty warbling song. *Nesting:* untidy globe of grass with long spout, lined with soft material and charcoal, in bush, tree or long grass; 4–6 white eggs. *Range:* common resident in grassy woodland in northern Aust from Broome, WA, to Mt Isa, Qld.

Black-throated Finch   *Poephila cincta*   10 cm
Black-bellied finch with blue-grey head, black throat, cinnamon underparts, white or black rump and short black tail. Two forms occur: white-rumped form (race *cincta*) in eastern Aust including Cape York, vanishing from southern range, now occurring only in isolated localities in north-eastern NSW, central and north-eastern Qld; and black-rumped form (race *atropygialis*) still locally common on Cape York. *Voice:* plaintive 'weet'. *Nesting:* grass dome with spout lined with soft material and sometimes charcoal, in tree, bush, grass, eagle's nest or hollow; 4–6 white eggs. *Range:* rare to uncommon nomad in woodland, riverine vegetation in eastern Aust from about Inverell, NSW, to Cape York, Qld.

Yellow-rumped Mannikin   *Lonchura flaviprymna*   10 cm
Grey-billed finch with pale grey head, pale buff underparts, yellowish rump and tail. *Juv:* very plain finch, indistinguishable from juv Chestnut-breasted, differs from juv Pictorella in dark undertail coverts and yellowish rump. Occurs in flocks sometimes large usually mixed with Chestnut-breasted. *Voice:* bell-like 'treet'. *Nesting:* large globe of flattened grass in long grass or bush; 5–6 white eggs. *Range:* locally common but decreasing nomad in swampy grassland in northern Aust from Kunanurra, WA, to Arnhem Land, NT.

Chestnut-breasted Mannikin   *Lonchura castaneothorax*   10 cm
Grey-billed finch with black face, chestnut breast, yellowish rump and tail. *Juv:* greyish-brown above, with yellowish rump and tail, paler greyish-brown below, similar to juv Yellow-rumped. Feeds mostly in seeding grass, usually in flocks, sometimes large in north. *Voice:* bell-like 'treet'. *Nesting:* large globe of flattened grass in long grass or bush; 5–6 white eggs. *Range:* common resident or nomad in grassland and grassy woodland, particularly near water, in east and north from Derby, WA, to Shoalhaven R, NSW; aviary escapees elsewhere.

Pictorella Mannikin   *Lonchura pectoralis*   11 cm
Grey-billed finch with black face, scalloped breast and small white spots on wing coverts. *Juv:* grey-brown, similar to juv Chestnut-breasted but with pale undertail coverts. Usually feeds in seeding grasses in small flocks, amalgamating towards end of dry season. *Voice:* reedy 'k-rt, k-rt'. *Nesting:* large globe of grass in long grass or low bush; 4–6 white eggs. *Range:* common nomad in wooded grassland in northern Aust, exploiting drier habitat than Chestnut-breasted, from Derby, WA, to Charters Towers, Qld.

FINCHES

MASKED

brown-eared form

LONG-TAILED

yellow-billed form

red-billed form

white-eared form

MANNIKINS

southern form

BLACK-THROATED

northern form

YELLOW-RUMPED

PICTORELLA

CHESTNUT-BREASTED

juv

ad

juv

ad

17 x 12	17 x 12	17 x 12	17 x 12	17 x 12	16 x 11
...ed	Long-tailed	Yellow-rumped	Black-throated	Chestnut-breasted	Pictorella

RB

y exotic finches and finchlike birds escape from captivity; only those that breed regularly in ural situations qualify for inclusion on Australian list.

pice Finch   *Lonchura punctulata*   11 cm
Dark-billed finch with chestnut-brown face, finely marbled underparts and golden-yellow rump. *Juv:* plain brown above, brownish-yellow below. Often occurs in same areas as Chestnut-breasted, and behaves similarly. *Voice:* high-pitched 'kit-teee'; sharp 'tret tret' rather like Chestnut-breasted; juvenile call like alarm note of Black-fronted Dotterel. *Nesting:* similar to Chestnut-breasted. *Range:* uncommon to locally common in long grasses mainly near towns and farms in coastal eastern Aust, mainly about Sydney and northern rivers, NSW, south-eastern and north-eastern Qld, but spreading.

Black-headed Mannikin   *Lonchura malacca*   11 cm
Grey-billed finch with black head, brown body and reddish-brown rump. Recorded as breeding near Sydney and at Nowra, NSW, but whether the populations are viable or are periodic escapees from captivity has not been established.

European Goldfinch   *Carduelis carduelis*   13 cm
White-billed finch with golden splashes in black wings, red face, pale brown body, white rump and black tail. Usually in pairs or small flocks, often seen feeding in thistles. *Voice:* pleasant tinkling 'swit-swit'. *Nesting:* open cup of twigs, grass and rootlets lined with down and fur, well hidden in leafy shrub or tree; 4–6 variably spotted pale blue eggs. *Range:* common nomad or resident mainly in urban and rural areas, particularly about pines and thistles, in south-east, Tas and south-west.

Greenfinch   *Carduelis chloris*   15 cm
Large white-billed finch with olive-green body, yellower on rump, yellow patches in wing and tail. *Juv:* dull brown body with darker streaks, yellow patches duller. Usually in pairs or small flocks; male sings from prominent perch. *Voice:* sharp 'swee-eee-e'; trilling downward 'brzzzt'. *Nesting:* open cup of twigs and hair, well lined with grass, feathers, down, etc, usually in pine or cypress; 4–6 spotted and streaked whitish to pale greenish-blue eggs. *Range:* uncommon resident in urban and rural areas in south-eastern Aust from Mt Lofty Ra to Canberra, Sydney and Tas.

Tree Sparrow   *Passer montanus*   14 cm
Black-billed finch with chestnut crown and nape, black face and throat, white cheek with small black patch (lacking in House Sparrow), streaked back. *Juv:* duller with less distinct markings. Similar to male House Sparrow and often overlooked in consequence. *Voice:* 'chick'; 'chick-up'; 'tek . . .' in flight; soft rapid twittering. *Nesting:* large untidy globe of grass and feathers in hollow in tree or building; 4–6 spotted pale brown eggs. *Range:* uncommon to common in south-east from Gilgandra, NSW, to Melbourne, Vic.

House Sparrow   *Passer domesticus*   15 cm
Male: crown grey, nape chestnut, face and throat black, cheek white (lacking black spot found in Tree Sparrow). *Female:* dull brown with buff eyebrow, streaked back, two pale bars on wing coverts. Possibly commonest bird in cities and towns of east and Tas, but so far kept out of WA. *Nesting:* untidy dome of grass lined with feathers in hollows in trees, buildings, etc, pre-empting many sites used by native birds; 4–6 spotted whitish eggs. *Range:* abundant in cities, towns, rural areas and bushland in eastern mainland and Tas.

White-winged Wydah   *Euplectes albonotatus*   15–18 cm
Male: black finch with gold and white patch in wing. *Male non-breeding:* similar to female but with gold and white wing patch. *Female:* rather like female House Sparrow, but more yellowish; back brown, streaked darker, without wing bars but with pale bases to wing feathers, pale buff underwing coverts; pale buff below, darker on breast. *Voice:* cheerful twittering. *Nesting:* woven globe of flat grass with large side entrance in long grass; 2–3 spotted greenish-blue eggs. Recorded in rank grasses near creeks and swamps near Windsor, NSW, until 1968.

Grenadier Weaver   *Euplectes orix*   11 cm
Male breeding: scarlet and black finch with black face, abdomen and tail, red breast, rump an undertail coverts, and puffy red nape and lower back. *Male non-breeding and female:* similar to fem House Sparrow; broadly streaked above with brown and black, buff eyebrow, below pale buff darker streaks on flanks. *Voice:* noisy chattering. *Nesting:* polygamous; woven oval of fine grass entrance near top; 2–4 pale turquoise-blue eggs. *Range:* colonies established in reedy river fla Murray Bridge at Woods Point, SA.

FINCHES AND SPARROWS

SPICE

ad

juv

BLACK-HEADED

GREEN-FINCH

GOLD-FINCH

TREE

HOUSE

♂ ♀

WHITE-WINGED

♂ ♀

GRENADIER WEAVER

♂ ♀

Sturnus vulgaris   20–22 cm

…ish and greenish black, with yellow bill and browner wings; in new plumage
…feathers pale buff giving plumage spotted appearance until spots wear off; bill
…n winter. *Juv:* dull brown with paler edges to wing feathers; paler underparts, whitish
…veniles moult directly into adult plumage, but some acquire heavily spotted immature
…y observed in large flocks; when flying singly may look like Dusky Woodswallow. *Voice:*
…whistling including mimicry. *Nesting:* untidy cup of grass, leaves, wool, feathers, etc, in
…n tree or building; 4–5 blue-white eggs. *Range:* introduced; well established in south-east and
…and spreading.

Metallic Starling   *Aplonis metallica*   18–24 cm

Adult: iridescent purplish and greenish black with large red eye and long
lorikeet-like tail. *Juv:* dark brown above, with pale edges to wing
feathers, white below streaked black on breast and flanks, eye dark
brown. *Imm:* head, back and tail iridescent black, wings brown,
underparts white streaked black, eye orange-red. In noisy flocks,
occasionally 'recorded' as black lorikeets. *Voice:* lorikeet-like chattering
and screeching; pretty canary-like song. *Nesting:* dozens of nests
clustered in high outer branches, often used year after year; hanging
woven dome of vine tendrils, palm fibres, lined with feathers; 2–4
spotted pale blue eggs. *Range:* common migrant (Jul–Apr) in rainforest,
mangrove and coastal towns in north-eastern Qld from Cape York to
Gladstone, occasionally farther south. (On Cape York and Torres
Strait islands, watch for Singing Starling *Aplonis cantoroides* 20 cm.
Similar to Metallic Starling, but has *square tail*. Recorded on islands
near New Guinea which are ecologically allied to New Guinea but
are politically Australian.)

Singing

Yellow Oriole   *Oriolus flavocinctus*   25–30 cm

Slender oriole with slender reddish bill, bright yellow-green above and
below, with pale yellow edges to black wing feathers and yellow tip to
green tail. *Juv:* more heavily streaked above with bright yellow edges to
wing feathers. Usually seen singly or in pairs, occasionally with
Figbirds. *Voice:* harsh sneezing 'snee-ach'; melodious monotonous
'cholonk-cholonk'. *Nesting:* deep cup slung in outer branch; 2 spotted
and blotched cream eggs. *Range:* uncommon nomad in wet forest and
mangrove from north of Derby, WA, to about Ingham, Qld.

Olive-backed Oriole   *Oriolus sagittatus*   25–28 cm

Slender oriole with slender pinkish bill, streaked white underparts and
narrow white tip to grey tail. *Male:* bright greenish-olive above,
greenish-grey throat. *Female:* dull olive above, streaked throat. *Juv:*
streaked brown above with buff edges to wing feathers. *Imm:* pale green
above with buff edges to wing feathers. Less sociable than Figbird but
often feeds and nests with them. *Voice:* loud sneezing 'chee-et'; low
whistling 'olly-oh'; beautiful varied whisper song including mimicry;
hollow chuckle; loud hiss like Satin Bowerbird. *Nesting:* bulky cup slung
in slender outer branch; 2–4 spotted blotched cream eggs. *Range:*
common nomad in north and east from south of Broome, WA, to
Adelaide, SA.

Figbird   *Sphecotheres viridis*   27–29 cm

Sociable stocky oriole with stout bill, red, pink or grey skin around eye,
broad white tips to outer tail feathers (male) or no white in tail (female
and juv). *Male:* head black with red or pink skin around eye. *Female:*
brown above, streaked below. *Juv:* streaked brown above and below, no
white tip to tail. *Imm:* paler above than female, less heavily streaked
below, some white in outer tail but less than other orioles. Two forms
occur, identifiable by males only: southern form (race *vieilloti*) with grey
breast, green abdomen; and northern form (race *flaviventris*) with
smoky-yellow underparts, with intergrades in between. *Voice:* loud
'chy-ock'; 't-cheer t-cheer'; pleasant warbling including mimicry.
Nesting: sparse cup slung in outer branches favouring paperbark or fig,
in loose colonies; 2–3 spotted or blotched greenish-brownish eggs.
Range: common nomad in fruit-bearing trees in eastern and northern
Aust from Prince Regent R, WA, to Shoalhaven R, NSW; vagrant
farther south.

COMMON
juv
imm
autumn and winter
bre[...]
and summ[...]

STAR[...]

METALLIC
ad
imm
juv

ORIOLES

YELLOW
juv
ad

juv
imm

OLIVE-BACKED
♀
♂

intergrade
northern form

FIGBIRD
southern form
♂
♀

[...]0 x 21   30 x 21   34 x 23   32 x 22

Metalic   Yellow   Olive-backed   Figbird 32 x 24

...s are related to starlings (p. 314); only one occurs in Aust, introduced from India. **Drongoes** are ... birds, usually with elongated tail feathers curling outwards, giving 'fish-tail' effect. One species ...Aust, distinguished by iridescent blue-green spangles on breast. Feeds on insects caught in flight, ...so chases small birds. Three birds build large cup-shaped nests from mud; two are related ...Apostlebird and White-winged Chough), the other, the Magpie-lark, which builds a similar nest though with more vertical walls, is classified in a different family. All feed on the ground but nest in trees.

Common Myna Acridotheres tristis 23–25 cm
Black-headed myna with brown body and large white 'windows' in wings, yellow bill and legs. Introduced in 1860s, spreading from Hobart, Adelaide, Melbourne, Sydney, Darling Downs and north-east, and now established in many country towns and rural areas.

Spangled Drongo Dicrurus megarhynchus 28–32 cm
Black 'fish-tailed drongo' with iridescent blue spangles on breast. *Adult:* eye red. *Juv:* more brownish with brown eye, white bars on undertail and underwing coverts. Noisy and obvious with loud 'scissoring' call. Usually seen on prominent perch frquently flicking tail open and shut or in erratic darting flight. *Voice:* harsh variable metallic 'grut-grut cris grut . . .'; harsh shattering and mimicry. *Nesting:* neat cup of twigs and tendrils slung at end of outer branch; 3–5 blotched pinkish to purplish eggs. *Range:* common resident, nomad or migrant in eastern and northern Aust, breeding south to north-eastern NSW; some migrate north after breeding, some southwards to about Jervis Bay (Apr–Sep).

Magpie-lark Grallina cyanoleuca 27 cm
Black and white mud-nest builder. *Male:* black throat, white eyebrow. *Female:* white face and throat, no white eyebrow. *Juv:* white throat, white eyebrow, dark eye and bill. Usually seen in pairs or small family parties, but in winter large flocks of young birds sometimes gather to feed in wet pastures and around lakes. *Voice:* 'pee-o-wit' often as duet; nasal 'clut'. *Nesting:* substantial mud nest with vertical sides, placed on horizontal branch, windmill or utility pole. *Range:* common to abundant resident or nomad throughout mainland Aust in most habitats except forest; rare in Tas.

White-winged Chough Corcorax melanorhamphos 45 cm
Sociable black mud-nest builder with white patch in wing obvious in flight. Usually in small flocks numbering 6–10 walking slowly and noisily or planing on white-windowed wings. More slender bill, longer tail than currawongs, which have yellow eyes, fly with looping wingbeats. *Voice:* harsh grating and piping. *Nesting:* large mud cup with sloping walls, usually on horizontal branch attended by family party; 3–5+ whitish spotted eggs. *Range:* common resident in woodland, dry sclerophyll forest and scrubland in southern and eastern Aust, from south-eastern WA to central Qld, mostly west of Divide.

Apostlebird Struthidea cinerea 29–32 cm
Sociable grey mud-nest builder with feathers of body and head tipped paler. Usually in small flocks numbering 6–20, walking on ground looking for food, commonly flushed from roadsides in inland eastern Aust. *Voice:* loud, harsh grating and scolding. *Nesting:* large mud cup with sloping sides, usually on horizontal branch, attended by family party; 2–5+ white or pale blue spotted, blotched or unmarked eggs. *Range:* common resident in woodlands and shrublands often in cypress and casuarina stands, mostly in eastern Aust, west of Divide, but spreading.

MYNA

COMMON

DRONG-

SPANGLED

juv

ad

MUD-NEST BUILDERS

juv

♀

♂

MAGPIE-LARK

WHITE-WINGED CHOUGH

APOSTLEBIRD

22	29 x 21	29 x 21	40 x 30		29 x 22
	Spangled	Magpie-Lark	White-winged		Apostlebird

OSWALLOWS

...ly-built birds with long wings and blue bills tipped black. Only passerine birds with powderdown ...hers, giving plumage soft compact look; only small passerines that habitually soar. Feed mainly on ...sects, some take nectar and may show yellow pollen on foreheads. Many cluster in cold weather; ...ome nest in loose colonies. Juveniles look like juv White-browed illustrated.

Little Woodswallow *Artamus minor* 12 cm
Very dark woodswallow with broad white tip in outer tail feathers and no white in wing quills. Similar to Dusky, which is larger, paler, has white outer quills in wing and ranges generally but not entirely more southerly. *Voice:* quiet 'check'; 'choo choo swit swit'; warbling whisper song including mimicry. *Nesting:* wispy cup of grass and leaves in hollow of dead tree or rock ledge; 3 spotted blotched white or pale buff eggs. *Range:* uncommon resident, nomad or migrant in woodland and shrubland in northern two-thirds of Aust.

Dusky Woodswallow *Artamus cyanopterus* 18 cm
Dark, with outer wing quills white. Usually in small groups flying and soaring or resting on prominent perches. *Voice:* loud 'check'. *Nesting:* sparse cup of twigs and grass in enclosed tree fork, upright hollow or behind loose bark; 3–4 spotted or blotched whitish or pale buff eggs. *Range:* common in forest and woodland particularly with grassy understorey in east, south-east, south-west and Tas, generally annual rainfall >500 mm.

Black-faced Woodswallow *Artamus cinereus* 18 cm
Smoky-grey, with small black face. Several forms occur: black-vented forms (race *melanops* and *tregallasi*) with black undertail, over most of Aust; and white-vented form (race *albiventris*) in north-eastern Qld from Cape York to about Rockhampton. *Voice:* twittering 'quet-quet'; excited chattering when defending territory. *Nesting:* untidy cup in bushy fork or hollow stump; 3–4 spotted blotched white or bluish eggs. *Range:* common resident in grassy woodlands generally <500 mm annual rainfall.

Masked Woodswallow *Artamus personatus* 19 cm
Highly nomadic pale grey woodswallow with extensive black or dusky face. *Male:* face black, underparts delicate grey. *Female:* face dusky-grey, underparts brownish-grey. *Imm and juv:* similar to imm and juv White-browed. *Voice:* similar to White-browed. *Nesting:* cup of green grass soon drying out in hollow stump or on broken branch in top of mulga sapling; 2–3 speckled greyish, greenish or brownish eggs. *Range:* common nomad (more common in west) mainly in drier areas; vagrant Tas.

White-browed Woodswallow *Artamus superciliosus* 19 cm
Highly nomadic dark grey woodswallow with white eyebrow. *Male:* head black with prominent white eyebrow, breast chestnut. *Female:* head dark grey with indistinct white eyebrow, breast cinnamon. *Imm:* wing feathers spotted and edged white. *Juv:* head and body streaked dark and light grey; wing feathers spotted and edged white. Usually in large flocks in east; less common in west. *Voice:* querulous nasal 'chirp' or 'check' often heard high overhead. *Nesting:* cup of green grass quickly drying out in upright tree stump, sometimes in mulga bush; 2–3 spotted blotched whitish or greyish green eggs. *Range:* common nomad mainly in drier areas; vagrant Tas.

White-breasted Woodswallow *Artamus leucorhynchus* 17 cm
Dark grey, with white breast and rump. Only woodswallow without white in tail tip. *Imm:* throat greyer, back browner with buff edges to feathers. Usually in small groups, favouring mangrove on coast and tree-lined watercourses inland. *Voice:* harsh 'eyeck'. *Nesting:* cup of grass in upright tree hollow; 3–4 spotted cream or pinkish eggs. *Range:* common resident or partial migrant in mangrove and woodland near water, northern central and eastern Aust and Tas.

WOODSWALLOWS

DUSKY

BLACK-FACED

black-vented form

LITTLE

imm

juv

WHITE-BROWED

♀

MASKED

♂

♀

ad

juv

WHITE-BREASTED

14

23 x 17	22 x 17	22 x 17	22 x 17	23 x 17
Dusky	Black-faced	Masked	White-browed	White-breasted

s are rather smaller than related magpies and currawongs and are more arboreal. Name
comes from habit of wedging large prey (insects, lizards, mice, small birds) into fork or
, it on broken branch to aid in tearing into smaller pieces. Beautiful songs and friendly
tion make them favourites in town and about homesteads (except during nesting period, when
may become aggressive). Generally they live in small family groups of adults and immatures, but
ch territorial activity in early autumn suggests adults may drive away offspring once they acquire
dult plumage at about 15 months of age after helping to rear the new season's brood.

Black-backed Butcherbird

Cracticus mentalis 25 cm

Adult: small white-throated pied butcherbird. *Juv:* black areas of adult replaced with brown. Grey is similar but has less white markings on wing, has grey back, and has different range. Usually in family parties up to six. *Voice:* rather like Grey but less penetrating. *Nesting:* shallow untidy cup of sticks and vines lined with finer twigs, placed in upright fork up to 25 m; 2–3 spotted and blotched greenish-grey to brownish eggs. *Range:* uncommon in woodland, rivercourses, pastoral holdings and about habitation, on Cape York north of about Cooktown.

Pied Butcherbird

Cracticus nigrogularis 32–35 cm

Adult: black-throated pied butcherbird. *Juv:* black areas of adult replaced with brown. Often in small family party consisting of two or three pied birds and several brown birds. Juveniles help with next season's nest, they moult to pied after breeding, then may disperse. *Voice:* most beautiful fluting piping and mimicry, loud 'zwit' of alarm. *Nesting:* untidy cup of sticks lined with grass usually placed in upright fork up to 15 m; 3–5 spotted olive to brown eggs. *Range:* common in woodland, farms, roadsides, towns, plains where there are enough trees for cover, over much of Aust; absent from highest rainfall areas, southern Vic and south-west.

Grey Butcherbird

Cracticus torquatus 28–32 cm

Adult: small white-throated butcherbird with grey back and greyish wash on underparts. *Juv:* similar in pattern to adult but black and grey areas replaced by brown and white areas washed olive. Four separate forms: large form (race *cinereus*) in Tas; grey-backed form (race *torquatus*) in more southern range, changing imperceptibly darker from east to west — old eastern birds often have white rather than grey underparts, and some south-western adults have freckled black bib; silver-backed form (race *argenteus* or *colletti*) in Top End, smaller, lacking white spot in front of eye; and Kimberley form (race *argenteus* or *latens*), larger, more pied-looking with silver back, white underparts, black necklace, more white on wing, tail-tip and rump, different song (like cross between Black Butcherbird and Yellow Oriole), looks like hybrid Pied/Grey, could be separate species. *Voice:* vigorous melodious piping, rattling cheep and soft mimicry (particularly when raining). *Nesting:* untidy cup of twigs lined with grass usually in upright fork in sapling 2–10 m; 3–5 speckled green to brownish eggs. *Range:* less common than Pied in woodland, forest and scrub, farms, towns wherever there are sufficient trees for cover but absent from rainforest (except margin); grey-backed form in Tas and southern two-thirds of Aust; silver-backed form in Top End, Kimberley form in north Kimberley.

Black Butcherbird

Cracticus quoyi 31–44 cm

Adult: large black butcherbird. *Juv:* dull black over most of range or rufous paler streaks on head and back on south-eastern Cape York. Several forms occur: two with dark juveniles (races *spaldingi* in NT and *quoyi* on northern Cape York); and one subspecies with rufous juvenile (race *rufescens*) in north-eastern Qld from Endeavour R to Rockhampton. *Voice:* melodious and flutelike, less sustained than other butcherbirds'. *Nesting:* large untidy cup-shaped nest of sticks in upright fork to 10 m; 4 spotted greyish-green to cream eggs. *Range:* common in mangrove, rainforest, riverine vegetation and nearby open woodland in four separate areas from Port Keats, NT, to about Rockhampton, Qld.

BUTCHERBIRDS

BLACK

rufous juv

ad

PIED

juv

ad

Grey 31 x 23   Black-backed 27 x 20

Pied 33 x 24   Black 35 x 25

juv

ad

BLACK-BACKED

Kimberley form

silver-backed form

western form

GREY

eastern form

ad

juv

BELL-MAGPIES (CURRAWONGS)

Currawongs are less terrestrial than magpies and have shorter legs, fly with characteristic looping wingbeats, often closing wings in flight; they are black or dark grey with white patches in wings, obvious in flight, and in the tail. Calls are loud and gong-like. Nests are large cups of sticks usually placed towards the top of an upright branch in a tree surrounded by other trees. After breeding, birds join nomadic flocks; Pied often visits habitation, becomes tame; Grey usually remains wary of humans.

Pied Currawong   *Strepera graculina*   41–51 cm

Black currawong with white in wing, rump, undertail coverts, base and tip of tail. Often seen in flocks, 100 in winter, but breeds in isolation. *Voice:* loud 'curra-wong' or 'caddow caddang'; whistling 'oo-oooo' like disapproving schoolgirl; 'kwok'. *Nesting:* large flat cup of sticks lined with soft material; 3 spotted and blotched brown eggs. *Range:* common nomad or altitudinal migrant in forest, woodland, urban and rural areas in eastern Aust from Cape York Peninsula to south-western Vic.

Grey Currawong   *Strepera versicolor*   45–53 cm

Dark-grey to black currawong with white in wing, undertail and tail tip, but no white on rump. Several forms occur: Clinking form (race *arguta*), in Tas, black with white undertail (Black Currawong has black undertail); grey form (race *versicolor*), in south-eastern, south-western and central Aust; brown form (race *intermedia*) on Yorke and Eyre Peninsulas, SA; and black-winged form (race *melanoptera*), in south-eastern mallee, brownish-black with no white in wing. *Voice:* loud gonglike call 'clang clang'; 'clink' or 'tew'; whistling 'eow'. *Nesting:* large shallow cup of twigs lined with soft material; 2–3 spotted and blotched buff eggs. *Range:* common to uncommon resident or nomad in dry forest, woodland and mallee in southern and central Aust, and central and eastern Tas.

Black Currawong   *Strepera fuliginosa*   46–48 cm

Black currawong with small white patch in wing and white tip to tail, confined to Tas. Clinking form of Grey Currawong has white undertail and large white patch in wing. *Voice:* loud 'kar-week, week car'. *Nesting:* large stick cup lined with softer material placed high in tree; 2–4 blotched buff eggs. *Range:* common nomad or altitudinal migrant in woodland and rural areas in Tas, mainly in mountain forest in summer and lowland in winter.

CURRAWONGS

BLACK

PIED

GREY

brown form

clinking form

widespread form

black-winged form

brown form

clinking form

40 x 29	42 x 30	42 x 30
Black	Pied	Grey

BELL-MAGPIES (AUSTRALIAN MAGPIE)

Bell-magpies are large pied or grey and white birds with loud melodious voices, variously known as magpies, currawongs and butcherbirds. Australian Magpie is found all over mainland Australia and Tasmania and has developed into a number of races that are sufficiently different in appearance to have been considered different species in past. Areas of hybridisation where some forms meet suggest that they form one species. However, the Tasmanian form may be separate species because it contradicts a universal trend known as Bergmann's Rule which states that individuals of a species are bigger in higher latitudes than in lower. In settled areas magpies are very tame; while nesting they might be aggressive towards humans to the point of occasionally inflicting damage; in natural bushland magpies are shy, in fact one of most difficult birds to approach. In natural bush, territories are well spaced; in agricultural, land territories are more concentrated, with considerable pressure to maintain boundaries. Small groups of birds varying from two to 20 birds (groups average bigger in west) occupy each territory. Unattached immature birds join non-breeding groups that occupy less-suitable ground, and may replace holders of productive territories as they die.

Australian Magpie *Gymnorhina tibicen* 34–44 cm

Sociable pied bell-magpie with white nape, white patch in wing, white rump and undertail, usually in groups 4–20, antagonistic towards other groups. Several forms occur; juvenile and immature plumages are basically similar. *Juv:* underparts dull black, back dull black with white edges to feathers, nape, shoulder, rump and base of tail ashy-white, bill small, leaden; moults at 3 months. *Imm:* underparts mottled grey with brownish tinge, back feathers grey with black subterminal fringe and white edges; bill longer, leaden grey. Subadult plumages are more like adults of various races, with bills becoming progressively whiter. Black-backed form (race *dorsalis*), over most of eastern and northern Aust, male and female basically similar in pattern but female has ashy cast to nape; Tasmanian form (race *hypoleuca*) like mainland white-backed form but smaller (according to Bergmann's Rule they should be larger); white-backed form (race *leuconota*) from lower south-east to central ranges, male with back white, female and subadult male with ashy backs (in areas of hybridisation with Black-backed form, variable amounts of white on back); western form (race *dorsalis*), in south-west, male with back white (with age amount of white in tail increases), female and subadult male white edges to back feathers (older females have less white edging, almost disappearing in very old birds); Papuan form (race *papuana*) in north-eastern NT and Groote Eylandt has white-backed male and black-backed immature. *Voice:* loud flutelike warbling; single 'sheow'; variable whisper song, including mimicry. *Nesting:* untidy cup of twigs, sometimes wire, lined with grass and soft material; 3 spotted and smeared brown or blue eggs. *Range:* common resident in woodland, urban and rural clearings throughout mainland and Tas.

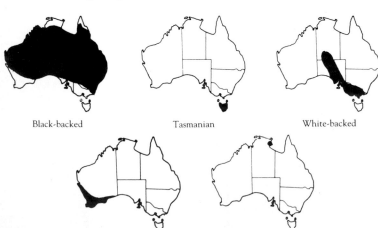

Black-backed   Tasmanian   White-backed

Western   Papuan

MAGPIES

PAPUAN

♂ ♀

imm juv

BLACK-
BACKED

WHITE-
BACKED

♀ ♂ ♀

WESTERN

old
♀

TASMANIAN

♂ ♀ ♀

38 x 27

Australian

BOWERBIRDS

Stocky stout-billed birds remarkable for the bowers built by males as focal points of territories where courting and mating occur. Three forms of bower: most simple is clearing strewn with large leaves made by Tooth-billed Catbird; Green Catbird may also lay leaves on forest floor. More complex is avenue bower consisting of two parallel rows of sticks or grass stuck into a platform of twigs, decorated with ornaments ranging from blue feathers, berries, flowers and cicada shell cases in Satin, bones and white stones in Spotted, Western and Great, and cicada cases and green berries in Regent and Fawn-breasted. Most 'paint' inside walls of bower with saliva, mixed with vegetable matter. Most complex is 'maypole' bower with sticks heaped about upright saplings; Golden's is less sophisticated than several related New Guinea species. Females build nest without help from males in upright fork of tree or sapling, usually in dense clump of leaves. Feed mainly on fruit and berries.

Green Catbird *Ailuroedus crassirostris* 29–32 cm

Southern catbird with plain ear coverts, lightly spotted head and underparts, loud miaowing call. Usually in pairs, but occurs in small flocks in winter. Female Satin Bowerbird somewhat similar, has pale rufous wings, blue eye, bill partly covered by feathers, duller green. *Voice:* cat-like miaowing; high-pitched clicking. *Nesting:* substantial deep cup of vines, twigs, leaves, moss lined with finer tendrils, etc, in vine, upright fork of slender sapling or in crown of treefern 3–6 m; 2 eggs. *Range:* common in or near rainforest from Mt Dromedary, NSW, to Kingaroy, Qld.

Spotted Catbird *Ailuroedus melanotis* 23 cm

Northern catbird with contrasting black ear coverts and heavily spotted plumage. *Imm:* duller. Usually attracts attention first by loud miaowing call in rainforest. Usually in pairs, but forms small flocks in winter, may visit gardens. *Voice:* loud 'wee-you-are' less catlike than Green; hissing and clocking notes. *Nesting:* open cup of vines and twigs decorated with moss in dense vine, open fork or on tree fern; 2 creamy eggs. *Range:* common resident in rainforest in north-east Qld from about Mt Spec near Townsville to Claudie R.

Spotted Bowerbird *Chlamydera maculata* 27–31 cm

Spotted bowerbird with whitish streak from gape and grey patch on lower hindneck, faintly spotted on head and neck. Lilac-pink crest on nape concealed except during display, large in male, lacking in immature. *Voice:* loud hiss; often mimics Brown Falcon or Whistling Kite when disturbed at bower. *Breeding:* avenue bower (described above). Nest is a sparse cup of twigs lined with leaves. *Range:* common in woodland, drier scrubs, central and western NSW and Qld; becoming rare in south, almost extinct on Murray R.

Western Bowerbird *Chlamydera guttata* 27–31 cm

Spotted bowerbird with heavily spotted head and neck. Lilac-pink crest hidden except during display. Spotted and Great are exclusive of range. *Voice:* loud hiss; variable calls and scolding, including mimicry during display. *Breeding:* avenue bower (described above). Nest is a sparse cup in upright fork; 2–3 eggs. *Range:* common in rocky hills where native figs grow, uncommon elsewhere in western interior from about Carnarvon, WA, to western edge of Simpson Desert east of Alice Springs, NT.

Great Bowerbird *Chlamydera nuchalis* 32–35 cm

Large bowerbird with grey head and breast without streaking or spotting. Lilac-pink crest hidden except during display; crest reduced to only a few feathers in some females. *Voice:* loud hiss; variable calls including mimicry during display. *Breeding:* bulky bower of sticks, often meeting over avenue, decorated. Nest is a sparse cup in upright fork; 2 eggs. *Range:* fairly common in dry sclerophyll woodland, thickets of limebush, figs, vines, bauhinias near watercourses in north from about Anna Plains (south of Broome), WA, to Mackay, Qld.

Fawn-breasted Bowerbird *Chlamydera cerviniventris* 25 cm

Spotted bowerbird with streaked head and breast and fawn abdomen. Lacks lilac-pink crest. *Voice:* hissing 'churr'; variable calls including mimicry at bower. *Breeding:* elaborate avenue bower with raised platform of sticks, decorated with green berries, leaves, shells and bones. Nest is a bulky shallow cup of twigs and sticks in tree or pandanus; 1 egg. *Range:* uncommon in mangroves, vine scrub, melaleuca thickets and rainforest glades on Cape York south to Jardine R and Chester R.

CATBIRDS

GREEN

SPOTTED

BOWERBIRDS

SPOTTED

WESTERN

GREAT

FAWN-
BREASTED

43 x 30	41 x 27	38 x 27	38 x 27	42 x 29	36 x 26
Green	Spotted Catbird	Spotted Bowerbird	Western	Great	Fawn-breasted

BOWERBIRDS

Regent Bowerbird *Sericulus chrysocephalus* 24–28 cm
Male: gold and black bowerbird. *Female:* black-crowned bowerbird with scalloped back, yellow eye and black bill. *Imm:* similar to female, young males gain yellow bill before changing to adult plumage. *Voice:* generally silent; low chattering in display, mimicry and soft warbling also recorded. *Breeding:* bower scanty avenue often consisting of only few twigs, decorated with green berries, leaves and cicada shells often stolen from Satin Bowerbirds' bowers. Nest is a loose saucer of twigs and grasses in dense bush or vine; 2 eggs. *Range:* fairly common in and near rainforest in eastern Aust from Eungella Ra, Qld, to Broken Bay and Cattai, NSW.

Tooth-billed Catbird *Ailuroedus dentirostris* 26–27 cm
Stout-billed olive bowerbird with coarse streaks on breast and abdomen. Large bill has cutting 'tooth' for collecting suitable leaves to decorate bower. *Voice:* most variable, vigorous and loud song at bower, making bowers easy to find. *Breeding:* arena bower of cleared forest floor decorated with large leaves laid with pale undersurface uppermost; male sings from perch above arena. Nest is a sparse shallow cup of twigs, 3–30 m above ground; 2 eggs. *Range:* common in upland rainforest 600–1400 m, occasionally lower, from Mt Amos to Mt Spec, Qld.

Golden Bowerbird *Amblyornis newtonianus* 23–25 cm
Male: olive and gold bowerbird. *Female:* small-billed, olive-backed bowerbird with ashy front and yellow eye. *Imm:* similar to female with dark eye. Female may be confused with female whistlers found in same habitat; olive back and yellow eye probably best characters. *Voice:* generally silent; rattling call followed by froglike croak, mimicry. *Breeding:* maypole bower of sticks joined by fallen branch which provides display perch; one pile of sticks is higher than other, may be 3 m in height (usually much less) decorated with lichen, seed pods, yellow or white flowers. Nest is a shallow cup in cavity in tree trunk or in gap between vine and trunk at 2–3 m; 2 eggs. *Range:* common in small areas of upland rainforest above 900 m (lower in winter) from Mt Cook to Mt Spec, Qld.

Satin Bowerbird *Ptilinorhynchus violaceus* 23–27 cm
Male: glossy blue-black bowerbird with violet eye and semi-covered greenish-yellow bill. *Female:* dull greenish and bluish-grey with pale rufous wings; breast and abdomen yellowish-white with darker scallops, eye blue. *Imm:* similar to female with dark eye becoming blue with age; young males in about 4th year have greenish unscalloped breast with fine white spots; bill becomes pale in about 5th year; blue patches appear in plumage, and in 7th year full blue plumage is acquired. Males in green plumage may build bowers but generally their blue ornaments are stolen and bower may be smashed by older birds. Two forms occur: smaller form (race *minor*) in north-eastern Qld; and larger form (race *violaceus*) in eastern and south-eastern Aust. *Voice:* loud 'wee-oo'; mechanical nasal racketing in display, mimicry. *Breeding:* avenue bower of twigs, painted with mixture of saliva (?) and hoop pine leaves, tobacco bush berries or charcoal, decorated mainly with blue items as well as cicada pupae cases, onion skins, flowers (blue or yellow) and berries; dominant males gather most ornaments from surrounding territories. Nest is a shallow cup of pliable twigs lined with leaves in tree, vine or mistletoe; 1–3 eggs. *Range:* common in rainforest and nearby eucalypt forest, moving into drier areas in winter (may raid fruit trees) in eastern Aust; smaller form from Atherton Tableland to Seaview Ra, Qld; larger form from Cooroy, Qld, to Otway Ra, Vic.

REGENT

BOWERBIRDS

♀

♂

TOOTH-BILLED

♂

♀ GOLDEN

♂ SATIN

♀

imm ♂

37 x 26	42 x 29	44 x 29	36 x 25
Regent	Tooth-billed	Satin	Golden

BIRDS OF PARADISE

Four birds of paradise inhabit various rainforests from Cape York to northern NSW. Male riflebirds have glistening blue-green gorgets; females have longer bill than males and are barred or spotted below. Some regard the two more southerly forms as races of one species. Much of their food is insects dug out of rotting wood using the long powerful bill; rain of debris falling to forest floor may alert observer. Fruit also eaten. Flight is noisy, sounding like rustling taffeta. Males display with outstretched wings on prominent branches. Nests are often decorated with snakeskins, eggs among most beautifully marked. Manucode is oily green and blue in plumage, New Guinea race has purple and blue iridescence. It draws attention with loud, gong-like call, uttered with aid of elongated windpipe. Feeds mainly on fruit. Male displays by spreading wings, erecting long hackles.

Magnificent Riflebird *Ptiloris magnificus* 26–33 cm
Male: large riflebird with plumes on flanks, found on northern Cape York. *Female and imm:* rufous above, grey below, profusely barred with black. *Voice:* two or three loud whistles, sometimes with additional indrawn 'hoo-oo'. *Nesting:* deep untidy cup of twigs and vines lined with fibrous materials usually in fork but also in pandanus, 1–10 m above ground; 2 streaked cream eggs. *Range:* fairly common in rainforest on northern Cape York north from Weipa on west coast and Chester R on east.

Victoria's Riflebird *Ptiloris victoriae* 23–25 cm
Male: small short-billed riflebird without plumes, found from Mt Amos to Seaview Ra, north-eastern Qld, and on nearby islands. *Female and imm:* greyish-brown above, buff below with darker spots on breast and flanks. *Voice:* loud rasping 'yass'. *Nesting:* shallow untidy cup of twigs and vines, lined with smaller twigs and tendrils, sometimes decorated with snakeskins or paper, placed in bushy vine or treetops at 25–30 m; 2 spotted and streaked pinkish eggs. *Range:* fairly common in rainforests in north-eastern Qld.

Paradise Riflebird *Ptiloris paradiseus* 28–30 cm
Male: medium-sized long-billed riflebird without plumes, found in south-eastern Qld and north-eastern NSW. *Female and imm:* brown above, pale buff below with 'horseshoe' markings. Normally found in rainforests but may move into sclerophyll forests in winter. *Voice:* loud 'yass'. *Nesting:* shallow cup of tendrils, sometimes decorated with snakeskin, placed in bushy vine or canopy at 25–30 m; 2 spotted and streaked pinkish eggs. *Range:* fairly common in rainforest in south-eastern Qld and north-eastern NSW, from Cooroy, Qld, to Barrington Tops, NSW.

Trumpet Manucode *Manucodia keraudrenii* 27–32 cm
Long-tailed black bird with oily green and blue sheen and with long hackles on neck. Other black birds in same habitat are sociable Shining Starling, which is much smaller, has graduated tail, Black Butcherbird, which is much bulkier, has pale bill and dark eye, and male Koel, which has longer tail, pale bill, lacks hackles. *Voice:* resonant 'grong' or 'grow'. *Nesting:* shallow sparse cup of vines and tendrils in top of tree or sapling at 6–20 m; 2 spotted pink eggs. *Range:* fairly common in rainforest and nearby woodland on tip of Cape York and between Claudie R and Chester R, Qld.

MAGNIFICENT

PARADISE

RIFLEBIRDS

VICTORIA'S

♂

♂

♂

MANUCODES

TRUMPET

MAGNIFICENT

♀

PARADISE

♀

VICTORIA'S

♀

34 x 23   36 x 24   Victoria's 34 x 23   37 x 24

Magnificent   Paradise   Victoria's   Trumpet

LYREBIRDS

Lyrebirds are basically terrestrial birds of paradise. Male's tail is specialised for display: outer feathers are broad and lyre-shaped, notched in Superb and plain in Albert's; other feathers are filamentous, dark on upper surface and silver or grey below, creating shimmering curtain when erected over back; as well there is a long plain feather which droops on each side as the bird dances on its arena, a small cleared platform in undergrowth. Normally displays difficult to observe, but where birds are tame, such as at Sherbrooke Forest, it is possible to see Superb more easily. Few people have seen full Albert's display. Song is extraordinary, incorporating variations on sounds heard in bush, mainly other birds' calls. Displays and peak of singing usually occur in autumn and winter, tailing off into spring when male moults. Female lacks plumes, has long pheasant-like tail with drooping tip. In winter she builds large domed nest of sticks placed in stump, in buttress or crevice in rocks, incubates egg and rears chick alone. Female also sings well, not as vigorously as male.

Superb Lyrebird *Menura novaehollandiae* 80–96 cm
A greyish-brown lyrebird with grey undertail coverts. *Male*: lyre-shaped outer tail feathers notched rufous, filamentous plumes blackish above and silver or grey below. *Female*: graduated tail long, dark brown. South-eastern form (race *novaehollandiae*) has silver undersides to plumes, found in mountain fern gullies; and Prince Edward's form (race *edwardi*) with grey undersides to plumes, found in open timber among granite outcrops. *Voice*: superb song, including mimicry. *Nesting*: large stick dome lined with grey feathers on rock ledge, tree stump or tree fern; 1 blotched brown egg. *Range*: common resident in fern gullies in rainforest, forest and woodlands; rugged sandstone and granite woodland in south-east and introduced to central and south-eastern Tas, increasing.

Albert's Lyrebird *Menura alberti* 65–90 cm
Reddish-brown lyrebird with rufous undertail coverts. *Male*: reduced lyre-shaped outer tail feathers, unnotched, dark brown above, grey below; filamentous plumes more wire-like than Superb. *Female*: graduated tail long, dark brown. *Voice*: superb songster, including mimicry equal to Superb. *Nesting*: large stick dome lined with rufous feathers; 1 blotched purplish-brown egg. *Range*: locally common in rainforest in south-east Qld and north-east NSW; range reduced by clearing; much of present range within national parks.

LYREBIRDS

SUPERB

♂

♀

65 x 45

Superb

♀

ALBERT'S

65 x 45

Albert's

♂

CROWS AND RAVENS

Torresian Crow *Corvus orru* 50 cm

Glossy-black corvid with white bases to neck feathers. Shuffles wings exaggeratedly on alighting; has wing-quivering greeting display. Distinguished from ravens by neck feather colour, unspecialised hackles, nasal quality to call, 'currawong' flight or 'missing a beat' while giving a single call note. Adults tend to remain in territories throughout year, defend them against other Torresians, Australian and Forest. Flocks of immatures in east, often seen in stubble and grain crops. *Voice:* high-pitched rapid nasal 'uk-uk-uk-uk . . .' or 'ok-ok-ok . . .' sometimes longer on last note(s); longer 'snarling' notes with dying fall on last note; deeper slower 'arr arr arr . . .' more like Little Raven but more nasal. Loud falsetto stuttering call in western birds. *Nesting:* large stick nest, smaller than Australian, more likely in outer branches; 3–6 eggs. *Range:* common resident in most habitats over northern two-thirds of Aust. Rare or absent in extreme desert areas.

Little Crow *Corvus bennetti* 48 cm

Glossy-black corvid with white bases to neck feathers. Outside breeding season in large flocks, fly high, often soar; performs spectacular tumbling aerobatics; performs 'currawong' flight less often than Torresian, while uttering two notes. Very confiding around inland towns and settlements. *Voice:* flat hoarse 'nark-nark-nark-nark . . .' lower than Torresian; occasionally high-pitched call with notes more rapid than Torresian. *Nesting:* stick cup with layer of mud below nest lining, in tree, telegraph pole or windmill; semi-colonial; 3–6 eggs. *Range:* common nomad in most habitats in drier areas west of Divide. Characteristic of arid zone.

House Crow *Corvus splendens* 43 cm

Vagrant dull-black corvid with dark eyes, ashy-brown collar on nape, upper back and breast. An unwelcome migrant, occasionally arriving at Aust ports on ships from India and Ceylon; report any sighting to Agricultural or Wildlife Departments (but do not confuse with Grey Currawong).

Forest Raven *Corvus tasmanicus* 52 cm

Glossy-black corvid with grey bases to neck feathers. Two forms occur: short-tailed Tasmanian form (race *tasmanicus*), only corvid in Tas and Flinders I, also on mainland at Wilson's Promontory, Otway Ra and in south-eastern SA; and longer-tailed form (race *boreus*) in NSW. Deep gravelly voice should separate from other corvids. Adults remain in territories throughout year, defend against Torresian and Australian. Immature flocks up to 100 in Tas, small flocks (< 10) usual in New England. *Voice:* rich deep gravelly 'korr-korr-korr-korr', powerful and far-carrying. *Nesting:* large stick nest in substantial fork; 3–6 eggs. *Range:* common resident (adult) or short-range nomad (immature) in most habitats in Tas, in Wilson's Promontory and Otway Ra; pine forest and pasture in south-east SA (flocks with Little and Australian); in eucalypt forest and farmland in north-eastern NSW; on beaches and heath, forest and paperbarks Port Stephens–Urunga, NSW.

Little Raven *Corvus mellori* 50 cm

Glossy-black corvid with grey bases to neck feathers. Outside breeding season in wandering flocks, fly in co-ordinated manner, has on occasions been recorded performing aerobatics like Little Crow; flips up wings with each call note, closing wings against body after each note; has neck-extending greeting ceremony. Small territory defended against other Littles. *Voice:* guttural, rapid, clipped 'kar-kar-kar-kar', flatter and less powerful than Forest. Occasional longer notes. *Nesting:* large stick nest, seldom higher than 10 m in tree; semi-colonial; 3–6 eggs. *Range:* common nomad in south-eastern Aust and King I (Tas).

Australian Raven *Corvus coronoides* 52 cm

Glossy-black corvid with grey bases to neck feathers. Has bare skin in inter-ramal area under bill, pinkish in juv; long hackles form prominent fan while calling or displaying. Adults remain in territories throughout year, defend against other Australians, Forest and Torresian; flocks of immatures usually small. *Voice:* high-pitched wailing 'aah-aah-aah-aaaahh'. *Nesting:* large stick nest in substantial fork of large tree, usually obvious, 10 m+; 3–6 eggs. *Range:* common resident (adult) or nomad (immature) in woodland and open habitats in eastern, central and south-western Aust, expanding with pastoral development.

(continued overleaf)

CROWS

TORRESIAN

LITTLE

RAVENS

FOREST

HOUSE

LITTLE

AUSTRALIAN

44 x 30	38 x 27	46 x 31	44 x 30	46 x 30
Torresian	Little Crow	Forest	Little Raven	Australian

CROWS AND RAVENS

The two species of crow and three ravens that occur in Aust are similar in appearance, with black glossy plumage, white eyes in adults and overlapping in size and behaviour, so there are some difficulties for the observer. Range will clearly eliminate some (only place where more than three occur is New England); areas where only one occurs are Tasmania, Flinders I, Otway Ranges and Wilson's Promontory, Vic, (all Forest Raven), and Cape York Peninsula, Top End and Kimberley, where all or nearly all are Torresian Crow. Where overlap occurs (see maps), identification should ideally be based on average appearance and behaviour and should encompass a variety of characteristics. First task is to separate crows from ravens. **Crows** have (a) white bases to neck feathers visible when ruffled by wind; (b) short hackles on neck (raised when calling in Torresian); (c) nasal quality to most-often-heard calls; (d) mostly western and northern in range. **Ravens** have (a) grey bases to neck feathers, less obvious when ruffled by wind; (b) medium-length (Little, Forest) to long (Australian) specialised hackles on throat, often distended when calling; (c) commonest calls lacking nasal quality; (d) mostly eastern and southern in range.

Calls

Each species has a wide variety of calls with different meanings; to learn them all requires an expert ear and much experience. However, some commonly uttered calls are diagnostic. Basically, Forest Raven is bass with gravelly quality, Little Raven and Little Crow are baritones, latter with nasal quality, Australian Raven and Torresian Crow are tenors, latter with nasal quality. Best for diagnostic purposes is territorial call, uttered with throat hackles fanned from prominent perches by Forest, Australian and Torresian, and uttered with throat hackles unfanned from variety of perches (including ground) by Little Raven and Little Crow:

Forest: rich, deep, slow 'korr-korr-korr-korrrr', last note often drawn out.

Australian: far-carrying, high-pitched wailing 'aah-aah-aah-aaaahh' with last note drawn out.

Little Raven: guttural 'kar-kar-kar-kar' or 'ark-ark-ark-ark', uttered from less prominent perches or on ground, wings flicked up at each note.

Little Crow: similar to Little Raven but with hoarse quality 'nark-nark-nark-nark', doesn't flick wings.

Torresian: notes usually shorter than Little Crow, nasal high-pitched clipped 'uk-uk-uk-uk-uk' or 'ok-ok-ok-ok-ok', sometimes finishing with longer notes.

Flight

Forest: wings rounded, tail short in southern form; wings rounder, tail broader than Australian in northern NSW form, soars, noisy wingbeat.

Australian: wings tapered, long slender rounded tail, noisy wingbeat.

Little Raven: wings tapered, more agile, rapid flight than Australian and Forest, often in large cohesive flocks, occasional aerobatics.

Little Crow: wings rounded, tail rounded, rapid agile flight, often in large cohesive flocks, performs spectacular tumbling aerobatics, soars.

Torresian: wings rounded more than Australian Raven but less than Little Crow (useful only in direct comparison), tail squarer than little Crow, in pairs or small (usually) flocks. Torresian often looks more compact in flight than Forest and Australian.

Diagnostic characteristics (found in one species only)

(a) Extensive loose skin under beak, characteristic of Australian; in juv noticeably pink. Juv Forest has pink edges to underside of beak, visible in field.

(b) 'Wing-quivering' greeting ceremony by Torresian: wings rapidly fluttered repeatedly above back.

(c) 'Wing-shuffling': when alighting, corvids shuffle wings to settle feathers; Torresian has exaggerated shuffling, often giving two or three shuffles.

(d) Bill length in Little Crow shorter than head length; equal in others. Tail shorter than folded wing in southern form of Forest, longer than folded wing in Little Crow, equal in others. Legs longest in Forest, shortest in Little Crow.

Other characteristics

(a) 'Currawong flight' or 'missing a beat': performed in flight by both crows, wings closed for space of one beat, calling at same time; performed by Torresian with single call-note and with throat hackles extended, less commonly by Little Crow, with double note.

(b) 'Winnowing flight' or 'reduced amplitude flight': flight display given in territory when one bird of pair is 'returning home': wings are 'winnowed' like typical kestrel flight with quick shallow wingbeats and descending call; Australian – high-pitched, tremulous wail; Forest – deeper and harsher; Little Raven – similar to Forest with creaky quality; Little Crow – similar to Little Raven; Torresian – more nasal than Australian not tremulous, sharply cut off.

(c) Throat hackles: most developed in Australian, with pointed tips; less developed in Forest and Little Raven with double tips; minimal development in either crow, look more like normal feathers. Erectile in Australian, Forest and Torresian.

(d) Nests are large, made of sticks; Australian, Forest and Torresian tend to nest in high situations, Little Raven and Little Crow more flexible and will nest in lower situations. Eggs of all species are similar, blue to blue-green with olive-brown blotches and spots, Little Crow smallest.

(e) Adult corvids are glossy black with white eyes; juveniles are less glossy, with eyes at first blue-grey turning brown before or shortly after leaving nest; immatures have eyes hazel (mottled brown and white), eyes become white at about 3 years in Forest and Australian, about 2 years in others.

Index